BARRON'S

SAT* SUBJECT TEST

CHEMISTRY

14TH EDITION

Mark C. Kernion, M.A.
Chemistry Teacher
Mount Lebanon High School, Pittsburgh, Pennsylvania

Joseph A. Mascetta, M.S., C.A.S.
Former Principal
Former Chemistry Teacher and Coordinator, Science Department
Mount Lebanon High School, Pittsburgh, Pennsylvania

Science Educational Consultant

BARRON'S

Dedication
To
my wife, Patty; and children, Emily, Sam, and Ben; graddaughter, Stella;
grandson, Nate; son-in-law, Rob; and daughter-in-law, Aly—M. Kernion
To
my wife, Jean; and daughters, Lisa, Linda, and Lori;
and their families,
who supported my efforts throughout the years—J. Mascetta

About the Authors

Mark Kernion has been teaching A.P. and Honors Chemistry for the past 25 years at Mt. Lebanon High School in Pittsburgh, Pennsylvania. He holds degrees and certifications from Penn State University, The University of Pittsburgh, and Duquesne University. In 2006 he was inducted into the Cum Laude Society, which is dedicated to honoring scholastic achievement in secondary schools. He was the recipient of the 2007 Yale Teaching Award. Prior to teaching he worked as a ceramic engineer for Harbison-Walker Refractories Company, and he holds a number of patents related to his research.

Joe Mascetta taught high school chemistry for twenty years. He was the science department coordinator and principal of Mt. Lebanon High School in Pittsburgh, Pennsylvania. He also served as a science consultant to the area schools and was a past-president of the Western Pennsylvania Association of Supervision and Curriculum Development (ASCD) and the State Advisory Committee of ASCD. He held degrees from the University of Pittsburgh, the University of Pennsylvania, and Harvard University, and was a participant in Harvard Project Physics, a General Electric Science Fellowship to Union College in Schenectady, New York, the Chemical Bond Approach Curriculum Study at Kenyon College, Ohio, and the Engineering Concepts Curriculum Project and Science Curriculum Supervision at the University of Colorado.

© Copyright 2018, 2016, 2014, 2012, 2010, 2008, 2006 by Kaplan, Inc., d/b/a Barron's Educational Series
Previous editions © Copyright 2002, 1998, 1994 by Kaplan, Inc., d/b/a Barron's Educational Series under the title *How to Prepare for the SAT II: Chemistry*.
Previous editions © Copyright 1990, 1986, 1981, 1969 by Kaplan, Inc., d/b/a Barron's Educational Series under the title *How to Prepare for College Board Achievement Test in Chemistry*.

Published by Kaplan, Inc., d/b/a Barron's Educational Series
750 Third Avenue
New York, NY 10017
www.barronseduc.com

Kaplan, Inc., d/b/a Barron's Educational Series print books are available at special quantity discounts to use for sales promotions, employee premiums, or educational purposes. For more information or to purchase books, please call the Simon & Schuster special sales department at 866-506-1949.

ISBN: 978-1-4380-1113-4

ISSN: 1559-3487

9 8 7 6 5 4 3

Contents

PRACTICE TESTS

APPENDIXES

Introduction: About the Test

The SAT Subject Tests are given in specific subject areas to assess your academic abilities for college. They are prepared by the College Board and give evidence about your readiness in specific academic areas. The test can complement and enhance your college admission credentials. The introduction of this book will give you all the basic information you need to know about the subject test in chemistry. To learn additional information about this testing program, go to the website *https://collegereadiness.collegeboard.org/sat-subject-tests/about/at-a-glance* and look under SAT Subject Tests.

All of the SAT Subject Tests are contained in the same test booklet. Each takes 1 hour of testing time, and you may choose any one, two, or three tests to take at one sitting.

Many colleges require or recommend one or more Subject Tests for admission or placement. The scores are used in conjunction with your high school record, results on the SAT, teacher recommendations, and other background information to provide a reliable measure of your academic achievements and a good predication of your future performance.

In addition to obtaining a standardized assessment of your achievement from your scores, some colleges use the test results for placement into their particular programs in the freshman year. At others, advisers use the results to guide freshmen in the selection of courses.

Is the SAT Subject Test in Chemistry Required?

The best information on whether SAT Subject Tests are required and, if so, which ones is found in the individual college catalogs or a directory of colleges. Some colleges specify which tests you must take, while others allow you to choose. Obviously, if you have a choice and you have done well in chemistry, you should pick the SAT Subject Test in Chemistry as one of your tests. Even if the test is not required by the colleges to which you are applying, you can add the result to your record to support your achievement level.

When Should You Take the Test?

You will undoubtedly do best if you take the test after completing the high school chemistry course or courses that you plan to take. At this time, the material will be fresh in your mind. Forgetting begins very quickly after you are past a topic or have finished the course. You should plan a review program for at least the last 6 weeks before the test date. (A plan is provided later in this book for such a review.) Careful review definitely helps—cramming just will not do if you want to get the best score of which you are capable!

Colleges that use SAT Subject Test results as part of the admissions process usually require that you take the test no later than December or January of your senior year. For early-decision programs, the latest test time is June of your junior year. The optimum time to take the test is the June test date of the year you take your last chemistry class in high school. At that time, the material you learned will be the easiest to recollect, and your preparations will likely correspond with the final exam in that course.

When Is the Test Offered?

The chemistry test is available every time the SAT Subject Tests are given, that is, on the first Saturday of October, November, December, May, and June. They are also given on the last Saturday of August. Be sure that the testing site for which you plan to register offers the SAT Subject Tests on each of these six times. Remember that you may choose to take one or two additional tests besides chemistry on any one test day. You do have to specify in advance which tests you plan to take on the test date you select; however, except for the Language Test with Listening, you may change your mind on the day of the test. Keep in mind that the SAT Test and the SAT Subject Test in Chemistry can't be taken on the same day.

How Do You Register?

You may get all of your registration information by going to *sat.collegeboard.org/register*. This is the quickest and easiest way to register for the test. This site will give you all the information you need to complete your registration. You can choose your test date and center as well as receive immediate registration confirmation. This website also gives you instructions for how to register by mail.

The deadline for registration is approximately one month before the test date.

How Should You Prepare for the Test?

Barron's SAT Subject Test in Chemistry will be very helpful. The more you know about the test, the more likely you are to get the best score possible for you. This book provides you with a diagnostic test, scoring information, four practice tests, and the equivalent of one more test incorporated with the chapter review tests that allow you to become familiar with the question types and the wording of directions. You will get a feeling for the degree of emphasis on particular topics and the ways in which information may be tested. Each of these aspects should be consciously pursued as you use this book.

ONLINE TESTS

In addition to the practice tests in this book, you can access two practice tests online. These tests will be scored automatically and will also provide you with complete answer explanations. Use these tests to further strengthen your ability to work through the problems and understand what it will take for you to ace the SAT Subject Test in Chemistry exam. You can access the online tests at *http://barronsbooks.com/tp/sat/chemistry/*.

What Topics Appear on the Test, and to What Extent?

The following charts show the content of the test and the levels of thinking skills tested:

	Topics	Percent of Test (Approx.)	Number of Questions (Approx.)
Structure of Matter		25	21
	I. <u>Atomic theory and structure</u>, energy levels, quantum numbers, orbitals, electron configurations, periodic trends II. <u>Molecular structures, shapes</u>, Lewis structures, polarity, three-dimensional shapes III. <u>Bonding</u> (ionic, covalent, metallic), relationships to properties and structures, intermolecular forces, hydrogen bonding, London dispersion forces, dipole-dipole forces		
States of Matter		16	14
	IV. <u>Gases</u>, kinetic molecular theory, gas law relationships, molar volume, density and related problems V. <u>Liquids and solids</u>, forces in these, types of solids, phase diagrams, phase changes VI. <u>Solutions</u>; molarity; percent by mass; solution preparation and related problems, solubility factors for solids, liquids, and gases; qualitative aspects of colligative properties		
Reaction Types		14	12
	VII. <u>Acids and bases, including Brønsted-Lowry theories</u>, strong and weak forms, pH, titration problems, indicators VIII. <u>Oxidation-reduction</u>, combustion, using oxidation numbers, use of activity series, <u>precipitation</u>, use of basic solubility rules		
Stoichiometry		14	12
	IX. <u>Mole concept</u>, molar mass, Avogadro's number, empirical and molecular formulas, <u>chemical equations</u>, balancing equations, solving related problems, determining percent yield, limiting factors		

	Topics	Percent of Test (Approx.)	Number of Questions (Approx.)
Equilibrium and Reaction Rates		5	4
	X. <u>Equilibrium systems</u>, factors affecting, Le Châtelier's Principle in gaseous and aqueous systems, equilibrium constants, expressions, <u>rates of reactions</u>, factors affecting rates, activation energies, reaction diagrams		
Thermochemistry		6	5
	XI. <u>Conservation of energy</u>, calorimetry, specific heat, thermal curves, enthalpy (heat) changes, entropy (randomness) changes		
Descriptive Chemistry		12	10
	XII. <u>Physical and chemical properties</u>, nomenclature of elements, compounds, ions, properties and trends related to the periodic table, reactivity of elements and prediction of chemical reactions, examples of basic organic compounds, environmental concerns		
Laboratory		8	7
	XIII. <u>Laboratory safety</u> rules, nomenclature, use of equipment, making measurements and observations, data to analyze, interpreting graphical data, drawing conclusions		
			Total Questions (85)

Note: Each test contains approximately five questions on equation balancing and/or predicting products of chemical reactions. These are distributed among the various content categories.

Thinking Skills Tested	Percent of Test (approx.)
Recalling fundamental concepts, specific pieces of information, and basic terminology (low-level skill)	20
Showing a *comprehension of the basics* and the *ability to apply this information* in a rather straightforward manner to questions, situations, and the solution of qualitative or quantitative problem-oriented questions (medium-level skill)	45
Using the ability to *analyze* quantitative and/or qualitative data and to *synthesize* the knowledge learned to *evaluate* how and what ideas or relationships should be used to draw conclusions or to solve problems (high-level skill)	35

The first chart gives you a general overview of the content of the test. Your knowledge of the topics and your skills in recalling, applying, and synthesizing this knowledge are evaluated through 85 multiple-choice questions. This material is that generally covered in an introductory course in chemistry at a level suitable for college preparation. While every test covers the topics listed, different aspects of each topic are stressed from year to year. Add to this the differences that exist in high school courses with respect to the percentage of time devoted to each major topic and to the specific subtopics covered, and you may find that there are questions on topics with which you have little or no familiarity.

Each of the sample tests in this book is constructed to match closely the distribution of topics shown in the preceding chart so that you will gain a feel for the makeup of the actual test. After each test, a chart will show you which questions relate to each topic. This will be very helpful to you in planning your review because you can identify the areas on which you need to concentrate in your studies. Another chart enables you to see which chapters correspond to the various topic areas.

What General Information Should You Have About the Test?

1. A periodic chart is provided in this test as a resource and as the source of atomic numbers and atomic masses of the elements.
2. You will *not* be allowed to use an electronic calculator during the test.
3. Mathematical calculations are limited to simple algebraic and numerical ones.
4. You should be familiar with the concepts of ratios and of direct and inverse proportions, scientific notation, and exponential functions.
5. Metric system units are used in this test.
6. The test is composed of three types of questions as explained in the next section.

What Types of Questions Appear on the Test?

There are three general types of questions on the SAT Subject Test in Chemistry—matching questions, true/false and relationship analysis questions, and general five-choice questions. This section will discuss each type and give specific examples of how to answer these questions. You should learn the directions for each type so that you will be familiar with them on the test day. The directions in this section are similar to those on the test.

TYPE 1. MATCHING QUESTIONS IN PART A. In each of these questions, you are given five lettered choices that you will use to answer all the questions in that set. The choices may be in the form of statements, pictures, graphs, experimental findings, equations, or specific situations. Answering a question may be as simple as recalling information or as difficult as analyzing the information given to establish what you need to do qualitatively or quantitatively to synthesize your answer. The directions for this type of question specifically state that a choice may be used once, more than once, or not at all in each set.

PART A

> **Directions:** Every set of the given choices below refers to the numbered statements or formulas immediately following it. Choose the one lettered choice that best fits each statement or formula and then fill in the corresponding oval on the answer sheet. Each choice may be used once, more than once, or not at all in each set.

➥ Example

<u>Questions 1–3</u> refer to the following graphs:

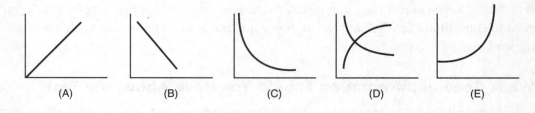

(A)   (B)   (C)   (D)   (E)

1. The graph that best shows the relationship of volume to temperature for an ideal gas while the pressure is held constant
2. The graph that best shows the relationship of volume to pressure for an ideal gas while the temperature is held constant
3. The graph that best shows the relationship of the number of grams of solute that is soluble in 100 grams of water at varying temperatures if the solubility begins as a small quantity and increases as the temperature is increased

These three questions require you to recall the basic gas laws and the graphic depiction of the relationship expressed in each law, as well as how solubility can be shown graphically.

To answer question 1, you must recognize that the relationship of gas volume to changes in temperature is a direct relationship that is depicted by graphing Charles's Law: $V_1/T_1 = V_2/T_2$. The only graph that shows that type of direct relationship with the appropriate slope is (A).

To answer question 2, you need to understand that Boyle's Law states that the pressure of a gas is inversely proportional to the volume at constant temperature. Mathematically, this means that pressure (P) times volume (V) is a constant, or $P_1V_1 = P_2V_2$. This inversely proportional relationship is accurately depicted as a hyperbola in (C). Although (B) shows the values on the x-axis increasing as the y-axis values decrease, it does not fit the graph for an inverse proportion.

Question 3 requires that you have knowledge about solubility curves and can apply the solubility relationship given in words to graph (E).

TYPE 2. TRUE/FALSE AND RELATIONSHIP ANALYSIS QUESTIONS IN PART B. On the actual SAT Subject Test in Chemistry, this type of question must be answered in a special section of your answer sheet labeled "chemistry." Type 2 questions are numbered beginning with 101. Each question consists of a statement or assertion in column I and, on the other side of the word BECAUSE, another statement or assertion in column II. Your first task is to determine whether each of the statements is true or false and to record your answer for each in the answer blocks for column I and column II in the answer grid by darkening either the ⓣ or the ⓕ oval. Here you must use your reasoning skills and your understanding of the topic to determine whether there is a causal relationship between the two statements.

Here are the directions and two examples of a relationship analysis question.

PART B

Directions: Every question below contains two statements, I in the left-hand column and II in the right-hand column. For each question, decide if statement I is true or false <u>and</u> if statement II is true or false and fill in the corresponding T or F ovals on your answer sheet. <u>Fill in oval CE only if statement II is a correct explanation of statement I.</u>

Sample Answer Grid:

CHEMISTRY * Fill in oval CE only if II is a correct explanation of I.

	I	II	CE*
101.	Ⓣ Ⓕ	Ⓣ Ⓕ	◯

➥ Example 1

101. When 2 liters of oxygen gas react completely with 2 liters of hydrogen gas, the limiting reactant is the oxygen   BECAUSE   the coefficients in the balanced equation of a gaseous reaction give the volume relationship of the reacting gases.

The reaction that takes place is

$$2H_2 + O_2 \rightarrow 2H_2O$$

The coefficients of this gaseous reaction show that 2 L of hydrogen react with 1 L of oxygen, leaving 1 L of oxygen unreacted, or in excess. The limiting reactant, then, is the quantity of hydrogen.

The ability to solve this quantitative relationship shows that statement I is not true. However, statement II does give a true statement of the relationship of coefficients in a balanced equation of gaseous chemical reaction. Therefore, the answer blocks would be completed like this:

	I	II	CE*
101.	Ⓣ **Ⓕ**	**Ⓣ** Ⓕ	◯

➥ Example 2

102. Water is generally a good solvent of ionic and polar molecular compounds   BECAUSE   the water molecule has polar properties due to the factors involved in the bonding of the hydrogen and oxygen atoms.

Statement I is true because water is such a good solvent for these types of compounds that, as you have probably learned, it is sometimes referred to as the universal solvent. This property is attributed mostly to its polar structure. The polar covalent bond between the oxygen and hydrogen atoms and the angular orientation of the hydrogens at 105 degrees between them contribute to the establishment of a permanent dipole moment in the water molecule. These properties combine to make water a powerful solvent for both polar and ionic compounds. Because of your familiarity with these concepts and the processes by which substances go into solution, you know that statement II not only is true but also is the reason that statement I is true. There is a causal relationship between the two statements. Therefore, the answer blocks would be marked like this:

	I		II		CE*
102.	Ⓣ	Ⓕ	Ⓣ	Ⓕ	⬭

As a test-taking tip, it should be noted that if statement I or II is false, the oval for CE should *never* be filled in. Therefore, the only time CE could *possibly* be filled in is if statements I and II are both marked true.

TYPE 3: GENERAL FIVE-CHOICE QUESTIONS IN PART C. The five-choice items in Part C are written usually as questions but sometimes as incomplete statements. You are given five suggested answers or completions. You must select the one that is best in each case and record your choice in the appropriate oval. In some questions you are asked to select the one inappropriate answer. Such questions contain a word in capital letters, such as NOT, LEAST, or EXCEPT.

In some of these questions, you may be asked to make an association between a graphic, pictorial, or mathematical representation and a stated explanation or problem. The solution may involve solving a scientific problem by correctly interpreting the representation. In some cases the same representation may be used for a series of two or more questions. In no case, however, is the correct answer to one question necessary for answering a subsequent question correctly. Each question in the set is independent of the others.

PART C

> **Directions:** Every question or incomplete statement below is followed by five suggested answers or completions. Choose the one that is best in each case and then fill in the corresponding oval on the answer sheet. Remember to return to the original part of the answer sheet.

➡ **Example 1** _____

40. In this graphic representation of a chemical reaction, which arrow depicts the activation energy?

(A) *A*
(B) *B*
(C) *C*
(D) *D*
(E) *E*

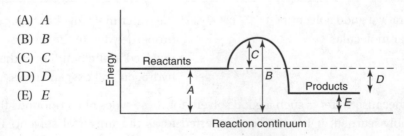

To answer this question, you need to know how to interpret the energy levels in this graphic representation of energy-level changes along the time continuum of the reaction. The activation energy is the minimum energy required for a chemical reaction to take place. The reactant molecules come together, and chemical bonds are stretched, broken, and formed in producing the products. During this process the energy of the system increases to a maximum, then decreases to the energy of the products. The activation energy is the difference between the maximum energy and the energy of the reactants. Choice (C) in the graphic depiction shows this energy barrier that has to be overcome for the reaction to proceed. The corresponding oval on the answer sheet should be darkened.

➡ **Example 2** _____

41. If the molar mass of NH_3 is 17 g/mol, what is the density of this compound at STP?

   (A) 0.25 g/L
   (B) 0.76 g/L
   (C) 1.25 g/L
   (D) 3.04 g/L
   (E) 9.11 g/L

$$D = \frac{M}{grams}$$

The solution of this quantitative problem depends on the application of several principles. One principle is that the molar mass of a gas expressed in grams/mole will occupy 22.4 L at standard temperature and pressure (STP). The other is that density is defined as the mass of a substance divided by the volume it occupies. Therefore, 17 g of ammonia (NH_3) will occupy 22.4 L at STP. So the density of the NH_3 is 17 g/22.4 L or 0.76 g/L. The correct answer is (B).

➡ **Example 3** _____

Some questions in this part are followed by three or four bits of information labeled by Roman numerals I through III or IV. One or more of these statements may correctly answer the question. You must select from the five lettered choices the one that best answers the question.

42. Which bond(s) is (are) considered predominantly ionic?

   I. H–Cl (g)
   II. S–Cl (g)
   III. Cs–F (g)

   (A) I only
   (B) III only
   (C) I and II only
   (D) II and III only
   (E) I, II, and III

To determine the type of bonding that exists in these three substances, you must use your knowledge of the way in which bonds are formed. You must also use your knowledge of the relationship of the electronegativity of an element and the position of that element in the Periodic Table. Compounds I and II are formed from elements that do not have enough difference in their respective electronegativities to cause the formation of an ionic bond. This can be inferred by checking the positions of the elements (H, Cl, and S) in the Periodic Table and noting how electronegativity varies with an element's position in the table. Compound III, cesium fluoride, consists of elements that appear in the lower right corner and the upper left corner, respectively, of the Periodic Table; therefore, the difference in their electronegativity values is sufficient so that an ionic bond can be predicted between them. Of the choices given, only (B) is a correct answer.

How Can You Use This Book to Prepare for the Test?

The best way to use this book is a two-stage approach, and the next sections are arranged accordingly. First, you should take the diagnostic test. This will give you a preliminary exposure to the type of test you are planning to take, as well as a measure of how well you achieve on each of the three parts. You will also become aware of the types of questions that the test includes. Use the test-scoring information following the diagnostic test to determine your raw score and your strengths and weaknesses in the specific areas of the test.

Having taken the diagnostic test, you should then follow a study program. A study plan covering the 6 weeks before the test has been developed for you and is given in detail on page 32. It requires a minimum of 1 or 2 hours per night on weekdays but leaves your weekends free.

Five Steps to Improve Your Problem-Solving Skills*

Chemistry is a subject that deals with many problem situations that you, the student, must be able to solve. Solving problems may seem to be a natural process when the degree of difficulty is not very great, and you may not need a structured method to attack these problems. However, for complex problems an orderly process is required.

The following is such a problem-solving process. Each step is vital to the next step and to the final solution of the problem.

STEP 1 Clarify the problem: to separate the problem into the facts, the conditions, and the questions that need to be answered, and to establish the goal.

STEP 2 Explore: to examine the sufficiency of the data, to organize the data, and to apply previously acquired knowledge, skills, and understanding.

STEP 3 Select a strategy: to choose an appropriate method to solve the problem.

STEP 4 Solve: to apply the skills needed to carry out the strategy chosen.

STEP 5 Review: to examine the reasonableness of the solution through estimation and to evaluate the effectiveness of the process.

The steps of the problem-solving process listed above should be followed in sequence. The subskills listed below for each step, however, are not in sequence. The order in which subskill patterns are used will differ with the nature of the problem and/or with the ways in which the individual problem solver thinks. Also, not every subskill need be employed in solving every problem.

Adapted with permission from Thinking Skills Resource Guide, *a noncopyrighted publication of Mount Lebanon School District, Pittsburgh, PA.*

1. **CLARIFY THE PROBLEM**
   a. Identify the facts. What is known about the problem?
   b. Identify the conditions. What is the current situation?
   c. Identify the questions. What needs to be answered before the problem can be solved?
   d. Visualize the problem.
      1. Make mental images of the problem.
      2. If desirable or necessary, draw a sketch or diagram, make an outline, write down symbols or equations that correspond to the mental images.
   e. Establish the goal. The goal defines the specific result to be accomplished through the problem-solving process. It defines the purpose or function the solution is expected to achieve and serves as the basis for evaluating the solution.

2. **EXPLORE**
   a. Review previously acquired knowledge, skills, and understanding. Determine whether the current problem is similar to a previously seen type.
   b. Estimate the sufficiency of the data. Does there seem to be enough information to solve the problem?
   c. Organize the data. There are many ways in which data can be organized. Some examples are outline, written symbols and equations, chart, table, graph, map, diagram, and drawing. Determine whether the data organized in the way(s) you have chosen will enable you to partially or completely solve the problem.
   d. Determine what new data, if any, need to be collected. What additional information may be needed to solve the problem? Can the existing data be reorganized to generate new information? Do other resources need to be consulted? This step may suggest possible strategies to be used to solve the problem.

3. **SELECT A STRATEGY**

   A strategy is a goal-directed sequence of mental operations. Selecting a strategy is the most important and also the most difficult step in the problem-solving process. Although there may be several strategies that will lead to the solution of a problem, the skilled problem solver uses the most efficient strategy. The choice of the most efficient strategy is based on knowledge and experience as well as a careful application of the clarify and explore steps of the problem-solving method. Some problems may require the use of a combination of strategies.

   The following search methods may help you to select a strategy. They do not represent all of the possible ways in which this can be done. Other methods of strategy selection are related to specific content areas.
   a. Trial-and-error search: Such a search either doesn't have or doesn't use information that indicates that one path is more likely to lead to the goal than any other path.

      Trial-and-error search comes in two forms, blind and systematic. In *blind search*, the searchers pick paths to explore blindly, without considering whether they have already explored these paths. A preferable method is *systematic search*, in which the searchers keep track of the paths they have already explored and do not duplicate them. Because this method avoids multiple searches, systematic search is usually twice as efficient as blind search.

b. Reduction method: This involves breaking the problem into a sequence of smaller parts by setting up subgoals. Subgoals make problem solving easier because they reduce the amount of search required to find the solution.

You can set up subgoals by working part way into a problem and then analyzing the partial goal to be achieved. In doing this, you can drop the problem restrictions that do not apply to the subgoal. By adding up all the subgoals, you can solve the "abstracted" problem.

c. Working backward: When you have trouble solving a problem head-on, it is often useful to try to work backward. Working backward involves a simple change in representation or point of view. Your new starting point is the original goal. Working backward can be helpful because problems are often easier to solve in one direction than another.

d. Knowledge-based method: This strategy uses information stored in the problem solver's memory, or newly acquired information, to guide the search for the solution. The problem solver may have solved a similar problem and can use this knowledge in a new situation. In other cases, problem solvers may have to acquire needed knowledge. For example, they may solve an auxiliary problem to learn how to solve the one they are having difficulty with.

Searching for analogous (similar) problems is a very powerful problem-solving technique. When you are having difficulty with a problem, try to pose a related, easier one and hope thereby to learn something that will help you solve the harder problem.

4. SOLVE

Use the strategy chosen to actually solve the problem. Executing the solution provides you with a very valuable check on the adequacy of your plan. Sometimes students will look at a problem and decide that, since they know how to solve it, they need not bother with the drudgery of actually executing the solution. Sometimes the students are right, but at other times they miss an excellent opportunity to discover that they were wrong.

5. REVIEW/VERIFY WITH ESTIMATION

a. Evaluation. The critical question in evaluation is this: "Does the answer I propose meet all of the goals and conditions set by the problem?" Thus, after the effort of finding a solution, you must turn back to the problem statement and check carefully to be sure your solution satisfies it.

With easy problems there is a strong temptation to skip evaluation because the probability of error seems small. In some cases, however, this can be costly. Evaluation may prove that errors were present.

b. Verification of the reasonableness of the answer. It is easy to become so involved with the process and mathematics of a problem that an answer is recorded that is totally illogical. To avoid this mistake, you should simplify the numbers involved and solve for an answer. Having done this, compare your estimated result with your answer to ensure that your answer is feasible.

For example, a problem requires the following operations:

$$5.12 \times 10^5 \times 3.98 \times 10^6 \text{ divided by } 910$$

And doing all the math, you get an answer of

$$0.02239 \times 10^{11} \text{ or } 2.24 \times 10^9$$

To estimate the answer, first simplify the numbers to one significant figure (significant figures are discussed in Chapter 1). This gives

$$5 \times 10^5 \times 4 \times 10^6 \text{ divided by } 9 \times 10^2$$

which is

$$20 \times 10^{11} \text{ divided by } 9 \times 10^2 = 2.2 \times 10^9$$

This is the estimated answer, which validates the answer above.

When you are dealing with test items that provide multiple-choice answers, you can often use estimation to arrive at the answer without doing the more complicated mathematics.

c. Consolidation. Here the basic question to be answered is: "What can I learn from the experience of solving this problem?" The following more specific questions may help you to answer this general one:

1. Why was this problem difficult?
2. Was it difficult to follow a plan?
3. Was it difficult to decide on a plan? If so, why?
4. Did I take the long way to the answer?
5. Can I use this plan again in similar problems?

The important thing is to reflect on the process that you used in order to make future problem solving easier.

Material in the following table may be useful in answering the questions in this examination.

Periodic Table of the Elements

1	2	3	4	5	6	7	8	9	10	11	12	13	14	15	16	17	18
1 **H** 1.0078																	**2** **He** 4.0026
3 **Li** 6.941	**4** **Be** 9.012											**5** **B** 10.811	**6** **C** 12.011	**7** **N** 14.007	**8** **O** 16.00	**9** **F** 19.00	**10** **Ne** 20.179
11 **Na** 22.99	**12** **Mg** 24.30											**13** **Al** 26.98	**14** **Si** 28.09	**15** **P** 30.974	**16** **S** 32.06	**17** **Cl** 35.453	**18** **Ar** 39.948
19 **K** 39.10	**20** **Ca** 40.08	**21** **Sc** 44.96	**22** **Ti** 47.90	**23** **V** 50.94	**24** **Cr** 52.00	**25** **Mn** 54.938	**26** **Fe** 55.85	**27** **Co** 58.93	**28** **Ni** 58.69	**29** **Cu** 63.55	**30** **Zn** 65.39	**31** **Ga** 69.72	**32** **Ge** 72.59	**33** **As** 74.92	**34** **Se** 78.96	**35** **Br** 79.90	**36** **Kr** 83.80
37 **Rb** 85.47	**38** **Sr** 87.62	**39** **Y** 88.91	**40** **Zr** 91.22	**41** **Nb** 92.91	**42** **Mo** 95.94	**43** **Tc** (98)	**44** **Ru** 101.1	**45** **Rh** 102.91	**46** **Pd** 106.42	**47** **Ag** 107.87	**48** **Cd** 112.41	**49** **In** 114.82	**50** **Sn** 118.71	**51** **Sb** 121.75	**52** **Te** 127.60	**53** **I** 126.91	**54** **Xe** 131.29
55 **Cs** 132.91	**56** **Ba** 137.33	**57** **La*** 138.91	**72** **Hf** 178.49	**73** **Ta** 180.95	**74** **W** 183.85	**75** **Re** 186.21	**76** **Os** 190.2	**77** **Ir** 192.2	**78** **Pt** 195.08	**79** **Au** 196.97	**80** **Hg** 200.59	**81** **Tl** 204.38	**82** **Pb** 207.2	**83** **Bi** 208.98	**84** **Po** (209)	**85** **At** (210)	**86** **Rn** (222)
87 **Fr** (223)	**88** **Ra** 226.02	**89** **Ac†** 227.03	**104** **Rf** (261)	**105** **Db** (262)	**106** **Sg** (266)	**107** **Bh** (264)	**108** **Hs** (277)	**109** **Mt** (268)	**110** **Ds** (281)	**111** **Rg** (272)	**112** **Cn** (285)	**113** **Nh** (286)	**114** **Fl** (289)	**115** **Mc** (289)	**116** **Lv** (293)	**117** **Ts** (294)	**118** **Og** (294)

*Lanthanide Series

58 **Ce** 140.12	**59** **Pr** 140.91	**60** **Nd** 144.24	**61** **Pm** (145)	**62** **Sm** 150.4	**63** **Eu** 151.97	**64** **Gd** 157.25	**65** **Tb** 158.93	**66** **Dy** 162.50	**67** **Ho** 164.93	**68** **Er** 167.26	**69** **Tm** 168.93	**70** **Yb** 173.04	**71** **Lu** 174.97

†Actinide Series

90 **Th** 232.04	**91** **Pa** 231.04	**92** **U** 238.03	**93** **Np** 237.05	**94** **Pu** (244)	**95** **Am** (243)	**96** **Cm** (247)	**97** **Bk** (247)	**98** **Cf** (251)	**99** **Es** (252)	**100** **Fm** (257)	**101** **Md** (258)	**102** **No** (259)	**103** **Lr** (262)

USE THIS PERIODIC TABLE WITH ALL THE PRACTICE TESTS.

PART 1
A Diagnostic Test

NOTE

In addition to the diagnostic test and the four practice tests at the end of this book, you have access to two online tests. You can find the link to the online tests on the inside front cover of this book.

Diagnostic Test

The following test of 85 questions is a sample of the actual test you will take to measure your chemistry achievement. It has basically the same distribution of topics, directions, and number and types of questions. Before taking this test, read the advice given in the section entitled "Final Preparation—The Day Before the Test" (p. 32). Use the answer sheet provided, and limit the test time to *1 hour.*

A Periodic Table of the Elements has been included for your use on problems requiring this source of information. Use this table also with the practice tests at the end of the book.

Turn now to the test.

ANSWER SHEET
Diagnostic Test

Determine the correct answer for each question. Then, using a No. 2 pencil, blacken completely the oval containing the letter of your choice.

1. Ⓐ Ⓑ Ⓒ Ⓓ Ⓔ
2. Ⓐ Ⓑ Ⓒ Ⓓ Ⓔ
3. Ⓐ Ⓑ Ⓒ Ⓓ Ⓔ
4. Ⓐ Ⓑ Ⓒ Ⓓ Ⓔ
5. Ⓐ Ⓑ Ⓒ Ⓓ Ⓔ
6. Ⓐ Ⓑ Ⓒ Ⓓ Ⓔ
7. Ⓐ Ⓑ Ⓒ Ⓓ Ⓔ
8. Ⓐ Ⓑ Ⓒ Ⓓ Ⓔ
9. Ⓐ Ⓑ Ⓒ Ⓓ Ⓔ
10. Ⓐ Ⓑ Ⓒ Ⓓ Ⓔ
11. Ⓐ Ⓑ Ⓒ Ⓓ Ⓔ
12. Ⓐ Ⓑ Ⓒ Ⓓ Ⓔ
13. Ⓐ Ⓑ Ⓒ Ⓓ Ⓔ
14. Ⓐ Ⓑ Ⓒ Ⓓ Ⓔ
15. Ⓐ Ⓑ Ⓒ Ⓓ Ⓔ
16. Ⓐ Ⓑ Ⓒ Ⓓ Ⓔ
17. Ⓐ Ⓑ Ⓒ Ⓓ Ⓔ

18. Ⓐ Ⓑ Ⓒ Ⓓ Ⓔ
19. Ⓐ Ⓑ Ⓒ Ⓓ Ⓔ
20. Ⓐ Ⓑ Ⓒ Ⓓ Ⓔ
21. Ⓐ Ⓑ Ⓒ Ⓓ Ⓔ
22. Ⓐ Ⓑ Ⓒ Ⓓ Ⓔ
23. Ⓐ Ⓑ Ⓒ Ⓓ Ⓔ
24. Ⓐ Ⓑ Ⓒ Ⓓ Ⓔ
25. Ⓐ Ⓑ Ⓒ Ⓓ Ⓔ

ON THE ACTUAL CHEMISTRY TEST, THE FOLLOWING TYPE OF QUESTION MUST BE ANSWERED ON A SPECIAL SECTION (LABELED "CHEMISTRY") AT THE LOWER LEFT-HAND CORNER OF PAGE 2 OF YOUR ANSWER SHEET. THESE QUESTIONS WILL BE NUMBERED BEGINNING WITH 101 AND MUST BE ANSWERED ACCORDING TO THE DIRECTIONS.

CHEMISTRY* Fill in oval CE only if II is a correct explanation of I.

	I	II	CE*
101.	Ⓣ Ⓕ	Ⓣ Ⓕ	◯
102.	Ⓣ Ⓕ	Ⓣ Ⓕ	◯
103.	Ⓣ Ⓕ	Ⓣ Ⓕ	◯
104.	Ⓣ Ⓕ	Ⓣ Ⓕ	◯
105.	Ⓣ Ⓕ	Ⓣ Ⓕ	◯
106.	Ⓣ Ⓕ	Ⓣ Ⓕ	◯
107.	Ⓣ Ⓕ	Ⓣ Ⓕ	◯
108.	Ⓣ Ⓕ	Ⓣ Ⓕ	◯
109.	Ⓣ Ⓕ	Ⓣ Ⓕ	◯
110.	Ⓣ Ⓕ	Ⓣ Ⓕ	◯
111.	Ⓣ Ⓕ	Ⓣ Ⓕ	◯
112.	Ⓣ Ⓕ	Ⓣ Ⓕ	◯
113.	Ⓣ Ⓕ	Ⓣ Ⓕ	◯
114.	Ⓣ Ⓕ	Ⓣ Ⓕ	◯
115.	Ⓣ Ⓕ	Ⓣ Ⓕ	◯

ANSWER SHEET
Diagnostic Test

ON THE ACTUAL CHEMISTRY TEST, THE REMAINING QUESTIONS MUST BE ANSWERED BY RETURNING TO THE SECTION OF YOUR ANSWER SHEET YOU STARTED FOR CHEMISTRY.

26. Ⓐ Ⓑ Ⓒ Ⓓ Ⓔ
27. Ⓐ Ⓑ Ⓒ Ⓓ Ⓔ
28. Ⓐ Ⓑ Ⓒ Ⓓ Ⓔ
29. Ⓐ Ⓑ Ⓒ Ⓓ Ⓔ
30. Ⓐ Ⓑ Ⓒ Ⓓ Ⓔ
31. Ⓐ Ⓑ Ⓒ Ⓓ Ⓔ
32. Ⓐ Ⓑ Ⓒ Ⓓ Ⓔ
33. Ⓐ Ⓑ Ⓒ Ⓓ Ⓔ
34. Ⓐ Ⓑ Ⓒ Ⓓ Ⓔ
35. Ⓐ Ⓑ Ⓒ Ⓓ Ⓔ
36. Ⓐ Ⓑ Ⓒ Ⓓ Ⓔ
37. Ⓐ Ⓑ Ⓒ Ⓓ Ⓔ
38. Ⓐ Ⓑ Ⓒ Ⓓ Ⓔ
39. Ⓐ Ⓑ Ⓒ Ⓓ Ⓔ
40. Ⓐ Ⓑ Ⓒ Ⓓ Ⓔ

41. Ⓐ Ⓑ Ⓒ Ⓓ Ⓔ
42. Ⓐ Ⓑ Ⓒ Ⓓ Ⓔ
43. Ⓐ Ⓑ Ⓒ Ⓓ Ⓔ
44. Ⓐ Ⓑ Ⓒ Ⓓ Ⓔ
45. Ⓐ Ⓑ Ⓒ Ⓓ Ⓔ
46. Ⓐ Ⓑ Ⓒ Ⓓ Ⓔ
47. Ⓐ Ⓑ Ⓒ Ⓓ Ⓔ
48. Ⓐ Ⓑ Ⓒ Ⓓ Ⓔ
49. Ⓐ Ⓑ Ⓒ Ⓓ Ⓔ
50. Ⓐ Ⓑ Ⓒ Ⓓ Ⓔ
51. Ⓐ Ⓑ Ⓒ Ⓓ Ⓔ
52. Ⓐ Ⓑ Ⓒ Ⓓ Ⓔ
53. Ⓐ Ⓑ Ⓒ Ⓓ Ⓔ
54. Ⓐ Ⓑ Ⓒ Ⓓ Ⓔ
55. Ⓐ Ⓑ Ⓒ Ⓓ Ⓔ

56. Ⓐ Ⓑ Ⓒ Ⓓ Ⓔ
57. Ⓐ Ⓑ Ⓒ Ⓓ Ⓔ
58. Ⓐ Ⓑ Ⓒ Ⓓ Ⓔ
59. Ⓐ Ⓑ Ⓒ Ⓓ Ⓔ
60. Ⓐ Ⓑ Ⓒ Ⓓ Ⓔ
61. Ⓐ Ⓑ Ⓒ Ⓓ Ⓔ
62. Ⓐ Ⓑ Ⓒ Ⓓ Ⓔ
63. Ⓐ Ⓑ Ⓒ Ⓓ Ⓔ
64. Ⓐ Ⓑ Ⓒ Ⓓ Ⓔ
65. Ⓐ Ⓑ Ⓒ Ⓓ Ⓔ
66. Ⓐ Ⓑ Ⓒ Ⓓ Ⓔ
67. Ⓐ Ⓑ Ⓒ Ⓓ Ⓔ
68. Ⓐ Ⓑ Ⓒ Ⓓ Ⓔ
69. Ⓐ Ⓑ Ⓒ Ⓓ Ⓔ
70. Ⓐ Ⓑ Ⓒ Ⓓ Ⓔ

THE DIAGNOSTIC TEST

Note: For all questions involving solutions, you should assume that the solvent is water unless otherwise noted. **Reminder: You may not use a calculator on this test!**

The following symbols have the meanings listed unless otherwise noted.

H	= enthalpy	g	= gram(s)
M	= molar	J	= joule(s)
n	= number of moles	kJ	= kilojoule(s)
P	= pressure	L	= liter(s)
R	= molar gas constant	mL	= milliliter(s)
S	= entropy	mol	= mole(s)
T	= temperature	mm	= millimeter(s)
V	= volume	V	= volt(s)
atm	= atmosphere(s)		

PART A

Directions: Every set of the given lettered choices below refers to the numbered statements or formulas that immediately follow it. Choose the one lettered choice that best fits each statement or formula; then fill in the corresponding oval on the answer sheet. Each choice may be used once, more than once, or not at all in each set.

Questions 1–4 refer to the following elements:

    (A) Fluorine
    (B) Chlorine
    (C) Bromine
    (D) Iodine
    (E) Astatine

1. The element that is most active chemically

2. The element with the smallest ionic radius

3. The element with the lowest first ionization energy

4. The element that, at room temperature and pressure, exists as a bluish-black solid

Questions 5–7 refer to the following sublevels:

    (A) $1s$
    (B) $2s$
    (C) $3s$
    (D) $3p$
    (E) $3d$

5. Contains up to 10 electrons

6. Contains one pair of electrons in the ground-state electron configuration of the lithium atom

7. Is exactly one-half filled in the ground-state electron configuration of the phosphorus atom

Questions 8–12 refer to the following:

    (A) Avogadro's number
    (B) Boyle's Law
    (C) Charles's Law
    (D) Dalton's Theory
    (E) Gay-Lussac's Law

8. Proposes basic postulates concerning elements and atoms

GO ON TO THE NEXT PAGE

9. Proposes a relationship between the combining volumes of gases with respect to the reactants and gaseous products

10. Proposes a temperature-volume relationship of gases

11. Proposes a concept regarding the number of particles in a mole

12. Proposes a volume-pressure relationship of gases

Questions 13–16 refer to the following structures:

(A) R–OH

(B) R–O–R*

(C) R–C$\overset{\nearrow O}{\searrow H}$

(D) R–C$\overset{\nearrow O}{\searrow OH}$

(E) $\overset{O}{\underset{\|}{R–C–O–R}}$*

(*Alkyl group that is not necessarily the same as R)

13. The organic structure that includes the functional group of an aldehyde

14. The organic structure that includes the functional group of an acid

15. The organic structure that includes the functional group of an ester

16. The organic structure that includes the functional group of an ether

Questions 17–21 refer to the following:

(A) $H_2(g)$

(B) $CO_2(g)$

(C) $2N_2O(g)$

(D) $2NaCl(aq)$

(E) $H_2SO_4(dilute\ aq)$

17. The expression that can be used to designate a linear nonpolar molecule that contains polar bonds

18. The expression that can be used to designate 2 moles of atoms

19. The expression that can be used to designate 3 moles of atoms

20. The expression that can be used to designate a maximum of 3 moles of ions

21. The expression that can be used to designate 6 moles of atoms

Questions 22–25 refer to the following pairs of substances:

(A) NH_3 and N_2H_4

(B) ^{16}O and ^{17}O

(C) NH_4Cl and NH_4NO_3

(D) CH_3OCH_3 and CH_3CH_2OH

(E) O_2 and O_3

22. Are isotopes

23. Have both ionic and covalent bonds

24. Are allotropes

25. Are strong electrolytes in aqueous solutions

GO ON TO THE NEXT PAGE

ON THE ACTUAL SAT SUBJECT TEST IN CHEMISTRY, THE FOLLOWING TYPE OF QUESTION MUST BE ANSWERED ON A SPECIAL SECTION (LABELED "CHEMISTRY") AT THE LOWER LEFT-HAND CORNER OF PAGE 2 OF YOUR ANSWER SHEET. THESE QUESTIONS ARE NUMBERED BEGINNING WITH 101 AND MUST BE ANSWERED ACCORDING TO THE FOLLOWING DIRECTIONS.

Directions: Every question below contains two statements, I in the left-hand column and II in the right-hand column. For each question, decide if statement I is true or false <u>and</u> whether statement II is true or false, and fill in the corresponding T or F ovals on your answer sheet. *<u>Fill in oval CE only if statement II is a correct explanation of statement I.</u>

Sample Answer Grid:

CHEMISTRY *Fill in oval CE only if II is a correct explanation of I.

	I	II	CE*
101.	T F	T F	◯

	I		II
101.	A catalyst can accelerate a chemical reaction	BECAUSE	a catalyst can decrease the activation energy required for the reaction to occur.
102.	Molten sodium chloride is a good electrical conductor	BECAUSE	sodium chloride in the molten state allows ions to move freely.
103.	Ice is less dense than liquid water	BECAUSE	water molecules are nonpolar.
104.	Two isotopes of the same element have the same mass number	BECAUSE	isotopes have the same number of protons.
105.	A 1.0 g sample of calcium citrate, $Ca_3(C_6H_5O_7)_2$ (molar mass 498 g/mol), contains more Ca than a 1.0 g sample of calcium carbonate, $CaCO_3$ (molar mass 100 g/mol)	BECAUSE	there are more Ca atoms in 1.0 mol of calcium carbonate than in 1.0 mol of calcium citrate.
106.	About two liters of CO_2 can be produced by 1 gram of carbon burning completely	BECAUSE	the amount of gas evolved in a chemical reaction can be determined by using the mole relationship of the coefficients in the balanced equation.
107.	A reaction is at equilibrium when it reaches completion	BECAUSE	the concentrations of the reactants in a state of equilibrium equal the concentrations of the products.

GO ON TO THE NEXT PAGE

108. The oxidation number of Mn in MnO_4^-, permanganate, is 8+ | BECAUSE | the oxidation number of oxygen in most substances is –2.

109. A solution with pH = 5 has a higher concentration of hydronium ions than a solution with a pH = 3 | BECAUSE | pH is defined as –log $[H_3O^+]$.

110. An endothermic reaction can be spontaneous | BECAUSE | both the enthalpy and the entropy changes affect the Gibbs free-energy change of the reaction.

111. Weak acids have small values for the equilibrium constant, K_a, | BECAUSE | the concentration of the hydronium ion is in the numerator of the K_a expression.

112. One mole of NaCl contains 2 moles of ions | BECAUSE | NaCl is a stable salt at room temperature.

113. A p orbital has a shape that can be described as two lobes along the x, y, or z axis in 3-D space | BECAUSE | each of the two lobes of a single p orbital can hold two electrons of opposite spin.

114. H_2S and H_2O have a significant difference in their boiling points | BECAUSE | hydrogen sulfide has a higher degree of hydrogen bonding than water.

115. Two miscible liquids, when shaken together, will form a suspension | BECAUSE | in a suspension, the mixed substances readily dissolve in each other.

PART C

Directions: Every question or incomplete statement below is followed by five suggested answers or completions. Choose the one that is best and then fill in the corresponding oval on the answer sheet.

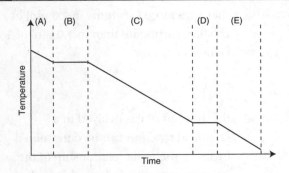

26. A thermometer is used to record the cooling of a confined pure substance over a period of time. During which interval on the cooling graph above is the system undergoing a change of state from a liquid to a solid?

27. If a principal energy level of an atom in the ground state contains 18 electrons, they will be arranged in orbitals according to the pattern

(A) $s^6p^6d^6$
(B) $s^2p^6d^{10}$
(C) $s^2d^6f^{10}$
(D) $s^2p^6f^{10}$
(E) $s^2p^2f^{14}$

GO ON TO THE NEXT PAGE

28. Which of the following molecules is a saturated hydrocarbon?

(A) C_3H_8
(B) C_2H_4
(C) C_4H_6
(D) CH_3OH
(E) CH_3COOH

29. A liter of hydrogen is at 5.0°C temperature and under 640. torr pressure. If the temperature were raised to 60.0°C and the pressure decreased to 320. torr, how would the liter volume be modified?

(A) $1\,L \times \dfrac{5.0}{60.} \times \dfrac{640.}{32.}$

(B) $1\,L \times \dfrac{60.}{5.0} \times \dfrac{320.}{640.}$

(C) $1\,L \times \dfrac{278.}{333.} \times \dfrac{640.}{320.}$

(D) $1\,L \times \dfrac{333.}{278.} \times \dfrac{640.}{320.}$

(E) $1\,L \times \dfrac{333.}{278.} \times \dfrac{320.}{640.}$

30. Of the following statements about the number of subatomic particles in an ion of $^{32}_{16}S^{2-}$, which is (are) true?

I. 16 protons
II. 14 neutrons
III. 18 electrons

(A) II only
(B) III only
(C) I and II only
(D) I and III only
(E) I, II, and III

31. The most reactive metallic elements are found in

(A) the upper right corner of the periodic chart
(B) the lower right corner of the periodic chart
(C) the upper left corner of the periodic chart
(D) the lower left corner of the periodic chart
(E) the middle of the periodic chart, just beyond the transition elements

32. If 1 mole of each of the following substances was dissolved in 1,000 grams of water, which solution would have the highest boiling point?

(A) NaCl
(B) KCl
(C) $CaCl_2$
(D) $C_6H_{10}O_5$
(E) $C_{12}H_{22}O_{11}$

33. A tetrahedral molecule, XY_4, would be formed if X were using the orbital hybridization

(A) p^2
(B) s^2
(C) sp
(D) sp^2
(E) sp^3

34. In the following reaction, how many liters of SO_2 at STP will result from the complete burning of pure sulfur in 8 liters of oxygen?

$$S(s) + O_2(g) \rightarrow SO_2(g)$$

(A) 1
(B) 4
(C) 8
(D) 16
(E) 32

GO ON TO THE NEXT PAGE

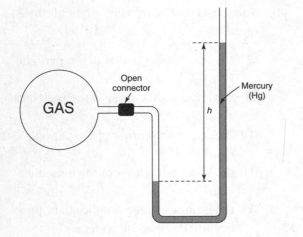

35. In the above laboratory setup to measure the pressure of the confined gas, what will be true concerning the calculated pressure on the gas?

    (A) The gas pressure will be the same as the atmospheric pressure.
    (B) The gas pressure will be less than the atmospheric pressure.
    (C) The gas pressure will be greater than the atmospheric pressure.
    (D) The difference in the height (h) of mercury levels is equal to the pressure of the gas.
    (E) The height (h) of mercury has no effect on the pressure calculation since the column of mercury is only used to enclose the gas volume.

36. Which of the following changes in the experiment shown in question 35 would cause the pressure in the glass container to vary from that shown?

    (A) Use a U-tube of a greater diameter and maintain the height of mercury.
    (B) Increase the temperature of gas in the tube.
    (C) Increase the length of the upper portion of the right side of tubing.
    (D) Use a U-tube of a smaller diameter and maintain the height of mercury.
    (E) Replace the flask with one that has the same volume but has a flat bottom.

37. Which of the following can be classified as amphoteric?

    (A) Na_3PO_4
    (B) HCl
    (C) $NaOH$
    (D) HSO_4^-
    (E) $C_2O_4^{2-}$

38. Standard conditions (STP) are

    (A) 0°C and 2 atm
    (B) 32°F and 76 torr
    (C) 273 K and 760 mm Hg
    (D) 4°C and 7.6 cm Hg
    (E) 0 K and 760 mm Hg

39. Laboratory results showed the composition of a compound to be 75% carbon and 25% hydrogen. What is the empirical formula of the compound?

    (A) CH_4
    (B) C_2H_{10}
    (C) C_2H_5
    (B) CH
    (E) C_4H

40. What is the percentage composition of sulfur in sulfur dioxide, SO_2?

    (A) 43%
    (B) 50%
    (C) 54%
    (D) 69%
    (E) 74%

41. How many moles of hydrogen gas can be produced from the following reaction if 65 grams of zinc and 36.5 grams of HCl are present in the reaction?

    $$Zn(s) + 2HCl(aq) \rightarrow ZnCl_2(aq) + H_2(g)$$

    (A) 0.50
    (B) 1.0
    (C) 3.6
    (D) 7.0
    (E) 58

GO ON TO THE NEXT PAGE

Questions 42 and 43 refer to the following setup.

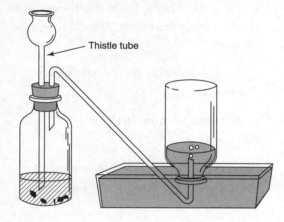

Thistle tube

42. The following statements were recorded while preparing carbon dioxide gas in the laboratory, as shown above. Which one involves an interpretation of the data rather than an observation?

(A) No liquid was transfered from the reaction bottle to the beaker.
(B) The quantity of solid minerals decreased.
(C) The cloudiness in the bottle of limewater on the right was caused by the product of the reaction of the colorless gas and the limewater.
(D) The bubbles of gas rising from the mineral remained colorless throughout the experiment.
(E) There was a 4°C rise in temperature in the reaction vessel during the experiment.

43. The previous laboratory setup can be used to prepare which of the following?

I. $CO_2(g)$
II. $H_2(g)$
III. $O_2(g)$

(A) I only
(B) III only
(C) I and III only
(D) II and III only
(E) I, II, and III

$$^2_1H + ^3_1H ® ^1_0n + \text{_____}$$

44. The missing product in the nuclear reaction represented above is

(A) ^1_1H
(B) ^3_2He
(C) ^4_2He
(D) ^4_3Li
(E) ^5_3Li

45. Which of the following is (are) true regarding the aqueous dissociation of HCN, $K_a = 4.9 \times 10^{-10}$, at 25°C?

I. At equilibrium, $[H^+] = [CN^-]$.
II. At equilibrium, $[H^+] > [CN^-]$.
III. HCN(aq) is a strong acid.

(A) I only
(B) II only
(C) I and II only
(D) I and III only
(E) I, II, and III

46. This question pertains to the reaction represented by the following equation:

$$2NO(g) + O_2(g) \rightleftharpoons 2NO_2(g) + 150 \text{ kJ}$$

Suppose that 0.8 mole of NO is converted to NO_2 in the above reaction. What amount of heat will be evolved?

(A) 30 kJ
(B) 60 kJ
(C) 80 kJ
(D) 130 kJ
(E) 150 kJ

47. How does a Brønsted-Lowry acid differ from its conjugate base?

(A) The acid has one more proton.
(B) The acid has one less proton.
(C) The acid has one more electron.
(D) The acid has one less electron.
(E) The acid has more than one additional proton.

GO ON TO THE NEXT PAGE

48. Two containers having 1 mole of hydrogen gas and 1 mole of oxygen gas, respectively, are opened. What will be the ratio of the rate of effusion of the hydrogen to that of the oxygen?

    (A) $\sqrt{2}:1$
    (B) $4:1$
    (C) $8:1$
    (D) $16:1$
    (E) $\sqrt{32}:1$

49. A molecule in which the electron configuration is a resonance hybrid is

    (A) SO_2
    (B) C_2H_6
    (C) Cl_2
    (D) HBr
    (E) NaCl

50. What is the pH of a solution in which the $[OH^-]$ is 1.0×10^{-4}?

    (A) -4
    (B) $+4$
    (C) $+7$
    (D) -10
    (E) $+10$

51. If 0.365 grams of hydrogen chloride is dissolved to make 1 liter of solution, the pH of the solution is

    (A) 0.00100
    (B) 0.0100
    (C) 1.00
    (D) 2.00
    (E) 12.000

52. In the laboratory, a sample of hydrated salt was heated at 110°C for 30 minutes until all the water was driven off. The data were as follows:

    Mass of the hydrate before heating = 250 grams

    Mass of the salt after heating = 160 grams

    From these data, what was the percent of water by mass in the original sample?

    (A) 26.5
    (B) 36
    (C) 47
    (D) 56
    (E) 90

53. Which of the following oxides dissolves in water to form an acidic solution?

    (A) Na_2O
    (B) CaO
    (C) Al_2O_3
    (D) ZnO
    (E) SO_3

54. In the laboratory, 20.0 milliliters of an aqueous solution of calcium hydroxide, $Ca(OH)_2$, was used in a titration. A drop of phenolphthalein was added to it to indicate the end point. The solution turned colorless after 20.0 milliliters of a standard solution of 0.050 M HCl solution was added. What was the molarity of the $Ca(OH)_2$?

    (A) 0.010 M
    (B) 0.025 M
    (C) 0.50 M
    (D) 0.75 M
    (E) 1.0 M

GO ON TO THE NEXT PAGE

55. Which of the following reactions will NOT occur spontaneously?

(A) $Zn(s) + 2HCl(aq) \rightarrow ZnCl_2(aq) + H_2(g)$
(B) $CaCO_3(s) + 2HCl(aq) \rightarrow CaCl_2(aq) + H_2O(aq) + CO_2(g)$
(C) $Ag^+(aq) + HCl(aq) \rightarrow AgCl(s) + H^+(aq)$
(D) $Cu(s) + 2H^+(aq) \rightarrow Cu^{2+}(aq) + H_2(g)$
(E) $H_2SO_4(aq) + 2NaOH(aq) \rightarrow Na_2SO_4(aq) + 2H_2O(aq)$

56. For a laboratory experiment, a student placed sodium hydroxide crystals on a watch glass, assembled the titration equipment, and prepared a solution of 0.10 M sulfuric acid. Then he weighed 4 grams of sodium hydroxide and added it to enough water to make 1 liter of solution. What might be a source of error in the results of the titration?

(A) Some sulfuric acid evaporated.
(B) The sulfuric acid became more concentrated.
(C) The NaOH solution gained weight, thus increasing its molarity.
(D) The NaOH crystals gained H_2O weight, thus making the solution less than 0.1 M.
(E) The evaporation of sulfuric acid solution countered the absorption of H_2O by the NaOH solution.

57. If 60. grams of NO is reacted with sufficient O_2 to form NO_2 that is removed during the reaction, how many grams of NO_2 can be produced? (Molar masses: NO = 30. g/mol, NO_2 = 46. g/mol)

(A) 46.
(B) 60.
(C) 92.
(D) 120
(E) 180

58. Based on the information shown, each of the following equations represents a reaction in which the change in entropy, ΔS, is positive EXCEPT

(A) $CaCO_3(s) \rightarrow CaO(s) + CO_2(g)$
(B) $Zn(s) + 2H^+(aq) \rightarrow H_2(g) + Zn^{2+}(aq)$
(C) $2C_2H_6(g) + 7O_2(g) \rightarrow 4CO_2(g) + 6H_2O(g)$
(D) $NaCl(s) \rightarrow Na^+(aq) + Cl^-(aq)$
(E) $N_2(g) + 3H_2(g) \rightarrow 2NH_3(g)$

59. $Cl_2(g) + 2Br^-(aq) \rightarrow$?

When 1 mole of chlorine gas reacts completely with excess KBr solution, as shown above, the products obtained are

(A) 1 mol of Cl^- ions and 1 mol of Br^-
(B) 1 mol of Cl^- ions and 2 mol of Br^-
(C) 1 mol of Cl^- ions and 1 mol of Br_2
(D) 2 mol of Cl^- ions and 1 mol of Br_2
(E) 2 mol of Cl^- ions and 2 mol of Br_2

60. Glucose and water are physically combined. Which of the following statements about the combination is most accurate?

(A) Glucose is a nonpolar solute and will cause the resulting solution to have a higher freezing point compared to the water by itself.
(B) Glucose is a nonpolar solute and will cause the resulting solution to have a lower freezing point compared to the water by itself.
(C) Glucose is a polar solute and will cause the resulting solution to have a higher freezing point compared to the water by itself.
(D) Glucose is a polar solute and will cause the resulting solution to have a lower freezing point compared to the water by itself.
(E) Glucose and water will not form a solution, and so the freezing point of the water will remain unchanged.

GO ON TO THE NEXT PAGE

61. Which K_a value indicates the strongest acid?

    (A) 1.3×10^{-2}
    (B) 6.7×10^{-5}
    (C) 5.7×10^{-10}
    (D) 4.4×10^{-7}
    (E) 1.8×10^{-16}

62. What mass of $CaCO_3$ is needed to produce 11.2 liters of CO_2 at STP when the calcium carbonate is reacted with an excess amount of hydrochloric acid? (Molar masses: $CaCO_3 = 100.$ g/mol, $HCl = 36.5$ g/mol, $CO_2 = 44.0$ g/mol)

    (A) 25.0 g
    (B) 44.0 g
    (C) 50.0 g
    (D) 100. g
    (E) None of the above

63. By experimentation it is found that a saturated solution of $BaSO_4$ at 25°C contains 4.0×10^{-5} mole/liter of Ba^{2+} ions. What is the K_{sp} of the $BaSO_4$?

    (A) 1.6×10^{-4}
    (B) 1.6×10^{-9}
    (C) 1.6×10^{-10}
    (D) 4.0×10^{-10}
    (E) 40×10^{-9}

64. What is the ΔH^0 value for the decomposition of sodium chlorate, given the following information?

    $$NaClO_3(s) \rightarrow NaCl(s) + \frac{3}{2}O_2(g)$$

    (ΔH^0_f values: $NaClO_3(s) = -358$ J/mol, $NaCl(s) = -410$ J/mol, $O_2(g) = 0$ kcal/mol)

    (A) 52 J
    (B) −52 J
    (C) 768 J
    (D) −768 J
    (E) $\frac{3}{2}(768 \text{ J})$

65. To the equilibrium reaction shown below:

    $$AgCl(s) \rightleftharpoons Ag^+(aq) + Cl^-(aq)$$

    a milliliter of concentrated HCl (12 M) is slowly added. Which is the best description of what will occur?

    (A) More salt will go into solution, and the K_{sp} will remain the same.
    (B) More salt will go into solution, and the K_{sp} will increase.
    (C) Salt will come out of the solution, and the K_{sp} will remain the same.
    (D) Salt will come out of the solution, and the K_{sp} will decrease.
    (E) No change in concentration will occur, and the K_{sp} will increase.

66. The oxidation state of manganese in the following reaction changes from what value in the reactants to what value in the products?

    $$HCl + KMnO_4 \rightarrow H_2O + KCl + MnCl_2 + Cl_2$$

    (A) +1 to +2
    (B) +1 to −2
    (C) +5 to −2
    (D) +7 to +2
    (E) +5 to +2

67. Each of the following systems is at equilibrium in a closed container. A decrease in the total volume of each container will increase the number of moles of product(s) for which system?

    (A) $2NH_3(g) \rightleftharpoons N_2(g) + 3H_2(g)$
    (B) $H_2(g) + Cl_2(g) \rightleftharpoons 2HCl(g)$
    (C) $2NO(g) + O_2(g) \rightleftharpoons 2NO_2(g)$
    (D) $CO(g) + H_2O(g) \rightleftharpoons CO_2(g) + H_2(g)$
    (E) $Fe_3O_4(s) + 4H_2(g) \rightleftharpoons 3Fe(s) + 4H_2O(g)$

GO ON TO THE NEXT PAGE

68. Which of the following is the correct and complete Lewis electron-dot diagram for PF_3?

(A) F:P:F
    F

(B) :F : P:F:
    :F :

(C) :F : P:F:
    :F :

(D) :F : P:F:
    :F :

(E) :F : P:F:
    :F :

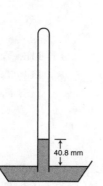

69. Hydrogen gas is collected in a eudiometer tube over water as shown above. The water level inside the tube is 40.8 millimeters higher than that outside. The barometric pressure is 730.0 millimeters Hg. The water vapor pressure at the room temperature of 29°C is found in a handbook to be 30.0 millimeters Hg. What is the pressure of the dry hydrogen?

(A) 659.2 mm Hg
(B) 689.2 mm Hg
(C) 697.0 mm Hg
(D) 740.8 mm Hg
(E) 800.8 mm Hg

70. Which substance is being oxidized in the following reaction?

$$HBr(aq) + Mg(s) \rightarrow MgBr_2(aq) + H_2(g)$$

(A) Hydrogen
(B) Magnesium
(C) Bromine
(D) Hydrogen bromide
(E) None of the substances are being oxidized; it is not a redox reaction.

If you finish before one hour is up, you may go back to check your work or complete unanswered questions.

DIAGNOSTIC TEST

1. **A**	8. **D**	15. **E**	22. **B**
2. **A**	9. **E**	16. **B**	23. **C**
3. **E**	10. **C**	17. **B**	24. **E**
4. **D**	11. **A**	18. **A**	25. **C**
5. **E**	12. **B**	19. **B**	
6. **A**	13. **C**	20. **E**	
7. **D**	14. **D**	21. **C**	

101. **T, T, CE**	105. **F, F**	109. **F, T**	113. **T, F**
102. **T, T, CE**	106. **T, T, CE**	110. **T, T, CE**	114. **T, F**
103. **T, F**	107. **F, F**	111. **T, T, CE**	115. **F, F**
104. **F, T**	108. **F, T**	112. **T, T**	

26. **D**	38. **C**	50. **E**	62. **C**
27. **B**	39. **A**	51. **D**	63. **B**
28. **A**	40. **B**	52. **B**	64. **B**
29. **D**	41. **A**	53. **E**	65. **C**
30. **D**	42. **C**	54. **B**	66. **D**
31. **D**	43. **E**	55. **D**	67. **C**
32. **C**	44. **C**	56. **D**	68. **E**
33. **E**	45. **A**	57. **C**	69. **C**
34. **C**	46. **B**	58. **E**	70. **B**
35. **C**	47. **A**	59. **D**	
36. **B**	48. **B**	60. **D**	
37. **D**	49. **A**	61. **A**	

ANSWERS EXPLAINED

1. **(A)** In the halogen family, the most active nonmetal would be the top element, fluorine, because fluorine replaces other halogens in compounds in single replacement reactions.

2. **(A)** As you proceed down a group, the ionic radius increases as additional energy levels are filled farther from the nucleus. Therefore fluorine, the top element, has the smallest ionic radius.

3. **(E)** Since astatine has the largest atomic radius and its outer electrons are shielded from the protons by a large number of interior electrons, it has the lowest first ionization energy.

4. **(D)** Because of its large molar mass (creating somewhat strong London dispersion forces between iodine molecules), iodine exists in the solid state at 25°C and 1 atmosphere of pressure. Elemental fluorine and chlorine are gases under these conditions while bromine is a liquid. Due to heat produced by its own radioactivity, a sample of elemental astatine has never been viewed.

5. **(E)** The *3d* sublevel has 5 orbitals, or 5 possible values of the quantum number m (-2, -1, 0, $+1$, and $+2$). The Pauli Exclusion Principle allows each orbital a maximum of 2 electrons with opposite spins. So the $3d$ sublevel can contain up to 10 electrons.

6. **(A)** The ground state is the lowest-energy, or most stable state, of an atom. The lithium atom has 3 electrons. In its stable state, the $1s$ sublevel, or lowest-energy orbital, contains a pair of electrons. The Pauli Exclusion Principle allows a maximum of 2 electrons with opposite spins, so the third electron resides in the $2s$ orbital.

7. **(D)** Phosphorus has 15 electrons. In its ground state, the lowest energy sublevels will be occupied. For phosphorus, the first 12 electrons fill the $1s$, $2s$, $2p$, and $3s$ orbitals, respectively. The remaining 3 electrons occupy each of the subsequent $3p$ orbitals (of which there are 3) with 1 electron each (with similar spins) due to Hund's Rule. Because 6 electrons are possible in the $3p$ sublevel, it is considered to be half-filled.

8. **(D)** John Dalton is credited with the basic postulates of atomic theory.

9. **(E)** Gay-Lussac is credited with the statement that, when gases combine, they do so in ratios of small whole numbers that are in relationship to the volumes of the reactants and the volumes of the products under the same conditions.

10. **(C)** Charles is credited with this temperature-volume relationship of gases: $\dfrac{V_1}{V_2} = \dfrac{T_1}{T_2}$ or $V_1 T_2 = V_2 T_1$ named in his honor. These equations show that the volume and temperature of gases are directly related.

11. **(A)** Named in honor of Amedeo Avogadro, Avogadro's number defines the number of particles in a mole, 6.02×10^{23}.

12. **(B)** Boyle is credited with the $P_1 V_1 = P_2 V_2$ relationship of gases that shows pressure and volume are inversely related.

13. **(C)** This includes the functional group of an aldehyde, in which an alkyl radical, R, is bonded to a carbon atom that is doubly bonded to an oxygen atom as well as singly bonded to a hydrogen atom.

14. **(D)** This includes the functional group of an organic acid, in which an alkyl radical, R, is bonded to a carbon atom that is also doubly bonded to an oxygen atom as well as singly bonded to an oxygen atom [that is bonded to a hydrogen atom].

15. **(E)** An ester is the equivalent of an organic salt since it is usually formed from an organic alcohol,

$$R\text{–}OH, \text{ plus an organic acid, } R^*\overset{\displaystyle O}{\overset{\|}{-C}}\text{–}OH.$$

The bonding is $R\text{–}O\underset{\cdot\cdot\cdot\cdot}{\overline{HHO}}\overset{\displaystyle O}{\overset{\|}{-C}}\text{–}R^*$, which gives

$$R\text{–}O\overset{\displaystyle O}{\overset{\|}{-C}}\text{–}R^* + H_2O.\ (R^* \text{ indicates that this alkyl radical need not be the same as R.})$$

16. **(B)** This includes the functional group of an ether. It can be formed by the dehydration of two alcohol molecules. The reaction is

$$R\text{–}O\overline{H + HO}\text{–}R^* \rightarrow R\text{–}O\text{–}R^* + H_2O.$$

17. **(B)** The molecular structure of carbon dioxide is O=C=O, where the oxygens are 180° apart and, although the bonding is polar to the carbon, counteract each other to constitute a nonpolar molecule.

18. **(A)** A hydrogen gas molecule is diatomic; it has 2 mol of atoms in each mole of molecules, represented by H_2.

19. **(B)** Each CO_2 has three atoms per molecule; hence the expression can represent 3 mol of atoms in 1 mol of molecules.

20. **(E)** With complete ionization $H_2SO_4 \rightarrow 2H^+ + SO_4^{2-}$, or 3 mol of ions per mole of H_2SO_4.

21. **(C)** The expression $2N_2O$ represents two triatomic molecules or 6 mol of atoms.

22. **(B)** The ^{16}O and ^{17}O are isotopes of the same element because they do not have the same number of neutrons. This difference in the number of neutrons shows up in the mass number (superscript to the left of the chemical symbol). The mass number of an element is the sum of the protons and neutrons in the nucleus.

23. **(C)** In NH_4Cl, the covalently bonded NH_4^+ ion is ionically bonded to the Cl^- ion. In NH_4NO_3, the covalently bonded NH_4^+ ion is ionically bonded to the covalently bonded NO_3^- ion. Each of these polyatomic ions contains covalent bonds within their structure where the electrons are shared equally.

24. **(E)** The O_2 and O_3 are allotropes because they are different molecular forms of the same element in the same state. They have different structures due to different bond arrangements.

25. **(C)** Strong electrolytes will almost completely dissociate in aqueous solutions. Both NH_4Cl and NH_4NO_3 are ionic compounds that dissociate to release the ammonium cation and either the Cl^- or the NO_3^- anion.

101. **(T, T, CE)** A catalyst can accelerate a chemical reaction by lowering the activation energy required for the reaction to occur.

102. **(T, T, CE)** Sodium chloride is an ionic substance and when molten is a good electrical conductor because in the molten state the ions are free to move and the property of electrical charge requires mobile charged particles.

103. **(T, F)** Ice is less dense than liquid water. Water molecules, however, are polar, not nonpolar, and water expands as these molecules arrange themselves into a crystal lattice.

104. **(F, T)** The statement that isotopes of the same element have the same mass number is false. Isotopes of the same element have the same number of protons but vary in the number of neutrons in the nucleus. Therefore, isotopes have the same atomic number but different mass numbers.

105. **(F, F)** The percent by mass of Ca in calcium citrate is about 24%. This is estimated from the fraction $\frac{120}{500}$, which is close to 25% (or $\frac{125}{500}$). The numerator part of the fraction is found by recognizing that the molar mass of calcium is 40 g and then tripling it due to the subscript of 3 on the calcium in the formula given. The denominator part of the fraction is found by rounding the molar mass of the compound, 498 g, to 500. By using similar thinking, the percent by mass of calcium in calcium carbonate is found to be 40%. Since the sample size is 1 gram for each compound, the mass of calcium in the calcium citrate is only 0.24 g. In the calcium carbonate, though, the mass of calcium is 0.40 g. These values make the first statement false. Additionally, there are fewer Ca atoms in 1 mole of calcium carbonate ($CaCO_3$) than in 1 mole of calcium citrate ($Ca_3(C_6H_5O_7)_2$) because each mole of calcium citrate contains 3 times more Ca than each mole of calcium carbonate. This makes statement II false as well.

106. **(T, T, CE)** The reaction is:

$$C + O_2 \rightarrow CO_2$$

and shows 1 mol or 12.0 g of carbon produces 1 mol or 22.4 L of CO_2 at STP. Then

$$2 \cancel{L CO}_2 \times \frac{12.0 \, g \, C}{22.4 \, \cancel{L CO}_2} = 1.07 \, g \, C$$

So the statement is correct, and the assertion explains it correctly.

107. **(F, F)** A reaction at equilibrium has reached a point where the forward and reverse reactions are occurring at equal rates. The concentrations of the reactants and products, however, are not necessarily equal and are described by the K_{eq} value at the temperature of the reaction.

108. **(F, T)** The oxidation state of oxygen in most situations is –2. The oxidation state of Mn in MnO_4^- then would be +7 because the addition of the oxidation states in polyatomic ions must equal the overall charge of the ion.

109. **(F, T)** The value pH = 5 can be expressed as

$$[H_3O^+] = 1 \times 10^{-5} \, \frac{mol}{L}$$

and pH = 3 as

$$[H_3O^+] = 1 \times 10^{-3} \, \frac{mol}{L}$$

Thus pH = 3 represents a larger concentration of hydronium ions, H_3O^+.

110. **(T, T, CE)** The first statement is true. The change in Gibbs free energy, ΔG, dictates the spontaneity of a process and depends on the enthalpy change, ΔH, and the entropy change, ΔS, from the equation $\Delta G = \Delta H - T\Delta S$. Thus, statement II is also true and explains the first statement.

111. **(T, T, CE)** In the expression for the equilibrium constant of an acid, $[H_3O^+]$ is in the numerator:

$$K_a = \frac{[H_3O^+][Y^-]}{[HY]}$$

As $[H_3O^+]$ decreases in the weak acids, the numerator becomes smaller as the denominator gets larger, therefore giving smaller K_a values.

112. **(T, T)** Both statements are true, but they are not related.

113. **(T, F)** A p orbital does have an "hourglass" or "dumbbell" shape and therefore can be described as having two "lobes." However, each p orbital can only hold a *total* of two electrons and not two electrons in *each* lobe, so the reason is false.

114. **(T, F)** The assertion that the boiling points of H_2S and H_2O are significantly different is true, but the reason is false. Water has the higher degree of hydrogen bonding.

115. **(F, F)** Substances that are miscible *will* dissolve in each other and form a solution, not a suspension. Substances that form suspensions do not dissolve readily in each other.

26. **(D)** In the graph, the first plateau must represent the condensation from gas to liquid because there is a second, lower plateau, which would represent the second change of state, from liquid to solid.

27. **(B)** If the electrons have the same principal energy level, they will fill the s^2, p^6, then the d^{10} level. This progression is from the lowest energy sublevel to the highest, to accommodate 18 electrons.

28. **(A)** When the chain or ring carries the full complement of hydrogen atoms, the hydrocarbon is said to be saturated. The general formula for a saturated chain molecule, called an alkane, is C_nH_{2n+2}. The chain is CH_3–CH_2–CH_3.

29. **(D)** Using the combined gas law equation,

$$\frac{V_1 P_1}{T_1} = \frac{V_2 P_2}{T_2}$$

and solving for V_2, you get

$$V_2 = V_1 \times \frac{T_2}{T_1} \times \frac{P_1}{P_2}$$

which is the answer. (Remember, temperatures must be changed to Kelvins.)

30. **(D)** The mass number, 32, is the total number of neutrons and protons. Since the atomic number, 16, gives the number of protons, $32 - 16 = 16$, or the number of neutrons, statement II is false. Since this ion has a charge of 2^-, it has two more electrons than protons, or 18 electrons. Statements I and III are true.

31. **(D)** The most reactive metallic elements are found in the lower left corner of the Periodic Table.

32. **(C)** The rise in the boiling point depends on the number of particles in solution. One mole of $CaCl_2$ gives 3 mol of ions, more than any other substance listed:

$$CaCl_2 \rightarrow Ca^{2+} + 2Cl^-$$

The number of moles of dissolved particles given by the other substances are as follows:

$$(A) = 2, (B) = 2, (D) = 1, (E) = 1$$

33. **(E)** The sp^3 hybridization produces a tetrahedral configuration. The sp^2 (D) is trigonal planar. The sp (C) is linear. The s (B) and p (A) are the usual orbital structures.

34. **(C)** The reaction is

$$\overset{8\,L}{S(s) + O_2(g)} \rightarrow \overset{x\,L}{SO_2(g)}$$

The given (8 L) and the unknown (x L) are shown above. Since the equation, according to Gay-Lussac's Law, shows that 1 volume of oxygen yields 1 volume of sulfur dioxide, then

$$8 \not{L}O_2 \times \frac{1\,L\,SO_2}{1\,\not{L}O_2} = 8\,L\,SO_2$$

35. **(C)** The fact that the mercury level in the U-tube is higher in the right side of the tube indicates that the pressure in the flask is higher than the atmospheric pressure exerted on the open end of the tube on the right side. If the pressure inside the flask were the same as the atmospheric pressure, the height of the mercury would be the same in both sides of the U-tube.

36. **(B)** The only change listed that would change the pressure of the gas inside the flask is to increase the temperature of the gas. This would cause the pressure to rise.

37. **(D)** An amphoteric substance must be able to be a proton, (H^+), donor, and a proton receiver. The bisulfate ion, HSO_4^-, is the only choice that can either accept a proton and become H_2SO_4, or lose a proton and become the sulfate ion, SO_4^{2-}.

38. **(C)** Standard conditions are 273 K and 760 mm Hg.

39. **(A)** Dividing the percentage value, expressed in grams, by the molar mass of each element in a compound and then comparing those values provides the simplest (or empirical) formula for a compound when expressed in whole numbers. Thus, 75 g $C \div 12$ g/mol ≈ 6 mol C, and 25 g $H \div 1$ g/mol ≈ 25 mol H. Comparison of the mole values discloses a ratio of about $1:4$ carbon to hydrogen or the empirical formula of CH_4.

40. **(B)** The percentage composition can be found by dividing each total atomic mass in the formula by the molar mass of the compound.

$$1S = 32 \text{ g} \quad 32 \text{ g} \div 64 \text{ g} \times 100\% = 50\% \text{ S}$$
$$2O = 32 \text{ g} \quad 32 \text{ g} \div 64 \text{ g} \times 100\% = 50\% \text{ O}$$

41. **(A)** The balanced equation shows that 1 mole of zinc requires 2 moles of hydrochloric acid to completely react. The masses provided show only 1.0 mole of hydrochloric acid (36.5 g) is present to react with the 1.0 mole of zinc (65 g), indicating that the hydrogen chloride is the limiting reactant. Because the balanced equation also shows that 1 mole of hydrogen is produced for every 2 moles of hydrogen chloride reacted, the number of moles of hydrogen gas that can be produced is 0.50.

42. **(C)** All the other statements represent observations because they merely record what was seen.

43. **(E)** The setup is appropriate for the collection of a nonsoluble gas by the displacement of water. All three gases fit this description.

44. **(C)** With the emission of a neutron, the mass number decreases by 1. However, the number of protons in the missing product is 2. The product is helium, ^4_2He.

45. **(A)** The small K_a indicates that this is a weak acid, so statement III is false. When HCN ionizes, it can be shown that $HCN \rightleftharpoons H^+ + CN^-$. This is a molar ratio of $1:1$, so statement I must be true.

46. **(B)** Because the equation shows that 2 mol of NO react to release 150 kJ, the solution is

$$0.8 \text{ mol NO} \times \frac{150 \text{ kJ}}{2 \text{ mol NO}} = 60 \text{ kJ}$$

This answer needs to be estimated as no calculator can be used on the test.

47. **(A)** In the Brønsted-Lowry acid-base theory, acids donate protons to bases to turn into their conjugates; therefore, acids have one more proton than their conjugate bases.

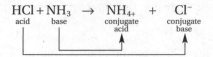

$$\underset{\text{acid}}{\text{HCl}} + \underset{\text{base}}{\text{NH}_3} \rightarrow \underset{\substack{\text{conjugate} \\ \text{acid}}}{\text{NH}_{4+}} + \underset{\substack{\text{conjugate} \\ \text{base}}}{\text{Cl}^-}$$

48. **(B)** According to Graham's Law, the rates of effusion of two gases are inversely proportional to the square roots of their molar masses. Since 1 mol O_2 = 32 g and 1 mol H_2 = 2 g

$$\frac{H_2 \text{ rate}}{O_2 \text{ rate}} = \frac{\sqrt{32}}{\sqrt{2}} \text{ (dividing the numerator by the denominator gives you)} = \sqrt{\frac{16}{1}} = \frac{4}{1}$$

Therefore, the effusion rate of hydrogen is four times faster than that of oxygen.

49. **(A)** The structure of SO_2 is a resonance hybrid, shown as these resonance structures:

$$S{\overset{\text{O}}{\underset{\text{O}}{}}} \leftrightarrow S{\overset{\text{O}}{\underset{\text{O}}{}}}$$

50. **(E)** The K_w of water is

$$K_w = [H^+][OH^-] = 10^{-14}$$
$$\text{If } [OH^-] = 1.0 \times 10^{-4}, \text{then}$$
$$[H^+] = \frac{10^{-14}}{10^{-4}} = 10^{-10}$$

and

$$pH = -\log [H^+] = 10$$

51. **(D)** $0.365 \text{ g } \cancel{HCl} \times \dfrac{1 \text{ mol HCl}}{36.5 \text{ g } \cancel{HCl}} = 0.0100 \text{ mol HCl}$

$0.0100 \text{ mol HCl} \rightarrow 0.0100 \text{ mol } H^+ + 0.0100 \text{ mol } Cl^-$

If $[H^+] = 0.0100 = 1.00 \times 10^{-2}$ mol/L, then pH = 2.00

52. **(B)**

$$\begin{array}{r} \text{Hydrate mass before heating} = 250.\,\text{g} \\ -\ \text{Salt mass after heating} = 160.\,\text{g} \\ \hline \text{Water loss} = \ \ 90.0\,\text{g} \end{array}$$

$$\frac{90.0 \text{ g mass loss}}{250. \text{ g original mass}} \times 100\% = 36\% \text{ by mass}$$

This value is easily estimated as "a little" less than 40% because 100 divided by 250 is 0.40 (or $\frac{2}{5}$). Remember that no calculator can be used while taking the test.

53. **(E)** Because metallic oxides are basic anhydrides, the only oxide that can form an acid is SO_3. The reaction is as follows

$$SO_3 + H_2O \rightarrow H_2SO_4 \text{ (sulfuric acid)}$$

54. **(B)** In the titration, the reaction is:

$$2HCl + Ca(OH)_2 \rightarrow CaCl_2 + 2H_2O$$

The acid to base ratio is $2:1$, or (moles acid used) = 2(moles base used), so $M_aV_a = 2M_bV_b$, where M is the molarity and V is the volume expressed in liters, then

$$M_b = \frac{M_aV_a}{2V_b}$$

$$M_b = \frac{0.05\,M \times 0.02\,L}{2(0.02\,L)} = 0.025\,M$$

55. **(D)** All the reactions are spontaneous except (D), which will not occur because copper is a less active metal than hydrogen.

56. **(D)** Sodium hydroxide is hygroscopic and will attract water to its surface. This water will influence its mass; consequently there will be less sodium hydroxide in the mass used.

57. **(C)** $\overset{60.\,g}{2NO} + O_2 \rightarrow \overset{x\,g}{2NO_2}$

This is a stoichiometry problem.

$$60.\ g\,NO \times \frac{1\ mol\ NO}{30\ g\ NO} = 2\ mol\ NO$$

Using the equation coefficients gives

$$2\ mol\ NO \times \frac{2\ mol\ NO_2}{2\ mol\ NO} = 2\ mol\ NO_2$$

then

$$2\ mol\ NO_2 \times \frac{46\ g\ NO_2}{1\ mol\ NO_2} = 92\ g\ NO_2$$

58. **(E)** Since in this reaction 4 volumes of gases are forming 2 volumes of product, the randomness of the system is decreasing. Therefore, entropy is decreasing and ΔS is negative. In all the other reactions randomness is increasing.

59. **(D)** The reaction is

$$Cl_2(g) + 2Br^-(aq) \rightarrow 2Cl^- + Br_2(aq)$$

Here 2 mol of chloride ions and 1 mol of Br_2 molecules are produced.

60. **(D)** Glucose, $C_6H_{12}O_6$, is a polar molecule (as are all sugars) and dissolves in water because water is also a polar molecule. A good rule of thumb is that "like dissolves like." Glucose is also a nonvolatile solid. So glucose lowers the vapor pressure of the solution which, in turn, lowers the freezing point of the solution. The freezing point is a colligative property.

61. **(A)** $HB \rightleftharpoons H^+ + B^-$

Since $[H^+]$ is in the numerator of K_a:

$$K_a = \frac{[H^+][B^-]}{[HB]}$$

The stronger the acid, the greater are $[H^+]$ and the K_a value. Of the choices given, 1.3×10^{-2} is the largest.

62. **(C)** $\overset{x\,g}{CaCO_3} + 2HCl \rightarrow CaCl_2 + H_2O + \overset{11.2\,L}{CO_2}$

This is a mass-volume problem.

Using dimensional analysis gives

$$11.2 \text{ L } CO_2 \times \frac{1 \text{ mol } CO_2}{22.4 \text{ L } CO_2} = 0.5 \text{ mol } CaCO_2$$

$$0.5 \text{ mol } CO_2 \times \frac{1 \text{ mol } CaCO_3}{1 \text{ mol } CO_2} = 0.5 \text{ mol } CaO_3$$

$$0.5 \text{ mol } CaCO_3 \times \frac{100 \text{ g } CaCO_3}{1 \text{ mol } CaCO_3} = 50 \text{ g } CaCO_3$$

63. **(B)** $K_{sp} = [Ba^{2+}][SO_4^{2-}]$

If $[Ba^{2+}] = 4.0 \times 10^{-5}$, then $[SO_4^{2-}]$ must also equal the same amount, so

$$K_{sp} = [4.0 \times 10^{-5}][4.0 \times 10^{-5}] = 16 \times 10^{-10} \text{ or } 1.6 \times 10^{-9}.$$

64. **(B)** $\Delta H^0_{reaction} = \Delta H^0_f \text{ (products)} - \Delta H^0_f \text{ (reactants)}$

$\Delta H^0_{reaction} = -410. \text{ J} - (-358. \text{ J})$

$\Delta H^0_{reaction} = -52. \text{ J}$

65. **(C)** Because the HCl solution will add to the chloride ion concentration, according to Le Châtelier's Principle the equilibrium will shift in the direction to reduce this disturbance, so the K_{sp} will remain the same but salt will come out of solution. This process:

$$AgCl \rightleftharpoons Ag^+ + Cl^-$$

will continue until the K_{sp} is reestablished. This phenomenon is called the "common ion effect."

66. **(D)** The oxidation state of Mn in MnO_4^- is +7 as the oxidation state of oxygen is generally −2. The oxidation state of Mn in $MnCl_2$ is +2 as the oxidation state of Cl is −1. Oxidation states of the elements in polyatomic ions or molecules must add up to the overall charge of that particle.

67. **(C)** A decrease in volume will cause the equilibrium to shift in the direction that has less volume(s) of gas(es). In every case except (C) this is the reverse reaction, which decreases the product. The coefficients give the volume relationships.

68. **(E)** The central P has 5 valence electrons. Of these, 2 are paired. The remaining 3 valence electrons each covalently bond with one of the 3 F atoms to fill the outer energy level. The Lewis electron-dot diagram is

$$\ddot{\underset{..}{F}}:\ddot{\underset{..}{P}}:\ddot{\underset{..}{F}}:$$
$$:\ddot{\underset{..}{F}}:$$

69. **(C)** Change the water height to the equivalent Hg height:

$$40.8 \; \cancel{mm \, H_2O} \times \frac{1 \; mm \, Hg}{13.6 \; \cancel{mm \, H_2O}}$$

$= 3.00$ mm Hg

Adjust for the difference in height to get the gas pressure. Pressure on the gas is

730.0 mm Hg – 3.00 mm Hg
= 727.0 mm Hg.

Vapor pressure of H_2O at 29°C accounts for 30.0 mm Hg pressure. Therefore,

727.0 mm Hg – 30.0 mm Hg
= 697.0 mm Hg

70. **(B)** The oxidation state of an element is zero. The oxidation state of a monatomic ion in a compound is the charge of the ion in the compound. Therefore, the oxidation state of magnesium in the reactants is 0 and the oxidation state of magnesium in the products is +2. Since the oxidation state of magnesium is increasing, magnesium is being oxidized.

CALCULATING YOUR SCORE

Your score on the diagnostic test can now be computed manually. The actual test will be scored by machine, but the same method is used to arrive at the raw score. You get one point for each correct answer. For each wrong answer, you lose one-fourth of a point. Questions that you omit or for which you have indicated more than one answer are not counted. On your answer sheet, mark all correct answers with a "C" and all incorrect answers with an "X."

Determining Your Raw Test Score

Total the number of correct answers you have recorded on your answer sheet. It should be the same as the total of all the numbers you place in the block in the lower left corner of each area of the Subject Area summary in the next section.

A. Enter the total number of correct answers here: _____
Now count the number of wrong answers you recorded on your answer sheet.
B. Enter the total number of wrong answers here: _____
Multiply the number of wrong answers in B by 0.25.
C. Enter that product here: _____
Subtract the result in C from the total number of right answers in A.
D. Enter the result of your subtraction here: _____
E. Round the result in D to the nearest whole number: _____.
This is your raw test score.

Conversion of Raw Scores to Scaled Scores

Your raw score is converted by the College Board into a scaled score. The College Board scores range from 200 to 800. This conversion is done to ensure that a score earned on any edition of a particular SAT Subject Test in Chemistry is comparable to the same scaled score earned on any other edition of the same test. Because some editions of the tests may be slightly easier or more difficult than others, scaled scores are adjusted so that they indicate the same level of performance regardless of the edition of the test taken and the ability of the group that takes it. Consequently, a specific raw score on one edition of a particular test will not necessarily translate to the same scaled score on another edition of the same test.

Since the practice tests in this book have no large population of scores with which they can be scaled, scaled scores can only be approximated.

Results from previous SAT Chemistry tests appear to indicate that the conversion of raw scores to scaled scores GENERALLY follows this pattern:

Raw Score	Scaled Score	Raw Score	Scaled Score
85–82	800–800	30–25	540–520
81–75	790–760	25–20	520–490
75–70	760–740	20–15	490–460
70–65	740–710	15–10	460–430
65–60	710–690	10–5	430–400
60–55	690–670	5–0	400–370
55–50	670–640	0 to –5	370–340
50–45	640–620	–5 to –10	340–310
45–40	620–590	–10 to –15	310–290
40–35	590–570	–15 to –20	290–270
35–30	570–540	–20 or lower	270–200

Note that this scale provides only a general idea of what a raw score may translate into on a scaled score range of 800–200. Scaling on every test is usually slightly different. Some students who have taken the SAT Subject Test in Chemistry after using this book have reported that they have scored slightly higher on the SAT test than on the practice tests in this book. They all reported that preparing well for the test paid off in a better score!

DIAGNOSING YOUR NEEDS

This section will help you to diagnose your need to review the various categories tested by the SAT Subject Test in Chemistry.

After taking the diagnostic test, check your answers against the correct ones. Then fill in the chart below. In the space under each question number, place a check (✓) if you answered that question correctly.

Next, total the checks for each section and insert the number in the designated block. Now do the arithmetic indicated, and insert your percentage for each area.

Subject Area*	(✔) Questions Answered Correctly										
I. **Atomic Theory and Structure,** including periodic relationships	1	2	3	5	6	7	8	27	104	30	44

☐ No. of checks ÷ 11 × 100 = _____%

Subject Area										
II. **Chemical Bonding and Molecular Structure**	17	20	22	23	24	113	114	33	68

☐ No. of checks ÷ 9 × 100 = _____%

III. **States of Matter and Kinetic Molecular Theory of Gases**	9	10	12	26	29	34	38	48

☐ No. of checks ÷ 8 × 100 = _____%

IV. **Solutions,** including concentration units, solubility, and colligative properties	105	115	32	54	60

☐ No. of checks ÷ 5 × 100 = _____%

V. **Acids and Bases**	109	111	37	47	50	51	53	61

☐ No. of checks ÷ 8 × 100 = _____%

VI. **Oxidation-Reduction**	102	108	55	59	66	70

☐ No. of checks ÷ 6 × 100 = _____%

VII. **Stoichiometry**	11	18	19	106	112	28	39	40	41	57	62

☐ No. of checks ÷ 11 × 100 = _____%

VIII. **Reaction Rates**	101	107

☐ No of checks ÷ 2 × 100 = _____%

IX. **Equilibrium**	45	63	65	67

☐ No. of checks ÷ 4 × 100 = _____%

X. **Thermodynamics:** energy changes in chemical reactions, randomness, and criteria for spontaneity	110	46	58	64

☐ No. of checks ÷ 4 × 100 = _____%

The subject areas have been expanded to identify specific areas in the text.

DIAGNOSTIC TEST

Subject Area*	(✔) Questions Answered Correctly							
XI. **Descriptive Chemistry:** physical and chemical properties of elements and their familiar compounds; organic chemistry; periodic properties	4	13	14	15	16	21	25	103
						31	36	49
☐ No. of checks ÷ 11 × 100 = _____%								
XII. **Laboratory:** equipment, procedures, observations, safety, calculations, and interpretation of results	35	42	43	52	56	69		
☐ No of checks ÷ 6 × 100 = _____%								

The subject areas have been expanded to identify specific areas in the text.

PLANNING YOUR STUDY

The percentages give you an idea of how you have done on the various major areas of the test. Because of the limited number of questions on some parts, these percentages may not be as reliable as the percentages for parts with larger numbers of questions. However, you should now have at least a rough idea of the areas in which you have done well and those in which you need more study. (There are four more practice tests in the back of this book, which may be used in a diagnostic manner as well.)

Start your study with the areas in which you are the weakest. The corresponding chapters are indicated in the table below.

Subject Area	Chapters to Review
I. Atomic Theory and Structure, including periodic relationships	2
II. Chemical Bonding and Molecular Structure	3, 4
III. States of Matter and Kinetic Molecular Theory of Gases	5, 7
IV. Solutions, including concentration units, solubility, and colligative properties	7
V. Acids and Bases	11
VI. Oxidation-Reduction	12
VII. Stoichiometry	5, 6
VIII. Reaction Rates	9
IX. Equilibrium	10
X. Thermodynamics: energy changes in chemical reactions, randomness, and criteria for spontaneity	8
XI. Descriptive Chemistry: physical and chemical properties of elements and their familiar compounds; organic chemistry; periodic properties	1, 2, 13, 14
XII. Laboratory: equipment, procedures, observations, safety, calculations, and interpretation of results	All lab diagrams, 15

After you have spent some time reviewing your weaker areas, plan a schedule of work that spans the 6 weeks before the test. Unless you set up a regular study pattern and goals, you probably will not prepare sufficiently.

The following schedule provides such a plan. Note that weekends are left free, and the time spans are held to 1- or 2-hour blocks. This will be time well spent!

	Monday	Tuesday	Wednesday	Thursday	Friday
First Week:	Ch. 1: 1 hr	Ch. 2: 2 hr	Ch 2: 1 hr.	Ch. 3: 2 hr	Ch. 3: 1 hr
Second Week:	Ch. 4: 1 hr	Ch. 4: 1 hr	Ch. 5: 2 hr	Ch. 5: 1 hr	Ch. 6: 1 hr
Third Week:	Ch. 6: 2 hr	Ch. 7: 2 hr	Ch. 8: 1 hr	Ch. 9: 2 hr	Ch. 10: 1 hr
Fourth Week:	Ch. 10: 1 hr	Ch. 11: 2 hr	Ch. 12: 2 hr	Ch. 13: 1 hr	Ch. 14: 2 hr
Fifth Week:	Ch. 16: 1 hr	Take Practice Test 1: 1 hr Study Ans.: 1 hr	Review weakest areas: 1 hr	Take Practice Test 2: 1 hr Study Ans.: 1 hr	Review weakest areas: 1 hr
Sixth Week:	Take Practice Test 3: 1 hr Study Ans.: 1 hr	Review weakest areas: 1 hr	Take Practice Test 4: 1 hr Study Ans.: 1 hr	Review weakest areas: 1 hr	Take 1 or 2 of the online practice tests: 1 to 2 hrs *Go to bed early.*

FINAL PREPARATION—THE DAY BEFORE THE TEST

The day before the test, review one of the practice tests you have already taken. Study again the directions for each type of question. Long hours of study at this point will probably only heighten your anxiety, so just look over the answer section of the practice test and refer to any chapter in the book if you need more information. This type of limited, relaxed review will probably make you feel more comfortable and better prepared.

Know the route to the test center and the entrance indicated. Then get together the materials that you will need. They are:

- Your admission ticket. Check the time your admission ticket specifies for arrival.
- Your identification. (You will not be admitted without some type of positive identification such as a student I.D. card with picture or a driver's license.)
- Two No. 2 pencils with erasers.
- Watch (without an audible alarm).

! Cell phone use is prohibited in both the test center and the testing room! If your cell phone is on, your scores will be canceled!

Note that calculator use is *not* allowed during the SAT Subject Test in Chemistry.
You should also go over this checklist:

A. Plan your activities so that you will have time for a good night's sleep.
B. Lay out comfortable clothes for the next day. You may want to bring a snack.

c. Review the following helpful tips about taking the test:

1. Read the directions carefully.

2. In each group of questions, answer first those that you know. Temporarily skip difficult questions, but mark them in the margin so you can go back if you have time. Keep in mind that an easy question answered correctly counts as much as a difficult one.

3. Avoid haphazard guessing since this will probably lower your score. Instead, guess smart! If you can eliminate one or more of the choices to a question, it will generally be to your advantage to guess which of the remaining answers is correct. Your score will be based on the number right minus a fraction of the number answered incorrectly.

4. Omit questions when you have no idea of how to answer them. You neither gain nor lose credit for questions you do not answer.

5. Keep in mind that you have 1 hour to complete the test, and pace yourself accordingly. If you finish early, go back to questions you skipped.

6. Mark the answer grid clearly and correctly. Be sure each answer is placed in the proper space and within the oval. *Erase all stray marks completely.*

7. Write as much as you like in the test booklet. Use it as a scratch pad. Only the answers on the answer sheet are scored for credit.

D. Set your alarm clock so as to allow plenty of time to dress, eat your usual (or even a better) breakfast, and reach the test center without haste or anxiety.

AFTER THE TEST

After several weeks, most scores will be reported online at *www.collegeboard.org*. A full report will be available to you online a few days later. You can request a paper report. Your score will also be mailed to your high school and to the colleges, universities, and programs that you indicated. The report includes your scores, percentiles, and interpretive information. You can also get your scores—for a fee—by telephone. Call customer service at 866-756-7346 in the United States. From outside the United States, call 212-713-7789.

If your scores are not reported by eight weeks after the test date, definitely contact customer service by telephone or e-mail (*sat@info.collegeboard.org*). The mailing address for comments or questions about the tests is:

The College Board SAT Program
P.O. Box 025505
Miami, FL 33102

PART 2
Review of Major Topics

Introduction to Chemistry

1

These skills are usually tested on the SAT Subject Test in Chemistry. You should be able to...

→ Distinguish types of matter: i.e., elements, mixtures, compounds, and pure substances.
→ Identify chemical and physical properties and changes.
→ Explain how energy is involved in these changes.
→ Identify and use the SI units of measurements.
→ Do mathematical calculations by using scientific notation, dimensional analysis, and proper significant figures.

This chapter will review and strengthen these skills. Be sure to do the Practice Exercises at the end of the chapter.

MATTER

Matter is defined as anything that occupies space and has mass. **Mass** is the quantity of matter that a substance possesses and, depending on the gravitational force acting on it, has a unit of weight assigned to it. Its formula is $w = mg$, where m is the mass of the substance and g is a gravitational constant. Although **weight** then can vary as the gravitational constant does, the mass of the body is a constant and can be measured by its resistance to a change of position or motion. This property of mass to resist a change of position or motion is called **inertia**. Since matter does occupy space, we can compare the masses of various substances that occupy a particular unit volume. This relationship of mass to a unit volume is called the

density of the substance. It can be shown in a mathematical formula as $D = \dfrac{m}{V}$. The unit of

mass (m) commonly used in chemistry is the gram (g), and of volume (V) is the cubic centimeter (cm³), milliliter (mL), or liter (L).

An example of how density varies can be shown by the difference in the volumes occupied by 1 gram of a metal, such as gold, and 1 gram of Styrofoam. Both have the same mass, 1 gram, but the volume occupied by the Styrofoam is much larger. Therefore, the density of the metal will be much larger than that of the Styrofoam. In chemistry, the standard units for density of gases are grams/liter at a standard temperature and pressure. This aspect of the density of gases is discussed in Chapter 6. Basically then, density can be defined as the mass per unit volume.

TIP

Matter occupies space and has mass.

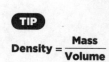

TIP

$$\text{Density} = \frac{\text{Mass}}{\text{Volume}}$$

States of Matter

Matter occurs in three states: solid, liquid, and gas. A **solid** has both a definite size and a definite shape. A **liquid** has a definite volume but takes the shape of the container, and a **gas** has neither a definite shape nor a definite volume. These states of matter can often be changed by the addition or subtraction of heat energy. An example is ice changing to liquid water and finally steam. Phases of matter are discussed in more depth in Chapter 7.

Composition of Matter

Matter can be subdivided into two general categories: pure substances and mixtures. A **pure substance** can be subdivided into the smallest particle that still has the properties of that substance. At that point, if the substance is made up of only one kind of atom, it is called an **element**. Atoms are considered to be the basic building blocks of matter that cannot be easily created nor destroyed. The word **atom** comes from the Greeks and means the smallest possible piece of something. Today, scientists recognize approximately 114 different kinds of atoms, each with its own unique composition. These atoms then are the building blocks of elements when only one kind of atom makes up the substance. If, however, two or more kinds of atoms join together in a definite grouping, this pure substance is called a **compound**. Compounds are made by combining atoms of two or more elements in a definite proportion (or ratio) by mass, according to the **Law of Definite Composition (or Proportions)**. The smallest naturally occurring unit of a compound is called a **molecule** of that compound. A molecule of a compound has a definite shape that is determined by how the atoms are bonded to or combined with each other, as described in Chapter 3. An example is the compound water: it always occurs in a two hydrogen atoms to one oxygen atom relationship. **Mixtures**, however, can vary in their composition.

In general, then:

TIP

Know how to separate mixtures by using their properties.

Mixtures	Pure Substances
1. Composition is indefinite (generally heterogeneous).* (Example: marble) 2. Properties of the constituents are retained. 3. Parts of the mixture react differently to changed conditions.	ELEMENTS 1. Composition is made up of one kind of atom. (Examples: nitrogen, gold, neon) 2. All parts are the same throughout (homogeneous).
*Solutions are mixtures, such as sugar in water, but since the substance, like sugar, is distributed evenly throughout the water, the mixture can be said to be homogeneous.	COMPOUNDS 1. Composition is definite (homogeneous). (Examples: water, carbon dioxide) 2. All parts react the same. 3. Properties of the compound are distinct and different from the properties of the individual elements that are combined in its makeup.

The following chart shows a classification scheme for matter.

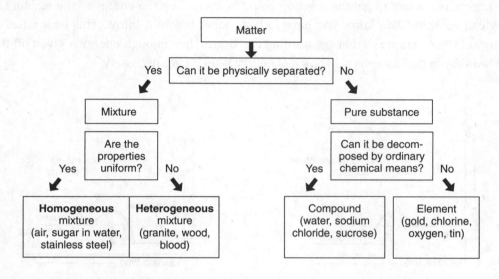

Chemical and Physical Properties

Physical properties of matter are those properties that can usually be observed with our senses. They include everything about a substance that can be noted when no change is occurring in the type of structure that makes up its smallest component. Some common examples are physical state, color, odor, solubility in water, density, melting point, taste, boiling point, and hardness.

Chemical properties are those properties that can be observed in regard to whether or not a substance changes chemically, often as a result of reacting with other substances. Some common examples are: iron rusts in moist air, nitrogen does not burn, gold does not rust, sodium reacts with water, silver does not react with water, and water can be decomposed by an electric current.

Chemical and Physical Changes

The changes matter undergoes are classified as either physical or chemical. In general, a **physical change** alters some aspect of the physical properties of matter, but the composition remains constant. The most often altered properties are form and state. Some examples of physical changes are breaking glass, cutting wood, melting ice, and magnetizing a piece of metal. In some cases, the process that caused the change can be easily reversed and the substance regains its original form. Water changing its state is a good example of physical changes. In the solid state, ice, water has a definite size and shape. As heat is added, it changes to the liquid state, where it has a definite volume but takes the shape of the container. When water is heated above its boiling point, it changes to steam. Steam, a gas, has neither a definite size, because it fills the containing space, nor shape, because it takes the shape of the container.

Chemical changes are changes in the composition and structure of a substance. They are always accompanied by energy changes. If the energy released in the formation of a new structure exceeds the chemical energy in the original substances, energy will be given off, usually in the form of heat or light or both. This is called an **exothermic reaction**. If, however, the new structure needs to absorb more energy than is available from the reactants, the result is an **endothermic reaction**. This can be shown graphically.

 TIP

Physical change does not alter the identity of the substance. Chemical change does.

Notice that in Figures 1 and 2 the term **activation energy** is used. The activation energy is the energy necessary to get the reaction going by increasing the energy of the reactants so they can combine. You know you have to heat paper before it burns. This heat raises the energy of the reactants so that the burning can begin; then enough energy is given off from the burning so that an external source of energy is no longer necessary.

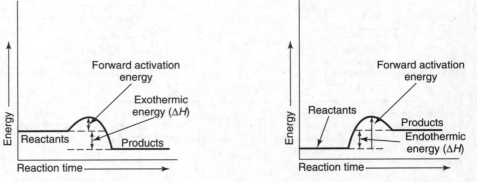

Figure 1. An Exothermic Reaction   **Figure 2.** An Endothermic Reaction

Conservation of Mass

When ordinary chemical changes occur, the mass of the reactants equals the mass of the products. This can be stated another way: In a chemical change, matter can neither be created nor destroyed, but only changed from one form to another. This is referred to as the **Law of Conservation of Matter** (Lavoisier—1785). This law is extended by the Einstein mass-energy relationship, which states that matter and energy are interchangeable (see page 42).

ENERGY

Definition of Energy

The concept of energy plays an important role in all of the sciences. In chemistry, all physical and chemical changes have energy considerations associated with them. To understand how and why these changes happen, an understanding of energy is required.

Energy is defined as the capacity to do work. Work is done whenever a force is applied over a distance. Therefore, anything that can force matter to move, to change speed, or to change direction has energy. The following example will help you understand this definition of energy. When you charge a battery with electricity, you are storing energy in the form of chemical energy. The charged battery has a capacity to do work. If you use the battery to operate a toy car, the stored energy is transformed into mechanical energy that exerts a force on the mechanism that turns the wheels and makes the car move. This process continues until the **charge** or stored energy is completely used. In its uncharged condition, the battery no longer has the capacity to do work.

Work itself is measured in **joules**, and so is energy. In some problems, however, energy may be expressed in **kilocalories**. The relationship between these two units is that 4.18×10^3 joules (J) equals 1 kilocalorie (kcal).

Forms of Energy

Energy may appear in a variety of forms. Most commonly, energy in reactions is evolved as **heat**. Some other forms of energy are **light**, **sound**, **mechanical energy**, **electrical energy**, and **chemical energy**. Energy can be converted from one form to another, as when the heat from burning fuel is used to vaporize water to steam. The energy of the steam is used to turn the wheels of a turbine to produce mechanical energy. The turbine turns the generator armature to produce electricity, which is then available in homes for use as light or heat, or in the operation of many modern appliances.

Two general classifications of energy are **potential energy** and **kinetic energy**. Potential energy is stored energy due to overcoming forces in nature. Kinetic energy is energy of motion. The difference can be illustrated by a boulder sitting on the side of a mountain. It has a high potential energy due to its position above the valley floor. If it falls, however, its potential energy is converted to kinetic energy. This illustration is very similar to the situation of electrons cascading to lower energy levels in the atomic model described in Chapter 2.

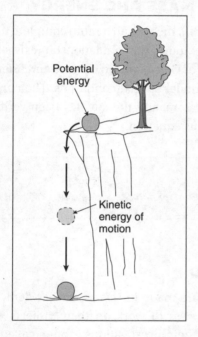

Types of Reactions (Exothermic Versus Endothermic)

When physical or chemical changes occur, energy changes are involved. Change of heat content can be designated as ΔH. The heat content (H) is sometimes referred to as the **enthalpy**. Every system has a certain amount of heat. This changes during the course of a physical or chemical change. The change in heat content, ΔH, is the difference between the heat content of the products and that of the reactants. The equation is:

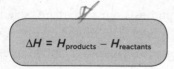

$$\Delta H = H_{products} - H_{reactants}$$

If the heat content of the products is greater than the heat content of the reactants, ΔH is a positive quantity ($\Delta H > 0$) and the reaction is **endothermic**. If, however, the heat content of the products is less than the heat content of the reactants, ΔH is a negative quantity ($\Delta H < 0$) and the reaction is **exothermic**. This relationship is shown graphically in Figures 1 and 2 on page 40. This topic is developed in detail in Chapter 8.

Conservation of Energy

Experiments have shown that energy is neither gained nor lost in physical or chemical changes. This principle is known as the **Law of Conservation of Energy** and is often stated as follows: Energy is neither created nor destroyed in ordinary physical and chemical changes. If the system under study loses energy, the reaction is exothermic and the ΔH is negative. Therefore, the system's surroundings must gain the energy that the system loses so that energy is conserved.

CONSERVATION OF MASS AND ENERGY

With the introduction of atomic theory and a more complete understanding of the nature of both mass and energy, it was found that a relationship exists between these two concepts. Einstein formulated the **Law of Conservation of Mass and Energy**. This states that mass and energy are interchangeable under special conditions. The conditions have been created in nuclear reactors and accelerators, and the law has been verified. This relationship can be expressed by Einstein's famous equation:

$$E = mc^2$$
$$\text{Energy} = \text{Mass} \times (\text{Velocity of light})^2$$

SCIENTIFIC METHOD

Although some discoveries are made in science by accident, in most cases, the scientists involved use an orderly process to work on their projects and discoveries. The process researchers use to carry out their investigations is often called the **scientific method**. It is a logical approach to solving problems by observing and collecting data, formulating a hypothesis, and constructing theories supported by the data. The formulating of a hypothesis consists of carefully studying the data collected and organized to see if a testable statement can be made with regard to the data. The hypothesis takes the form of an "if . . . then" statement. If certain data are true, then a prediction can be made concerning the outcome. The next step is to test the prediction to see if it withstands the experimentation. The hypothesis can go through several revisions as the process continues. If the data from experimentation show that the predictions of the hypothesis are successful, scientists usually try to explain the phenomena by constructing a **model**. The model can be a visual, verbal, or mathematical means of explaining how the data is related to the phenomena.

The stages of this process can be illustrated by the diagram shown below:

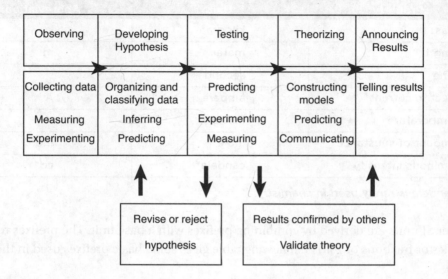

Observing	Developing Hypothesis	Testing	Theorizing	Announcing Results
Collecting data	Organizing and classifying data	Predicting	Constructing models	Telling results
Measuring	Inferring	Experimenting	Predicting	
Experimenting	Predicting	Measuring	Communicating	

Revise or reject hypothesis

Results confirmed by others
Validate theory

MEASUREMENTS AND CALCULATIONS

The student of chemistry must be able to make good observations. Observations are either qualitative or quantitative. Qualitative observations involve descriptions of the nature of the substances under investigation. Quantitative observations involve making measurements to describe the substances under observation. The chemistry student must also be able to use correct measurement terms and the required mathematical skills to solve the problems. The following sections review these topics.

Metric System

It is important that scientists around the world use the same units when communicating information. For this reason, scientists use the modernized metric system, designated in 1960 by the General Conference on Weights and Measures as the International System of Units. This is commonly known as **SI**, an abbreviation for the French name Le Système International d'Unités. It is now the most common system of measurement in the world. There are minor differences between the SI and metric systems. For the most part, the quantities are interchangeable.

Only metric units are used on the SAT test.

The reason SI is so widely accepted is twofold. First, it uses the decimal system as its base. Second, many units for various quantities are defined in terms of units for simpler quantities.

There are seven basic units that can be used to express the fundamental properties of measurement. These are called the SI base units and are shown in the table that follows.

SI Base Units

Property	Unit	Abbreviation
mass	kilogram	kg
length	meter	m
time	second	s
electric current	ampere	A
temperature	kelvin	K
amount of substance	mole	mol
luminous intensity	candela*	cd

*The candela is rarely used in chemistry.

Other SI units are derived by combining prefixes with a base unit. The prefixes represent multiples or fractions of 10. The following table gives some basic prefixes used in the metric system.

Prefixes Used with SI Units

Prefix	Symbol	Meaning	Exponential Notation
exa-	E	1,000,000,000,000,000,000	10^{18}
peta-	P	1,000,000,000,000,000	10^{15}
tera-	T	1,000,000,000,000	10^{12}
giga-	G	1,000,000,000	10^{9}
mega-	M	1,000,000	10^{6}
kilo-	k	1,000	10^{3}
hecto-	h	100	10^{2}
deka-	da	10	10^{1}
—	—	1	10^{0}
deci-	d	0.1	10^{-1}
centi-	c	0.01	10^{-2}
milli-	m	0.001	10^{-3}
micro-	μ	0.000 001	10^{-6}
nano-	n	0.000 000 001	10^{-9}
pico-	p	0.000 000 000 001	10^{-12}
femto-	f	0.000 000 000 000 001	10^{-15}
atto-	a	0.000 000 000 000 000 001	10^{-18}

Commonly Used Prefixes

For an example of how a prefix works in conjunction with the base word, consider the term *kilometer*. The prefix *kilo-* means "multiply the root word by 1,000," so a kilometer is 1,000 meters. By the same reasoning, a millimeter is 1/1,000 of a meter.

Because of the prefix system, all units and quantities can be easily related by some factor of 10. Here is a brief table of some metric unit equivalents.

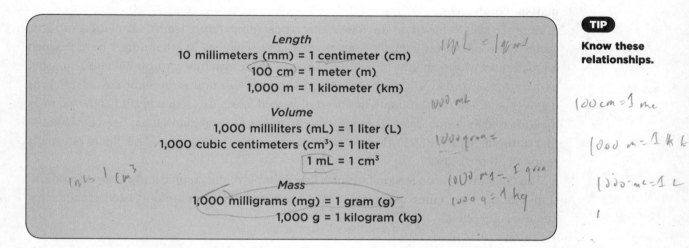

Length
10 millimeters (mm) = 1 centimeter (cm)
100 cm = 1 meter (m)
1,000 m = 1 kilometer (km)

Volume
1,000 milliliters (mL) = 1 liter (L)
1,000 cubic centimeters (cm³) = 1 liter
1 mL = 1 cm³

Mass
1,000 milligrams (mg) = 1 gram (g)
1,000 g = 1 kilogram (kg)

TIP

Know these relationships.

A unit of length, used especially in expressing the length of light waves, is the nanometer, abbreviated as nm and equal to 10^{-9} meter.

The metric system standards were chosen as natural standards. The meter was once described as 1/10,000,000 of the distance between the equator and the North Pole but now is defined as the length of the path traveled by light in a vacuum during a time interval of $1/2.99792458 \times 10^8$ second.

There are some interesting relationships between volume and mass units in the metric system. Because water is most dense at 4°C, the gram was intended to be 1 cubic centimeter of water at this temperature. This means, then, that:

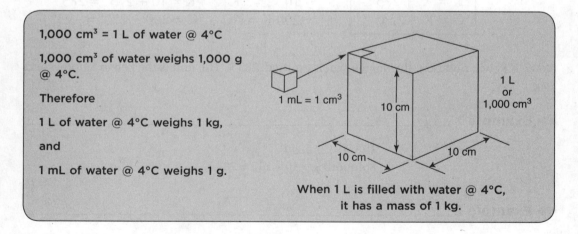

1,000 cm³ = 1 L of water @ 4°C

1,000 cm³ of water weighs 1,000 g @ 4°C.

Therefore

1 L of water @ 4°C weighs 1 kg,

and

1 mL of water @ 4°C weighs 1 g.

1 mL = 1 cm³
10 cm
10 cm
10 cm
1 L or 1,000 cm³

When 1 L is filled with water @ 4°C, it has a mass of 1 kg.

Temperature Measurements

The most commonly used temperature scale in scientific work is the **Celsius** scale. It gets its name from the Swedish astronomer Anders Celsius and dates back to 1742. For a long time it was called the centigrade scale because it is based on the concept of dividing the distance on a thermometer between the freezing point of water and its boiling point into 100 equal markings or degrees.

Another scale is based on the lowest theoretical temperature (called absolute zero). This temperature has never actually been reached, but scientists in laboratories have reached temperatures within about a billionth of a degree above absolute zero. Sir William Thomson, also known as Lord Kelvin, proposed this scale on which a unit is the same size as a Celsius degree but where the zero mark has been displaced lower. Consequently, it is referred to as the **Kelvin** temperature scale. Through experiments and calculations, it has been determined that absolute zero is 273.15 degrees below zero on the Celsius scale. This figure is usually rounded off to $-273°C$.

The diagram and conversion formulas that follow give the graphic and algebraic relationships among three temperature scales: the Celsius and Kelvin, commonly used in chemistry, and the Fahrenheit.

TIP

Kelvin and °C units are used on the SAT test.

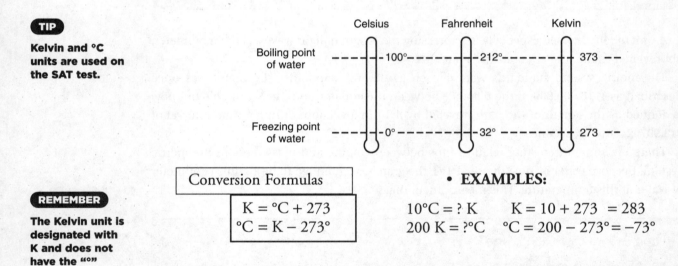

REMEMBER

The Kelvin unit is designated with K and does not have the "°" symbol.

Conversion Formulas
$K = °C + 273$
$°C = K - 273°$

• EXAMPLES:

$10°C = ? \ K \qquad K = 10 + 273 = 283$

$200 \ K = ?°C \qquad °C = 200 - 273° = -73°$

Note: In Kelvin notation, the degree sign is omitted: 283 K. The unit is the kelvin, abbreviated as K.

➡ **Example 1** _____

$10°C = $ _____ K

Solution: $K = 10 + 273 = 283 \ K$

➡ **Example 2** _____

$200 \ K = $ _____ °C

Solution: $°C = 200 - 273 = -73°C$

Heat Measurements

Heat energy (or just *heat*) is a form of energy that transfers among particles in a substance (or system) by means of the kinetic energy of those particles. In other words, under kinetic theory, heat is transferred by particles bouncing into each other.

The scales above are used to measure the degree of heat. A pail and a thimble can both be filled with water at 100° Celsius. The water in both measures the same degree of heat. However, the pail of water has a greater quantity of heat. This could be easily demonstrated by the amount of ice that could be melted by the water in these two containers. Obviously, the pail of water at 100° Celsius will melt more ice than will a thimble full of water at the same temperature. Therefore, the pail of water contains a greater number of **calories** of heat. The calorie unit is used to measure the quantity of heat. It is defined as the amount of heat needed to raise the temperature of 1 gram of water by 1 degree on the Celsius scale. This is a rather small unit to measure the quantities of heat involved in most chemical reactions. Therefore, the **kilocalorie** is more often used. The kilocalorie equals 1,000 calories. It is the quantity of heat that will increase the temperature of 1 kilogram of water by 1 degree on the Celsius scale. Although the calorie is commonly used, the SI unit for heat energy is the **joule**. It is abbreviated as J and, because it is a rather small unit, it is commonly given in kilojoules (kJ). The relationship between the calorie and the joule is that 1 calorie equals 4.18 joules.

TIP

Because the joule is rather small, kJ is used most often.

Scientific Notation

When students must do mathematical operations with numerical figures, **scientific notation** is very useful. Basically this system uses an exponential means of expressing figures. With large numbers, such as 3,630,000., move the decimal point to the left until only one digit remains to the left (3.630000) and then indicate the number of moves of the decimal point as the exponent of 10 (3.63×10^6). With a very small number such as 0.000000123, move the decimal point to the right until only one digit is to the left (0000001.23) and then express the number of moves as the negative exponent of 10 (1.23×10^{-7}).

TIP

Scientific notation is based on exponents of 10.

With numbers expressed in this exponential form, you can now use your knowledge of exponents in mathematical operations. An important fact to remember is that in multiplication you add the exponents of 10, and in division you subtract the exponents. Addition and subtraction of two numbers expressed in scientific notation can be performed only if the numbers have the same exponent.

REMEMBER

Only one digit can be to the left of the decimal point.

➡ Examples _____

Multiplication:

$(2.3 \times 10^5)(5.0 \times 10^{-12})$. Multiplying the first numbers, you get 11.5, and addition of the exponents gives 10^{-7}. Now, changing to a number with only one digit to the left of the decimal point gives you 1.15×10^{-6} for the answer.

Try these:
$(5.1 \times 10^{-6})(2 \times 10^{-3}) = 10.2 \times 10^{-9} = 1.02 \times 10^{-8}$
$(3 \times 10^5)(6 \times 10^3) = 18 \times 10^8 = 1.8 \times 10^9$

Division:
$(1.5 \times 10^3) \div (5.0 \times 10^{-2}) = 0.3 \times 10^5 = 3 \times 10^4$
$(2.1 \times 10^{-2}) \div (7.0 \times 10^{-3}) = 0.3 \times 10^1 = 3$
(Notice that in division the exponents of 10 are subtracted.)

Addition and subtraction:

$(4.2 \times 10^4 \text{ kg}) + (7.9 \times 10^3 \text{ kg}) =$

$(4.2 \times 10^4 \text{ kg}) + (0.79 \times 10^4 \text{ kg})$ (note that the exponents of 10 are now the same) $= 4.99 \times 10^4 \text{ kg}$
This can be rounded to $5.0 \times 10^4 \text{ kg}$.

$(6.02 \times 10^{-3}) - (2.41 \times 10^{-4}) = (6.02 \times 10^{-3}) - (.241 \times 10^{-3})$ (note that the exponents of 10 are now the same) $= 5.779 \times 10^{-3}$ or 5.8×10^{-3} when rounded to two significant figures.

Dimensional Analysis (Factor-Label Method of Conversion)

When you are working problems that involve numbers with units of measurement, it is convenient to use this method so that you do not become confused in the operations of multiplication or division. For example, if you are changing 0.001 kilogram to milligrams, you set up each conversion as a fraction so that all the units will factor out except the one you want in the answer.

$$1 \times 10^{-3} \ \cancel{\text{kg}} \times \frac{1 \times 10^3 \ \cancel{\text{g}}}{1 \ \cancel{\text{kg}}} \times \frac{1 \times 10^3 \ \text{mg}}{1 \ \cancel{\text{g}}} = 1 \times 10^3 \ \text{mg}$$

Notice that the kilogram is made the denominator in the first fraction to be factored with the original kilogram unit. The numerator is equal to the denominator except that the numerator is expressed in smaller units. The second fraction has the gram unit in the denominator to be factored with the gram unit in the preceding fraction. The answer is in milligrams because this is the only unit remaining and it assures you that the correct operations have been performed in the conversion.

Precision, Accuracy, and Uncertainty

Two other factors to consider in measurement are **precision** and **accuracy**. Precision indicates the reliability or reproducibility of a measurement. Accuracy indicates how close a measurement is to its known or accepted value.

For example, suppose you were taking a reading of the boiling point of pure water at sea level. Using the same thermometer in three trials, you record 96.8, 96.9, and 97.0 degrees Celsius. Since these figures show a high reproducibility, you can say that they are precise. However, the values are considerably off from the accepted value of 100 degrees Celsius, so we say they are not accurate. In this example we probably would suspect that the inaccuracy was the fault of the thermometer.

Regardless of precision and accuracy, all measurements have a degree of **uncertainty**. This is usually dependent on one or both of two factors—the limitation of the measuring instrument and the skill of the person making the measurement. Uncertainty can best be shown by example.

The graduated cylinder in the illustration contains a quantity of water to be measured. It is obvious that the quantity is betwen 30 and 40 milliliters because the meniscus lies between these two marked quantities. Now, checking to see where the bottom of the **meniscus** lies with reference to the ten intervening subdivisions, we see that it is between the fourth and fifth. This means that the volume lies between 34 and 35 milliliters. The next step introduces the uncertainty. We have to guess how far the reading is between these two markings. We can make an approximate guess, or estimate, that the level is more than 0.2 but less than 0.4 of the distance. We therefore report the volume as 34.3 milliliters. The last digit in any measurement is an estimate of this kind and is uncertain.

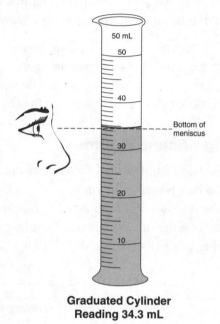

**Graduated Cylinder
Reading 34.3 mL**

Significant Figures

Any time a measurement is recorded, it includes all the digits that are certain plus one uncertain digit. These certain digits plus the one uncertain digit are referred to as **significant figures**. The more digits you are able to record in a measurement, the less relative uncertainty there is in the measurement. The following table summarizes the rules of significant figures.

Rule	Example	Number of Significant Figures
All digits other than zeros are significant.	25 g	2
	5.471 g	4
Zeros between nonzero digits are significant.	309 g	3
	40.06 g	4
Final zeros to the right of the decimal point are significant.	6.00 mL	3
	2.350 mL	4
In numbers smaller than 1, zeros to the left or directly to the right of the decimal point are not significant.	0.05 cm	1 The zeros merely mark the position of the decimal point.
	0.060 cm	2 The first two zeros mark the position of the decimal point. The final zero is significant.

One last rule deals with final zeros in a whole number. These zeros may or may not be significant, depending on the measuring instrument. For instance, if an instrument that measures to the nearest mile is used, the number 3,000 miles has four significant figures. If, however, the instrument in question records miles to the nearest thousands, there is only one significant figure. The number of significant figures in 3,000 could be one, two, three, or four, depending on the limitation of the measuring device.

This problem can be avoided by using the system of scientific notation. For this example, the following notations would indicate the numbers of significant figures:

$$3 \times 10^3 \qquad \text{one significant figure}$$
$$3.0 \times 10^3 \qquad \text{two significant figures}$$
$$3.00 \times 10^3 \qquad \text{three significant figures}$$
$$3.000 \times 10^3 \qquad \text{four significant figures}$$

Calculations with Significant Figures

When you do calculations involving numbers that do not have the same number of significant figures in each, keep the following two rules in mind.

First, in multiplication and division, the number of significant figures in a product or a quotient of measured quantities is the same as the number of significant figures in the quantity having the smaller number of significant figures.

 Example 1

Problem	Unrounded answer	Answer rounded to the correct number of significant figures
$4.29 \text{ cm} \times 3.24 \text{ cm} =$	$13.8996 \text{ cm}^2 =$	13.9 cm^2

Explanation: Both measured quantities have three significant figures. Therefore, the answer should be rounded to three significant figures.

 Example 2

Problem	Unrounded answer	Answer rounded to the correct number of significant figures
$4.29 \text{ cm} \times 3.2 \text{ cm} =$	$13.728 \text{ cm}^2 =$	14 cm^2

Explanation: One of the measured quantities has only two significant figures. Therefore, the answer should be rounded to two significant figures.

 Example 3

Problem	Unrounded answer	Answer rounded to the correct number of significant figures
$8.47 \text{ cm}^2/4.26 \text{ cm} =$	$1.9882629 \text{ cm} =$	1.99 cm

Explanation: Both measured quantities have three significant figures. Therefore, the answer should be rounded to three significant figures.

Second, when adding or subtracting measured quantities, the sum or difference should be rounded to the same number of decimal places as the quantity having the fewest decimal places.

➡ **Example 1** _____

Problem	Unrounded answer	Answer rounded to the correct number of significant figures
3.56 cm		
2.6 cm		
+ 6.12 cm		
Total =	12.28 cm =	12.3 cm

Explanation: One of the quantities added has only one decimal place. Therefore, the answer should be rounded to only one decimal place.

➡ **Example 2** _____

Problem	Unrounded answer	Answer rounded to the correct number of significant figures
3.514 cm		
−2.13 cm		
Difference =	1.384 cm =	1.38 cm

Explanation: One of the quantities has only two decimal places. Therefore, the answer should be rounded to only two decimal places.

CHAPTER SUMMARY

The following terms summarize all the concepts and ideas that were introduced in this chapter. You should be able to explain their meaning and how you would use them in chemistry. They appear in boldface type in this chapter to draw your attention to them. The boldface type also makes it easier for you to look them up if you need to. You could also use Internet search engines like *google.com* on your computer to get a quick and expanded explanation of these terms, laws, and formulas.

accuracy	gas	matter
activation energy	heat energy	meniscus
calorie	heterogeneous	mixture
Celsius	homogeneous	physical change
chemical change	inertia	physical property
chemical property	joule	potential energy
compound	Kelvin	precision
density	kilocalorie	significant figures
element	kilojoule	SI units
endothermic	kinetic energy	solid
exothermic	liquid	uncertainty
exponential notation	mass	

Law of Conservation of Energy
Law of Conservation of Matter
Law of Conservation of Mass and Energy
Law of Definite Composition or Proportion

INTERNET RESOURCES

Online content that reinforces the major concepts discussed in this chapter can be found at the following Internet addresses if they are still available. *Some may have been changed or deleted.*

The Classification of Matter
http://www.prenhall.com/wps/media/objects/165/169061/blb9ch0102.html
This site explains the differences among types of matter and provides exercises to check your understanding.

The Scientific Method
http://en.wikipedia.org/wiki/Scientific_method
This site gives an historical overview of examples of using the scientific method.

The SI System
http://en.wikipedia.org/wiki/International_System_of_Units
This site gives an overview of the SI System as well as an historical and a global view of the system.

1. 1.2 mg = _.0012_ g

2. 6.3 cm = _63_ mm

3. 5.12 m = _512_ cm

4. 32°C = _305_ K

5. 6.111 mL = _.006111_ L

6. 1 km = _1,000,000_ mm

7. 1.03 kg = _1030_ g

8. 0.003 g = _.000003_ kg

9. 22.4 L = _22400_ mL

10. 10,013 cm = _.10013_ km

11. The density of CCl₄ (carbon tetrachloride) is 1.58 grams/milliliter. What would be the mass of 100. milliliters of CCl₄?

158 gram

12. The mass of a piece of sulfur is 227 grams. When it was submerged in a graduated cylinder containing 50.0 milliliters of H₂O, the level rose to 150. milliliters. What is the density (g/mL) of the sulfur? _2.27 g/mL_

13. (a) A box 20.0 centimeters × 20.0 centimeters × 5.08 inches has what volume in cubic centimeters? _5160 cm³_
    (b) What mass, in grams, of H₂O @ 4°C will the box hold? _5160 grams_

14. Set up the following using *dimensional analysis:*

$$\frac{5\,cm}{s} = \frac{?\,km}{h}$$

15. How many significant figures are in each of the following?

    (a) 1.01 cg 3
    (b) 200.0 cg 4
    (c) 0.0021 cg 2
    (d) 0.0230 cg 3

16. A baking powder can carries the statement, "Ingredients: corn starch, sodium bicarbonate, calcium hydrogen phosphate, and sodium aluminum sulfate." Therefore, this baking powder is

    (A) a compound
    (B) a mixture
    (C) a molecule
    (D) a mixture of elements

17. Which of the following is a physical property of sugar?

    (A) It decomposes readily.
    (B) Its composition is carbon, hydrogen, and oxygen.
    (C) It turns black with concentrated H₂SO₄.
    (D) It can be decomposed with heat.
    (E) It is a white crystalline solid.

18. A substance that can be further simplified using ordinary means may be either

    (A) an element or a compound
    (B) an element or a mixture
    (C) a mixture or a compound
    (D) a mixture or an atom

19. Chemical action may involve all of the following EXCEPT

    (A) combining of atoms of elements to form a molecule
    (B) separation of the molecules in a mixture
    (C) breaking down compounds into elements
    (D) reacting a compound and an element to form a new compound and a new element

20. The energy of a system can be

   (A) easily changed to mass
   (B) transformed into a different form
   (C) measured only as potential energy
   (D) measured only as kinetic energy

21. If the ΔH of a reaction is a negative quantity, the reaction is definitely

   (A) endothermic
   (B) unstable
   (C) exothermic
   (D) reversible

22. Write E for each element, C for each compound, and M for each mixture in the following list:

Water C	Aluminum oxide C	Hydrochloric acid C M
Wine M	Hydrogen E	Nitrogen E
Soil M	Carbon dioxide C	Tin E
Silver E	Air M	Potassium chloride C

The following questions are in the format that is used on the **SAT Subject Test in Chemistry**. If you are not familiar with these types of questions, study pages xiii–xviii before doing the remainder of the review questions.

23. If the graphic representation of the energy levels of the reactants and products in a chemical reaction looks like this:

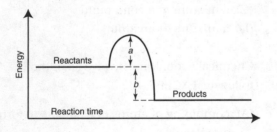

Which of the following statements are true?

   I. The activation energy for the forward reaction is represented by the "a" portion.
   II. The activation energy for the forward reaction is represented by the "b" portion of the graph.
   III. The "a" portion is the energy given off in the forward reaction.

   (A) I only
   (B) II only
   (C) I and III only
   (D) II and III only
   (E) I, II, and III

24. A substance that can be further simplified by ordinary chemical means may be which of the following?

   I. An element or a compound
   II. A mixture or a compound
   III. An element of a mixture

   (A) I only
   (B) II only
   (C) III only
   (D) I and II only
   (E) I, II, and III

Questions 25–29 refer to the following terms.

   (A) Density
   (B) A solid
   (C) Volume
   (D) Weight
   (E) Matter

25. Gives the mass per unit volume

26. Has mass and a definite size and shape

27. Gives the space occupied

28. Has mass and occupies space

29. Defined as a measure of the mass times the gravitational force

Directions:

(See the explanation for this type of question and the examples on pages xv–xviii before attempting questions 30–32.)

Every question below contains two statements, I in the left-hand column and II in the right-hand column. For each question, decide if statement I is true or false <u>and</u> whether statement II is true or false, and fill in the corresponding T or F ovals in the answer spaces. *<u>Fill in oval CE only if statement II is a correct explanation of statement I.</u>

	I		II
30.	A substance composed of two or more elements chemically combined is called a mixture	BECAUSE	the properties of the constituents of a mixture are retained.
31.	A chemical change involves change in the composition and molecular structure of the reactants	BECAUSE	in a chemical reaction bonds are broken and new substances and new bonds are formed.
32.	The burning of paper is a physical change	BECAUSE	when a chemical change occurs energy is either gained or lost by the reactants.

*Fill in oval CE only if II is a correct explanation of I.

	I	II	CE*
30.	T F	T F	○
31.	T F	T F	●
32.	T F	T F	●

F, T
T, T, CE
F, T

Answers and Explanations

1. 0.0012 g

$$1.2 \ \cancel{mg} \times \frac{1 \ g}{1,000 \ \cancel{mg}} = 0.0012 \ g$$

2. 63 mm

$$6.3 \ \cancel{cm} \times \frac{10 \ mm}{1 \ \cancel{cm}} = 63 \ mm$$

3. 512 cm

$$5.12 \ \cancel{m} \times \frac{100 \ cm}{1 \ \cancel{m}} = 512 \ cm$$

4. 305 K

$$32°C + 273 = 305 \ K$$

5. 0.006111 L

$$6.111 \ \cancel{mL} \times \frac{1 \ L}{1,000 \ \cancel{mL}} = .006111 \ L$$

6. 1,000,000 or 1×10^6 mm

$$1 \ \cancel{km} \times \frac{1,000 \ \cancel{m}}{1 \ \cancel{km}} \times \frac{1,000 \ mm}{1 \ \cancel{m}} = 1,000,000 \ or \ 1 \times 10^6 \ mm$$

7. 1.03×10^3 kg

$$1.03 \ \cancel{kg} \times \frac{1,000 \ g}{\cancel{kg}} = 1030 \ g \ or \ 1.03 \times 10^3 \ g$$

8. .000003 kg or 3×10^{-6} kg

$$.003 \ \cancel{g} \times \frac{1 \ kg}{1,000 \ \cancel{g}} = .000003 \ kg \ or \ 3 \times 10^{-6} \ kg$$

9. 22,400 or 2.24×10^4 mL

$$22.4 \ \cancel{L} \times \frac{1,000 \ mL}{1 \ \cancel{L}} = 22,400 \ or \ 2.24 \times 10^4 \ mL$$

10. .10013 km

$$10,013 \ \cancel{cm} \times \frac{1 \ \cancel{m}}{100 \ \cancel{cm}} \times \frac{1 \ km}{1000 \ \cancel{m}} = 0.10013 \ km$$

11. 158 g

Because the density is 1.58 g/mL and you want the weight of 100 mL, you use the formula that density × volume = mass.

Inserting the values gives

$$1.58 \ g/\cancel{mL} \times 100 \ \cancel{mL} = 158 \ g$$

12. 2.27 g/mL

   To find the volume of 227 g of sulfur, subtract the volume of water before from the volume after immersion.

$$150. \text{ mL} - 50. \text{ mL} = 100. \text{ mL}$$

   Then

$$\frac{227 \text{ g}}{100. \text{ mL}} = 2.27 \text{ g/mL}$$

13. (a) $5,160 \text{ cm}^3$ (b) $5,160$ g

   (a) Convert 5.08 in. to centimeters:

$$5.08 \text{ in.} \times \frac{2.54 \text{ cm}}{1 \text{ in.}} = 12.9 \text{ cm}$$

   Then $20.0 \text{ cm} \times 20.0 \text{ cm} \times 12.9 \text{ cm} = \underline{5,160} \text{ cm}^3$

   (b) Since 1.00 cm^3 of water at 4°C has a mass of 1.00 g, then $5,160 \text{ cm}^3 = \underline{5,160}$ g

14. $\dfrac{5 \text{ cm}}{s} \times \dfrac{1 \text{ m}}{100 \text{ cm}} \times \dfrac{1 \text{ km}}{1,000 \text{ m}} \times \dfrac{60 \text{ s}}{1 \text{ min}} \times \dfrac{60 \text{ min}}{1 \text{ h}} =$

15. (a) 3 (b) 4 (c) 2 (d) 3

16. **(B)** Because all the substances are compounds.

17. **(E)** Because all the other statements refer to chemical properties of sugar.

18. **(C)** Neither an element nor an atom can be simplified into anything else by ordinary means.

19. **(B)** Separating molecules in a mixture (e.g., evaporating water from a sugar solution) does not involve a chemical change action.

20. **(B)** Because the other choices are not true. Transforming energy into different forms can be demonstrated by batteries in a flashlight, changing chemical energy into electrical energy and then into light energy.

21. **(C)** Anytime the ΔH of a reaction is negative, the reaction gives off energy and it is classified as exothermic.

22.

C	C	M
M	E	E
M	C	E

E	M	C

23. **(A)** In the diagram part of question 23

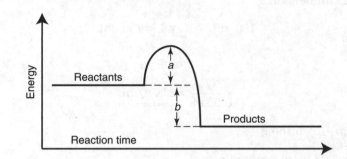

"a" is the activation energy for the forward reaction. The sum of "a" and "b" is the activation energy for the reverse reaction, and "b" alone is the energy given off by the forward reaction.

24. **(B)** Because an element is a basic building block in the periodic table, it cannot be further simplified by ordinary chemical means. A compound and a mixture can be simplified by chemical and physical means, respectively.

25. **(A)** The mass per unit volume is the definition of density.

26. **(B)** Matter that has mass and a definite size and shape is called a solid.

27. **(C)** The space occupied by a substance is called its volume.

28. **(E)** That a substance occupies space and mass is the basic definition of all matter.

29. **(D)** Weight is the product of mass times a gravitational force. In outer space where there is no gravitational force, we say a mass is weightless.

30. **(F, T)** The column I statement is false, because a mixture is elements and/or compounds in physical contact (without bonds) with one another.

31. **(T, T, CE)** Chemical reactions involve the changing of molecular structures by the breaking of bonds and the formation of new bonds.

32. **(F, T)** Burning paper is a chemical reaction not a physical one. The statement in column II is true, because a chemical change always involves an energy change.

Atomic Structure and the Periodic Table of the Elements

2

These skills are usually tested on the SAT Subject Test in Chemistry. You should be able to...

→ Describe the history of the development of atomic theory.
→ Explain the structure of atoms, their main energy levels, sublevels, orbital configuration, and the rules that govern how they are filled.
→ Place atoms in groups and periods based on their atomic structure.
→ Write formulas and names of compounds.
→ Explain how chemical and physical properties are related to positions in the Periodic Table, including atomic size, ionic size, electronegativity, acid-forming properties, and base-forming properties.
→ Explain the nature of radioactivity, the types and characteristics of each, and the inherent dangers.
→ Identify the changes that occur in a decay series.
→ Do the mathematical calculations to determine the age of a substance using its half-life.

This chapter will review and strengthen these skills. Be sure to do the Practice Exercises at the end of the chapter.

The idea of small, invisible particles being the building blocks of matter can be traced back more than 2,000 years to the Greek philosophers Democritus and Leucippus. These particles, considered to be so small and indestructible that they could not be divided into smaller particles, were called **atoms**, the Greek word for indivisible. The English word **atom** comes from this Greek word. This early concept of atoms was not based upon experimental evidence but was simply a result of thinking and reasoning on the part of the philosophers. It was not until the eighteenth century that experimental evidence in favor of the atomic hypothesis began to accumulate. Finally, around 1805, John Dalton proposed some basic assumptions about atoms based on what was known through scientific experimentation and observation at that time. These assumptions are very closely related to what scientists presently know about atoms. For this reason, Dalton is often referred to as the father of modern atomic theory. Some of these basic ideas were:

1. All matter is made up of very small, discrete particles called atoms.
2. All atoms of an element are alike in weight, and this weight is different from that of any other kind of atom.
3. Atoms cannot be subdivided, created, or destroyed.
4. Atoms of different elements combine in simple whole-number ratios to form chemical compounds.
5. In chemical reactions, atoms are combined, separated, or rearranged.

 TIP

Know Dalton's five basic ideas about atoms.

By the second half of the 1800s, many scientists believed that all the major discoveries related to the elements had been made. The only thing left for young scientists to do was to refine what was already known. This came to a suprising halt when J. J. Thomson discovered the electron beam in a cathode ray tube in 1897. Soon afterward, Henri Becquerel announced his work with radioactivity, and Marie Curie and her husband, Pierre, set about trying to isolate the source of radioactivity in their laboratory in France.

During the late nineteenth and early twentieth centuries, more and more physicists turned their attention to the structure of the atom. In 1913 the Danish physicist Niels Bohr published a theory explaining the line spectrum of hydrogen. He proposed a planetary model that quantized the energy of electrons to specific orbits. The work of Louis de Broglie and others in the 1920s and 1930s showed that quantum theory described a more probabilistic model of where the electrons could be found that resulted in the theory of orbitals.

ELECTRIC NATURE OF ATOMS

From around the beginning of the twentieth century, scientists have been gathering evidence about the structure of atoms and fitting the information into a model of the atomic structure.

Basic Electric Charges

TIP

The electron was discovered by J. J. Thomson.

The discovery of the electron as the first subatomic particle is credited to **J. J. Thomson** (England, 1897). He used an evacuated tube connected to a spark coil as shown in Figure 3. As the voltage across the tube was increased, a beam became visible. This was referred to as a cathode ray. Thomson found that the beam was deflected by both electrical and magnetic fields. Therefore, he concluded that cathode rays are made up of very small, negatively charged particles, which became known as **electrons**.

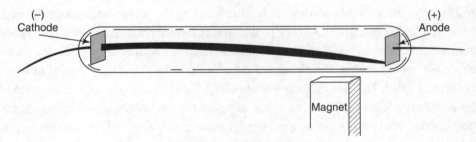

Figure 3. Cathode Ray Tube

Further experimentation led Thomson to find the ratio of the electrical charge of the electron to its mass. This was a major step toward understanding the nature of the particle. He was awarded a Nobel Prize in 1906 for his accomplishment.

It was an American scientist, **Robert Millikan**, who in 1909 was able to measure the charge on an electron using the apparatus pictured in Figure 4.

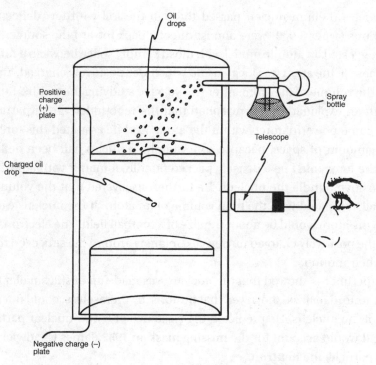

Figure 4. Millikan Oil Drop Experiment

Oil droplets were sprayed into the chamber and, in the process, became randomly charged by gaining or losing electrons. The electric field was adjusted so that a negatively charged drop would move slowly upward in front of the grid in the telescope. Knowing the rate at which the drop was rising, the strength of the field, and the mass of the drop, Millikan was able to calculate the charge on the drop. Combining the information with the results of Thomson, he could calculate a value for the mass of a single electron. Eventually, this number was found to be 9.11×10^{-28} gram.

Ernest Rutherford (England, 1911) performed a gold foil experiment (Figure 5) that had tremendous implications for atomic structure.

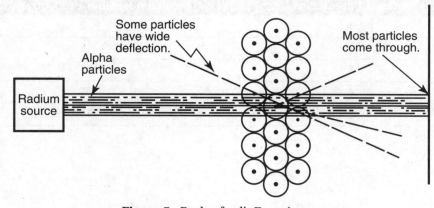

Figure 5. Rutherford's Experiment

Alpha particles (helium nuclei) passed through the foil with few deflections. However, some deflections (1 per 8,000) were almost directly back toward the source. This was unexpected and suggested an atomic model with mostly empty space between a **nucleus**, in which most of the mass of the atom was located and which was positively charged, and the electrons that defined the volume of the atom. After two years of studying the results, Rutherford finally came up with an explanation. He reasoned that the rebounded alpha particles must have experienced some powerful force within the atom. And he assumed this force must occupy a very small amount of space, because so few alpha particles had been deflected. He concluded that the force must be a densely packed bundle of matter with a positive charge. He called this positive bundle the nucleus. He further discovered that the volume of a nucleus was very small compared with the total volume of an atom. If the nucleus were the size of a marble, then the atom would be about the size of a football field. The electrons, he suggested, surrounded the positively charged nucleus like planets around the sun, even though he could not explain their motion.

Further experiments showed that the nucleus was made up of still smaller particles called **protons**. Rutherford realized, however, that protons, by themselves, could not account for the entire mass of the nucleus. He predicted the existence of a new nuclear particle that would be neutral and would account for the missing mass. In 1932, James Chadwick (England) discovered this particle, the **neutron**.

Bohr Model of the Atom

In 1913, **Niels Bohr** (Denmark) proposed his model of the atom. This pictured the atom as having a dense, positively charged nucleus and negatively charged electrons in specific spherical orbits, also called energy levels or shells, around this nucleus. These energy levels are arranged concentrically around the nucleus, and each level is designated by a number: 1, 2, 3, ... The closer to the nucleus, the less energy an electron needs in one of these levels, but it has to gain energy to go from one level to another that is farther away from the nucleus.

Because of its simplicity and general ability to explain chemical change, the Bohr model still has some usefulness today.

TIP

Bohr's electron distribution to principal energy levels has the formula $2n^2$.

Principal Energy Level	Maximum Number of Electrons ($2n^2$)
1	2
2	8
3	18
4	32
5	50

Components of Atomic Structure

The chart below lists the basic particles of the atom and important information about them.

Particle	Charge	Symbol	Actual Mass	Relative Mass Compared to Proton	Discovery
Electron	$- (e^-)$	$_{-1}^{0}e$	9.109×10^{-28} g	1/1,837	J. J. Thomson–1897
Proton	$+ (p^+)$	$_{1}^{1}H$	1.673×10^{-24} g	1	—— early 1900s
Neutron	$0 (n^0)$	$_{0}^{1}n$	1.675×10^{-24} g	1	J. C. Chadwick–1932

When these components are used in the model, the protons and neutrons are shown in the nucleus. These particles are known as **nucleons**. The electrons are shown outside the nucleus.

The number of protons in the nucleus of an atom determines the **atomic number**. All atoms of the same element have the same number of protons and therefore the same atomic number; atoms of different elements have different atomic numbers. Thus, the atomic number identifies the element. An English scientist, **Henry Moseley**, first determined the atomic numbers of the elements through the use of X-rays.

The sum of the number of protons and the number of neutrons in the nucleus is called the **mass number**.

Table 1 (see page 64) summarizes the relationships just discussed. Notice that the outermost energy level can contain no more than eight electrons. The explanation of this is given in the next section.

In some cases, different types of atoms of the same element have different masses. For example, three types of hydrogen atoms are known. The most common type of hydrogen, sometimes called protium, accounts for 99.985% of the hydrogen atoms found on Earth. The nucleus of a protium atom contains one proton only, and it has one electron moving about it. The second form of hydrogen, known as deuterium, accounts for 0.015% of Earth's hydrogen atoms. Each deuterium atom has a nucleus containing one proton and one neutron. The third form of hydrogen, tritium, is radioactive. It exists in very small amounts in nature, but it can be prepared artificially. Each tritium atom contains one proton, two neutrons, and one electron.

Protium, deuterium, and tritium are isotopes of hydrogen. **Isotopes** are atoms of the same element that have different masses. The isotopes of a particular element all have the same number of protons and electrons but different numbers of neutrons. In all three isotopes of hydrogen, the positive charge of the single proton is balanced by the negative charge of the electron. Most elements consist of mixtures of isotopes. Tin, for example, has ten stable isotopes, the most of any element.

The percentage of each isotope in the naturally occurring element on Earth is nearly always the same, no matter where the element is found. The percentage at which each of an element's isotopes occurs in nature is taken into account when calculating the element's average atomic mass. **Average atomic mass** is the weighted average of the atomic masses of the naturally occurring isotopes of an element.

TIP

Isotopes have the same atomic number but a different atomic mass. This means they differ in the number of neutrons, not protons.

Table 1. Table of the First 21 Elements*

Element	Atomic No.	Mass No.	Number of Protons	Number of Neutrons	Number of Electrons	Electrons in PELs** 1 2 3 4
Hydrogen	1	1	1	0	1	1
Helium	2	4	2	2	2	2
Lithium	3	7	3	4	3	2 1
Beryllium	4	9	4	5	4	2 2
Boron	5	11	5	6	5	2 3
Carbon	6	12	6	6	6	2 4
Nitrogen	7	14	7	7	7	2 5
Oxygen	8	16	8	8	8	2 6
Fluorine	9	19	9	10	9	2 7
Neon	10	20	10	10	10	2 8
Sodium	11	23	11	12	11	2 8 1
Magnesium	12	24	12	12	12	2 8 2
Aluminum	13	27	13	14	13	2 8 3
Silicon	14	28	14	14	14	2 8 4
Phosphorus	15	31	15	16	15	2 8 5
Sulfur	16	32	16	16	16	2 8 6
Chlorine	17	35	17	18	17	2 8 7
Argon	18	40	18	22	18	2 8 8
Potassium	19	39	19	20	19	2 8 8 1
Calcium	20	40	20	20	20	2 8 8 2
Scandium	21	45	21	24	21	2 8 9 2

*The particular atom shown is the most abundant in the group of isotopes for that element.
**PEL is used to represent the principal energy levels.

Calculating Average Atomic Mass

The average atomic mass of an element depends on both the mass and the relative abundance of each of the element's isotopes. For example, naturally occurring copper consists of 69.17% copper-63, which has an atomic mass of 62.919 598 amu, and 30.83% copper-65, which has an atomic mass of 64.927 793 amu. The average atomic mass of copper can be calculated by multiplying the atomic mass of each isotope by its relative abundance (expressed in decimal form) and adding the results.

$$0.6917 \times 62.919\ 598\ \text{amu} + 0.3083 \times 64.927\ 793\ \text{amu} = 63.55\ \text{amu}$$

Therefore, the calculated average atomic mass of naturally occurring copper is 63.55 amu. Average atomic masses of the elements listed in the Periodic Table, rounded to one decimal place for use in calculations and also in full to four decimal places, are given in the Chemical Elements table in the Tables for Reference section at the back of the book.

Valence Electrons

Each atom attempts to have its outer energy level complete and accomplishes this by borrowing, lending, or sharing its electrons. The electrons found in the outermost energy level are called **valence electrons**. The remainder of the electrons are called core electrons. The absolute number of electrons gained, lost, or borrowed is referred to as the valence of the atom.

➡ **EXAMPLE:** _____

$$_{17}Cl = \overset{\text{nucleus}}{\bullet} \quad)2 \quad)8 \quad)7 \leftarrow \text{valence electrons}$$

This picture can be simplified to ·C̈l:, showing only the valence electrons as dots in an electron dot notation. This is called the **Lewis structure** of the atom. To complete its outer orbit to eight electrons, chlorine must borrow an electron from another atom. Its valence number then is 1. As stated above, when electrons are gained, we assign a – sign to this number, so the oxidation number of chlorine is –1 (more on pages 75 and 76).

➡ **ANOTHER EXAMPLE:** _____

$$_{11}Na = \overset{\text{nucleus}}{\bullet} \quad)2 \quad)8 \quad)1 \leftarrow \text{valence electrons}$$

Na • (Lewis dot structure)

Since sodium tends to lose this electron, its oxidation number is +1.

TIP

A Lewis structure shows the atomic symbol to represent the nucleus and inner shell electrons. It shows dots to represent the valence electrons.

ATOMIC SPECTRA

The Bohr model was based on a simple postulate. Bohr applied to the hydrogen atom the concept that the electron can exist only in certain energy levels without an energy change but that, when the electron changes its state, it must absorb or emit the exact amount of energy that will bring it from the initial state to the final state. The **ground state** is the lowest energy state available to the electron. The **excited state** is any level higher than the ground state. The formula for a change in energy (ΔE) is:

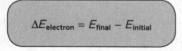

$$\Delta E_{electron} = E_{final} - E_{initial}$$

When an electron moves from the ground state to an excited state, it must absorb energy. When it moves from an excited state to the ground state, it emits energy. This release of energy is the basis for **atomic spectra**. (See Figure 6.)

The energy values shown were calculated from Bohr's equation.

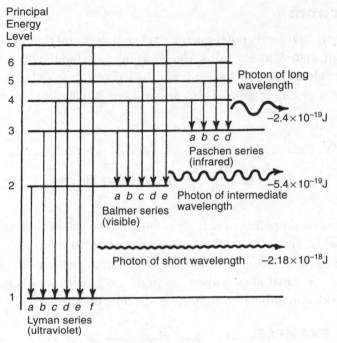

Figure 6. Atomic Spectra Chart

When energy is released in the "allowed" values, it is released in the form of discrete radiant energy called **photons**. Each of the first three levels has a particular name associated with the emissions that occur when an electron reaches its ground state on that level. The emissions, consisting of ultraviolet radiation, that occur when an electron cascades from a level higher than the first level down to $n = 1$ are known as the **Lyman series**. Note in Figure 6 that the next two higher levels have the names **Balmer** (for $n = 2$) and **Paschen** ($n = 3$) **series**, respectively.

Spectroscopy

When the light emitted by energized atoms is examined with an instrument called a **spectroscope**, the prism or diffraction grating in the spectroscope disperses the light to allow an examination of the **spectra** or distinct colored lines. Since only particular energy jumps are available in each type of atom, each element has its own unique emission spectra made up of only the lines of specific wavelength that correspond to its atomic structure. The relationship of wavelength to frequency is shown below.

Visible Light Spectra Wavelengths

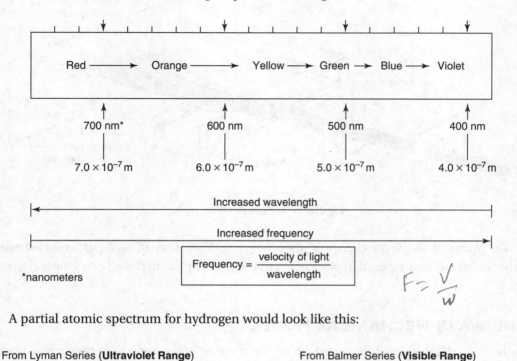

Red ⟶ Orange ⟶ Yellow ⟶ Green ⟶ Blue ⟶ Violet

700 nm* 600 nm 500 nm 400 nm

7.0×10^{-7} m 6.0×10^{-7} m 5.0×10^{-7} m 4.0×10^{-7} m

Increased wavelength

Increased frequency

$$\text{Frequency} = \frac{\text{velocity of light}}{\text{wavelength}}$$

*nanometers

$F = \dfrac{V}{W}$

A partial atomic spectrum for hydrogen would look like this:

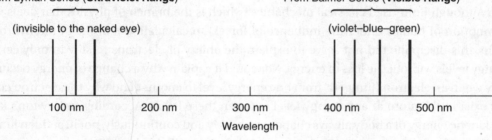

From Lyman Series (**Ultraviolet Range**) From Balmer Series (**Visible Range**)

(invisible to the naked eye) (violet–blue–green)

100 nm 200 nm 300 nm 400 nm 500 nm

Wavelength

The right-hand group is in the visible range and is part of the Balmer series. The left-hand group is in the ultraviolet region and belongs to the Lyman series.

Spectral lines like these can be used in the identification of unknown specimens.

Mass Spectroscopy

Another tool used to identify specific atomic structures is mass spectroscopy, which is based on the concept that differences in mass cause differences in the degree of bending that occurs in a beam of ions passing through a magnetic field. This is shown in Figure 7.

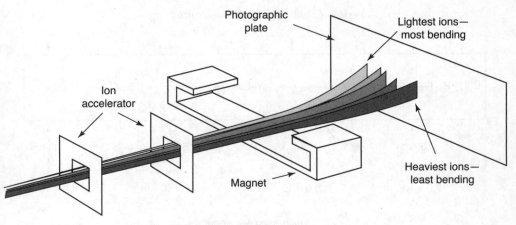

Figure 7. Mass Spectroscope

The intensity on the photographic plate indicates the amount of each particular **isotope**. Other collectors may be used in place of the photographic plate to collect and interpret these data.

THE WAVE-MECHANICAL MODEL

In the early 1920s, some difficulties with the Bohr model of the atom were becoming apparent. Although Bohr used classical mechanics (which is the branch of physics that deals with the motion of bodies under the influence of forces) to calculate the orbits of the hydrogen atom, this discipline did not serve to explain the ability of electrons to stay in only certain energy levels without the loss of energy. Nor could it explain why a change of energy occurred only when an electron "jumped" from one energy level to another and why the electron could not exist in the atom at any energy level between these levels. According to Newton's laws, the kinetic energy of a body always changes smoothly and continuously, not in sudden jumps. The idea of only certain **quantized** energy levels being available in the Bohr atom was a very important one. The energy levels explained the existence of atomic spectra, described in the preceding sections.

Another difficulty with the Bohr model was that it worked well only for the hydrogen atom with its single electron. It did not work with atoms that had more electrons. A new approach to the laws governing the behavior of electrons inside the atom was needed, and such an approach was developed in the 1920s by the combined work of many scientists. Their work dealt with a more mathematical model usually referred to as **quantum mechanics** or **wave mechanics**. By this time, Albert Einstein had already proposed a relativity mechanics model to deal with the relative nature of mass as its speed approaches the speed of light. In the same manner, a quantum/wave mechanics model was now needed to fit the data of the atomic model. **Max Planck** suggested in his quantum theory of light that light has both particlelike properties and wavelike characteristics. In 1924, **Louis de Broglie**, a young French physicist, suggested that, if light can have both wavelike and particlelike characteristics as Planck had suggested, then perhaps particles can also have wavelike characteristics. In 1927, de Broglie's ideas were verified experimentally when investigators showed that electrons could produce diffraction patterns, a property associated with waves. Diffraction patterns are produced by waves as they pass through small holes or narrow slits.

In 1927, **Werner Heisenberg** stated what is now called the **uncertainty principle**. This principle states that it is impossible to know both the precise location and precise velocity of a subatomic particle at the same time. Heisenberg, in conjunction with the Austrian physicist Erwin Schrödinger, agreed with the de Broglie concept that the electron is bound to the nucleus in a manner similar to a standing wave. They developed the complex equations that describe the **wave-mechanical model** of the atom. The solution of these equations gives specific wave functions called **orbitals**. These are not related at all to the Bohr orbits. The electron does not move in a circular orbit in this model. Rather, the orbital is a three-dimensional region around the nucleus that indicates the probable location of an electron but gives no information about its pathway. The drawings in Figures 8a and 8b are only probability distribution representations of where electrons in these orbitals might be found.

Quantum Numbers and the Pauli Exclusion Principle

Each electron orbital of an atom may be described by a set of four quantum numbers in the wave-mechanical model. These numbers give the position with respect to the nucleus, the shape of the orbital, its spatial orientation, and the spin of the electron in the orbital.

Principal quantum number (n)
  1, 2, 3, 4, 5, etc.
The values of $n = 1, 2, 3, \ldots$

This number refers to average distance of the orbital from the nucleus. 1 is closest to the nucleus and has the least energy. The numbers correspond to the orbits in the Bohr model. They are called energy levels.

TIP

The principal quantum number refers to the principal energy level: 1, 2, 3, and so on. The angular momentum quantum number refers to shape.

Angular momentum (ℓ)
  quantum number
s, p, d, f
(in order of increasing energy)

The value of ℓ can $= 0, 1, \ldots,$
  $(n - 1)$
$\ell = 0$ indicates a spherical-shaped *s* orbital
$\ell = 1$, indicates a dumbbell-shaped *p* orbital
$\ell = 2$, indicates a five orbital orientation *d* orbital

This number refers to the shape of the orbital. The number of possible shapes is limited by the principal quantum number. The first energy level has only one possible shape, the *s* orbital because $n = 1$ and the limit of $\ell = (n - 1) = 0$. The second has two possible shapes, the *s* and *p*. See Figures 8a and 8b for representations of these shapes.

Magnetic quantum number (m_ℓ)
$s = 1$ space-oriented orbital
$p = 3$ space-oriented orbitals
$d = 5$ space-oriented orbitals
$f = 7$ space-oriented orbitals
The value of m, can equal
$-\ell, \ldots, 0, \ldots, + \ell.$
Spin quantum number (m_s)
$+$ spin $-$ spin

The value of $m = +\dfrac{1}{2}$ or $-\dfrac{1}{2}$

The drawings in Figure 8a show the *s*-orbital shape, which is a sphere, and the *p* orbitals, which have dumbbell shapes with three possible orientations on the axis shown. The number of spatial orientations of orbitals is referred to as the magnetic quantum number. The possible orientations are listed. Figure 8b represents the *d* orbitals. Electrons are assigned one more quantum number called the spin quantum number. This describes the spin in either of two possible directions. Each orbital can be filled by only two electrons with opposite spins. The main significance of electron spin is explained by the postulate of Wolfgang Pauli. It states that in a given atom no two electrons can have the same set of four quantum numbers (n, ℓ, m_ℓ, and m_s). This is referred to as the **Pauli Exclusion Principle**. Therefore, each orbital in Figures 8a and 8b can hold only two electrons.

TIP

Pauli Exclusion Principle: No two electrons can have the same four quantum numbers.

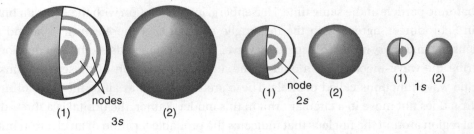

Two representations of the hydrogen 1*s*, 2*s*, and 3*s* orbitals. (1) The electron probability distribution; the nodes indicate regions of zero probability. (2) The surface that contains 90% of the total electron probability (the size of the orbital, by definition).

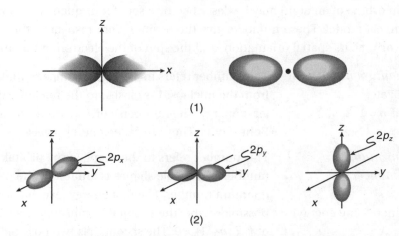

Representation of the 2*p* orbitals. (1) The electron probability distribution and (2) the boundary surface representations of all three orbitals.

Figure 8a. Representations of *s* and *p* orbitals

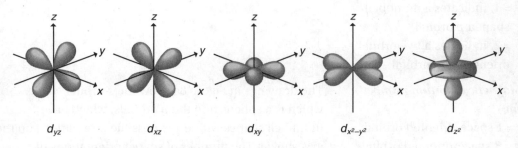

(Indicates orbitals in *y* and *z* planes)
Representations of the 3*d* orbitals in terms of their boundary surfaces.
The subscripts of the first four orbitals indicate the planes in which the four lobes are centered.

Figure 8b. Representations of *d* orbitals

Quantum numbers are summarized in the table below.

Summary of Quantum Numbers for the First Four Levels of Orbitals in the Hydrogen Atom

Principal Quantum No., n	Angular Momentum Quantum No., ℓ	Orbital Shape Designation	Magnetic Quantum No., m_ℓ	Number of Orbitals	Total Electrons
1	0	1s	0	1	2
2	0	2s	0	1	2
	1	2p	–1, 0, +1	3	6
3	0	3s	0	1	2
	1	3p	–1, 0, +1	3	6
	2	3d	–2, –1, 0, +1, +2	5	10
4	0	4s	0	1	2
	1	4p	–1, 0, +1	3	6
	2	4d	–2, –1, 0, +1, +2	5	10
	3	4f	–3, –2, –1, 0, +1, +2, +3	7	14

Limits of Quantum Numbers		
$n = 1, 2, 3, \ldots$	$\ell = 0, 1, \ldots, (n - 1)$	$m_\ell = -\ell, \ldots, 0, \ldots, +\ell$

Hund's Rule of Maximum Multiplicity and the Aufbau Principle

It is important to remember that, when there is more than one orbital at a particular energy level, such as three p orbitals or five d orbitals, only one electron will fill each orbital until each has one electron. This principle, that an electron occupies the lowest energy orbital that can receive it, is called the **Aufbau Principle.** After this, pairing will occur with the addition of one more electron to each orbital. This principle, called **Hund's Rule of Maximum Multiplicity**, is shown in Table 2, where each arrowhead indicates an electron (⬆).

TIP

Know the Aufbau Principle and Hund's Rule of Maximum Multiplicity.

Table 2. Orbital Notations

Chemical Symbol	Atomic No.	Orbital Notation 1s	Orbital Notation 2s	Orbital Notation 2p	Electron Orbital Notation
H	1	⊡	☐	☐☐☐	$1s^1$
He	2	⊡⊡	☐	☐☐☐	$1s^2$
Li	3	⊡⊡	⊡	☐☐☐	$1s^2 2s^1$
Be	4	⊡⊡	⊡⊡	☐☐☐	$1s^2 2s^2$
B	5	⊡⊡	⊡⊡	⊡☐☐	$1s^2 2s^2 2p^1$
C	6	⊡⊡	⊡⊡	⊡⊡☐	$1s^2 2s^2 2p^2$
N	7	⊡⊡	⊡⊡	⊡⊡⊡	$1s^2 2s^2 2p^3$
O	8	⊡⊡	⊡⊡	⊡⊡⊡⊡	$1s^2 2s^2 2p^4$
F	9	⊡⊡	⊡⊡	⊡⊡⊡⊡⊡	$1s^2 2s^2 2p^5$
Ne	10	⊡⊡	⊡⊡	⊡⊡⊡⊡⊡⊡	$1s^2 2s^2 2p^6$

Maximum electrons in orbitals at a particular sublevel:

s = 2 (one orbital)
p = 6 (three orbitals)
d = 10 (five orbitals)
f = 14 (seven orbitals)

If each orbital is indicated in an energy diagram as a square (☐), we can show relative energies in a chart such as Figure 9. If this drawing represented a ravine with the energy levels as ledges onto which stones could come to rest only in numbers equal to the squares for orbitals, then pushing stones into the ravine would cause the stones to lose their potential energy as they dropped to the lowest potential level available to them. Much the same is true for electrons.

SUBLEVELS AND ELECTRON CONFIGURATION
Order of Filling and Notation

The sublevels do not fill up in numerical order, and the pattern of filling is shown on the right side of the approximate relative energy levels chart (Figure 9). In the first instance of failure to follow numerical order, the $4s$ fills before the $3d$. (Study Figure 9 carefully before going on.)

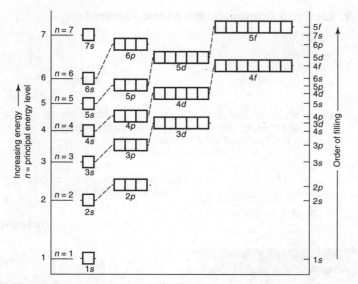

Figure 9. Approximate Relative Energy Level of Subshells

$_{19}$K $\qquad$ $1s^2 2s^2 2p^6 3s^2 3p^6 4s^1$

$_{20}$Ca $\qquad$ $1s^2 2s^2 2p^6 3s^2 3p^6 4s^2$

$_{21}$Sc $\qquad$ $1s^2 2s^2 2p^6 3s^2 3p^6 4s^2 3d^1$ (note $4s$ filled before $3d$)

There is a more stable configuration to a half-filled or filled sublevel, so at atomic number 24 the $3d$ sublevel becomes half-filled by taking a $4s$ electron;

$_{24}$Cr $\qquad$ $1s^2 2s^2 2p^6 3s^2 3p^6 3d^5 4s^1$

and at atomic number 29 the $3d$ becomes filled by taking a $4s$ electron:

$_{29}$Cu $\qquad$ $1s^2 2s^2 2p^6 3s^2 3p^6 3d^{10} 4s^1$

Table 3 shows the electron configurations of the elements. A triangular mark indicates an outer-level electron dropping back to a lower unfilled orbital. These phenomena are exceptions to the Aufbau Principle. By following the atomic numbers throughout this chart, you will get the same order of filling as shown in Figure 9.

Table 3. Electron Configuration of the Elements

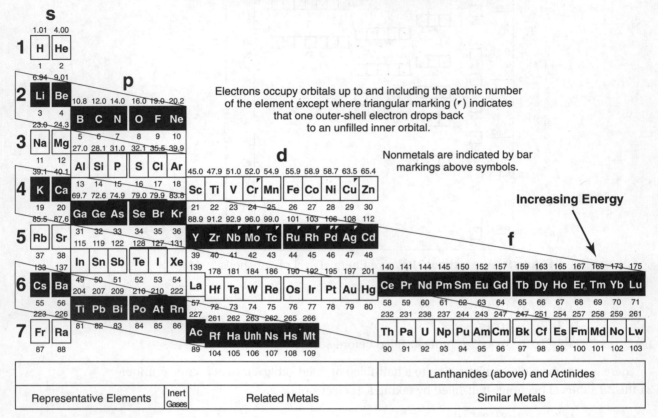

Note: Follow order of the atomic numbers to ascertain the order of filling.

By following the atomic numbers in numerical order in Table 3 you can plot the order of filling of the orbitals for every element shown.

A simplified method of showing the order in which the orbitals are filled is to use the following diagram. It works for all the naturally occurring elements through radium, atomic number 88.

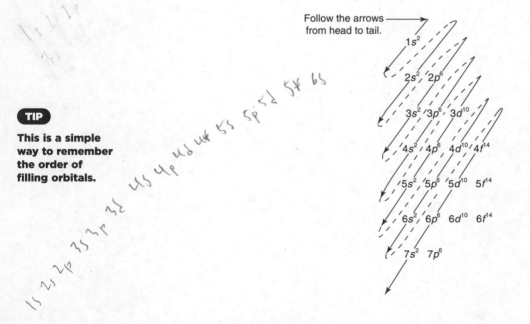

TIP

This is a simple way to remember the order of filling orbitals.

Start by drawing the diagonal arrows through the diagram as shown. The order of filling can be charted by following each arrow from tail to head and then to the tail of the next one. In this way you get the same order of filling as is shown in Figure 9 and Table 3:

$$1s^2 2s^2 2p^6 3s^2 3p^6 4s^2 3d^{10} 4p^6 5s^2 4d^{10} 5p^6 6s^2 4f^{14} 5d^{10} 6p^6 7s^2$$

Lewis Structures (Electron Dot Notation)

In 1916 **G. N. Lewis** devised the electron dot notation, which may be used in place of the electron configuration notation. The electron dot notation shows only the chemical symbol surrounded by dots to represent the electrons in the incomplete outer level. Examples are:

$$K, \cdot \ddot{As} \cdot, \quad Sr, \quad :\ddot{I}:, \quad \text{and} \quad :\ddot{Rn}:$$

The symbol denotes the nucleus and all electrons except the valence electrons. The dots are arranged at the four sides of the symbol and are paired when appropriate. In the examples above, the depicted electrons are the valence electrons found in the outer energy level orbitals.

$4s^1$ is shown for potassium (K)

$4s^2 4p^3$ are shown for arsenic (As)

$5s^2$ is shown for strontium (Sr)

$5s^2 5p^5$ are shown for iodine (I)

$6s^2 6p^6$ are shown for radon (Rn)

Noble Gas Notation

Another method of simplifying the electron distribution to the orbitals is called the noble gas notation. In this method you represent all of the lower filled orbitals up to the closest noble gas. By enclosing its symbol in brackets, it represents all of the complete noble gas configuration. Then the remaining orbitals are written in the usual way. An example of this can be shown by using the third period of elements. By using neon as the noble gas, you write [Ne] to represent its orbital structure, which is $1s^2 2s^2 2p^6$. This allows you to write an element like sodium as [Ne] $3s^1$, which is called sodium's noble gas notation. The table in the next section shows the noble gas notations of some of the transition elements in the fourth period of elements. Notice that the base structure of argon is used and represented as [Ar].

TRANSITION ELEMENTS

The elements involved with the filling of a d sublevel with electrons after two electrons are in the s sublevel of the next principal energy level are often referred to as the **transition elements**. The first examples of these are the elements between calcium, atomic number 20, and gallium, atomic number 31. Their electron configurations are the same in the $1s$, $2s$, $2p$, $3s$, and $3p$ sublevels. It is the filling of the $3d$ and changes in the $4s$ sublevels that are of interest, as shown in the following table.

Electron Configuration of Some Elements in the Fourth Period

Element	Symbol	Atomic No.	1s²	2s²	2p⁶	3s²	3p⁶	3d¹⁰	4s	Noble Gas Notation
Scandium	Sc	21						1	2	[Ar] $d^1 s^2$
Titanium	Ti	22	All					2	2	[Ar] $d^2 s^2$
Vanadium	V	23						3	2	[Ar] $d^3 s^2$
Chromium	Cr	24			the			5	1*	[Ar] $d^5 s^1$
Manganese	Mn	25						5	2	[Ar] $d^5 s^2$
Iron	Fe	26					same	6	2	[Ar] $d^6 s^2$
Cobalt	Co	27						7	2	[Ar] $d^7 s^2$
Nickel	Ni	28						8	2	[Ar] $d^8 s^2$
Copper	Cu	29						10	1*	[Ar] $d^{10} s^1$
Zinc	Zn	30						10	2	[Ar] $d^{10} s^2$

The asterisk (*) shows where a $4s$ electron is promoted into the $3d$ sublevel. This is because the $3d$ and $4s$ sublevels are very close in energy and that there is a state of greater stability in half-filled and filled sublevels. Therefore, chromium gains stability by the movement of an electron from the $4s$ sublevel into the $3d$ sublevel to give a half-filled $3d$ sublevel. It then has one electron in each of the five orbitals of the $3d$ sublevel. In copper, the movement of one $4s$ electron into the $3d$ sublevel gives the $3d$ sublevel a completely filled configuration.

The fact that the electrons in the $3d$ and $4s$ sublevels are so close in energy levels leads to the possibility of some or all the $3d$ electrons being involved in chemical bonding. With the variable number of electrons available for bonding, it is not surprising that transition elements can exhibit variable oxidation numbers. An example is manganese with possible oxidation numbers of +2, +3, +4, +6, and +7, which correspond, respectively, to the use of no, one, two, four, and five electrons from the $3d$ sublevel.

The transition elements in the other periods of the table show this same type of anomaly, as they have d sublevels filling in the same manner.

Transition elements have several common characteristic properties.

- They often form colored compounds.
- They can have a variety of oxidation states.
- At least one of their compounds has an incomplete d electron subshell.
- They are often good catalysts.
- They are silvery blue at room temperature (except copper and gold).
- They are solids at room temperature (except mercury).
- They form complex ions.
- They are often paramagnetic due to unpaired electrons.

PERIODIC TABLE OF THE ELEMENTS

History

The history of the development of a systematic pattern for the elements includes the work of a number of scientists such as John Newlands, who, in 1863, proposed the idea of repeating octaves of properties.

Dimitri I. Mendeleev in 1869 proposed a table containing 17 columns and is usually given credit for the first periodic table since he arranged elements in groups according to their atomic weights and properties. It is interesting to note that Lothar Meyer proposed a similar arrangement about the same time. In 1871 Mendeleev rearranged some elements and proposed a table of eight columns, obtained by splitting each of the long periods across into a period of seven elements, an eighth group containing the three central elements (such as Fe, Co, Ni), and a second period of seven elements. The first and second periods of seven across were later distinguished by use of the letters A and B attached to the group symbols, which were Roman numerals. This nomenclature of periods (IA, IIA, etc.) has been revised in the present periodic table, even in the extended form of assigning Arabic numbers from 1–18 as shown in Table 4.

TIP

Mendeleev is given credit for the first Periodic Table. It was based on placement by properties.

Table 4. Periodic Table Properties

TIP

Periods are the horizontal rows 1–7. Groups are the vertical columns 1–18.

Period / Light metals (Hydrogen doesn't fit well anywhere.)

1   2   Group Number

Noble gases   18   O

Nonmetals   13   14   15   16   17

Period	1	2											13	14	15	16	17	18
1	H			Transition Metals														He
				brittle					ductile			Low melting						
2	Li	Be	3	4	5	6	7	8	9	10	11	12	B	C	N	O	F	Ne
3	Na	Mg											Al	Si	P	S	Cl	Ar
4	K	Ca	Sc	Ti	V	Cr	Mn	Fe	Co	Ni	Cu	Zn	Ga	Ge	As	Se	Br	Kr
5	Rb	Sr	Y	Zr	Nb	Mo	Tc	Ru	Rh	Pd	Ag	Cd	In	Sn	Sb	Te	I	Xe
6	Cs	Ba	rare earth	Hf	Ta	W	Re	Os	Ir	Pt	Au	Hg	Tl	Pb	Bi	Po	At	Rn
7	Fr	Ra																

Amphoteric elements along this line are called metalloids.

←— *s* subshell —→ ←————— *d* subshell —————→ ←— *p* subshell —→

——————————————— Acid properties increase ————————————→

←—— Base properties increase ——————————————

——————————— Atomic radii decrease ———————————→

———————— Ionization energy increases ————————→

———————— Nonmetallic properties increase ————————→

TIP

Know these relationships across the table.

Mendeleev's table had the elements arranged by atomic weights with recurring properties in a periodic manner. Where atomic weight placement disagreed with the properties that should occur in a particular spot in the table, Mendeleev gave preference to the element with the correct properties. He even predicted elements for places that were not yet occupied in the table. These predictions proved to be amazingly accurate and led to wide acceptance of his table.

Periodic Law

Henry Moseley stated, after his work with X-ray spectra in the early 1900s, that the properties of elements are a periodic function of their atomic numbers, thus changing the basis of the periodic law from atomic weight to atomic number. This is the present statement of the **periodic law**.

The Table

The horizontal rows of the periodic table are called **periods** or **rows**. There are seven periods, each of which begins with an atom having only one valence electron and ends with a complete outer shell structure of an inert gas. The first three periods are short, consisting of 2, 8, and 8 elements, respectively. Periods 4 and 5 are longer, with 18 each, while period 6 has 32 elements, and period 7 is incomplete with 22 elements, most of which are radioactive and do not occur in nature.

In Table 4, you should note the relationship of the length of the periods to the orbital structure of the elements. In the first period, the $1s^2$ orbital is filled with the noble gas helium, He. The second period begins with the $2s^1$ orbital and ends with the filling of the $2p^6$ orbital, again with a noble gas, neon, Ne. The same pattern is repeated in period three, going from $3s^1$ to $3p^6$. The eight elements from sodium, Na, to argon, Ar, complete the filling of the $n = 3$ energy level with $3s^2$ and $3p^6$. In the fourth period, the first two elements fill the $4s^2$ orbital. Beyond calcium, Ca, the pattern becomes more complicated. As discussed in the section "Order of Filling and Notation," the next orbitals to be filled are the five $3d$ orbitals whose elements represent transition elements. Then the three $4p$ orbitals are filled, ending with the noble gas krypton, Kr. The fifth period is similar to the fourth period. The $5s^2$ orbital filling is represented by rubidium, Rb, and strontium, Sr, both of which resemble the elements directly above them on the table. Next come the transition elements that fill the five $4d$ orbitals before the next group of elements, from indium, In, to xenon, Xe, complete the three $5p$ orbitals. (Table 3 should be consulted for the irregularities that occur as the d orbitals fill.) The sixth period follows much the same pattern and has the filling order $6s^2$, $4f^{14}$, $5d^{10}$, $6p^6$. Here, again, irregularities occur and can best be followed by using Table 3.

The vertical columns of the Periodic Table are called **groups** or **families**. The elements in a group exhibit similar or related properties. In 1984 the IUPAC agreed that the groups would be numbered 1 through 18.

PROPERTIES RELATED TO THE PERIODIC TABLE

Metals are found on the left of the chart (see Table 4) with the most active metal in the lower left corner. Nonmetals are found on the right side with the most active nonmetal in the upper right-hand corner. The noble or inert gases are on the far right. Since the most active metals react with water to form bases, the Group 1 metals are called alkali metals. As you proceed to the right, the base-forming property decreases and the acid-forming properties increase. The metals in the first two groups are the light metals, and those toward the center are heavy metals.

The elements found along the dark line in the Periodic Table (Table 4) are called **metalloids**. These elements have certain characteristics of metals and other characteristics of nonmetals. Some examples of metalloids are boron, silicon, arsenic, and tellurium.

Here are some important general summary statements about the Periodic Table:

- Acid-forming properties increase from left to right on the table.
- Base-forming properties are high on the left side and decrease to the right.
- The atomic radii of elements decrease from left to right across a period.
- First ionization energies increase from left to right across a period.
- Metallic properties are greatest on the left side of the table and decrease to the right.
- Nonmetallic properties are greatest on the right side of the table and decrease to the left.

TIP

These are important trends to remember.

Study Table 4 carefully because it summarizes many of these properties. For a more detailed description of metals, alloys, and metalloids, see pages 274–276 in Chapter 13.

Radii of Atoms

The size of an atom is difficult to describe because atoms have no definite outer boundary. Unlike a volleyball, an atom does not have a definite circumference.

To overcome this problem, the size of an atom is estimated by describing its radius. In metals, this is done by measuring the distance between two nuclei in the solid state and dividing this distance by 2. Such measurements can be made with X-ray diffraction. For a nonmetallic element that exists in pure form as a molecule, such as chlorine, measurements can be made of the distance between nuclei for two atoms covalently bonded together. Half of this distance is referred to as the **covalent radius**. The method for finding the covalent radius of the chlorine atom is illustrated in the following diagram.

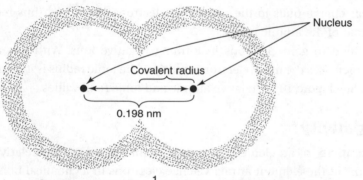

Covalent radius of Cl = $\frac{1}{2}$ (0.198) nm = 0.099 nm

Figure 10 shows the relative **atomic and ionic radii** for some elements. As you review this chart, you should note two trends:

1. Atomic radii decrease from left to right across a period in the Periodic Table (until the noble gases).
2. Atomic radii increase from top to bottom in a group or family.

The reason for these trends will become clear in the following discussions.

TIP

Know these trends in atomic radii.

Atomic Radii in Periods

Since the number of electrons in the outer principal energy level increases as you go from left to right in each period, the corresponding increase in the nuclear charge because of the additional protons pulls the electrons more tightly around the nucleus. This attraction more than balances the repulsion between the added electrons and the other electrons, and the radius is generally reduced. The inert gas at the end of the period has a slight increase in radius because of the electron repulsion in the filled outer principal energy level. For example, lithium's atomic radius in Figure 10 is 0.152 nm at the one end of period 2 whereas fluorine has a radius of only 0.064 nm at the far end of the period. This trend can be seen in Figure 10 across every period.

Atomic Radii in Groups

For a group of elements, the atoms of each successive member have another outer principal energy level in the electron configuration, and the electrons there are held less tightly by the nucleus. This is so because of their increased distance from the nuclear positive charge and the shielding of this positive charge by all the core electrons. Therefore, the atomic radius increases down a group. For example, oxygen's atomic radius in Figure 10 is 0.066 nm at the top of group 16, whereas polonium has a radius of 0.167 nm at the bottom of the same group. This trend can be seen in Figure 10 down every group.

Ionic Radius Compared with Atomic Radius

Metals tend to lose electrons in forming positive ions. With this loss of negative charge, the positive nuclear charge pulls in the remaining electrons closer and thus reduces the ionic radius below that of the atomic radius.

Nonmetals tend to gain electrons in forming negative ions. With this added negative charge, which increases the inner electron repulsion, the ionic radius is increased beyond the atomic radius. See Figure 10 for relative atomic and ionic radii values.

Electronegativity

The **electronegativity** of an element is a number that measures the relative strength with which the atoms of the element attract valence electrons in a chemical bond. This electronegativity number is based on an arbitrary scale going from 0 to 4. In general, a value of less than 2 indicates a metal.

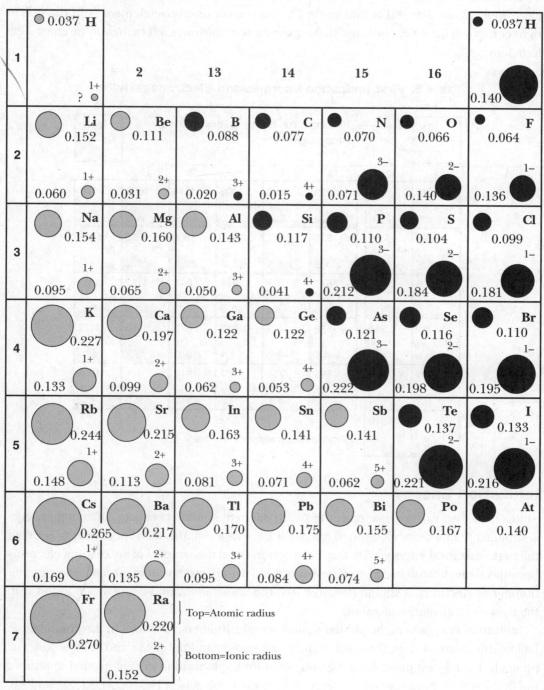

Figure 10. Radii of Some Atoms and Ions (in nanometers)

Notes: The atomic radius is usually given for metal atoms, which are shown in gray, and the covalent radius is usually given for atoms of nonmetals, which are shown in black.

Notice in Table 5 that the electronegativity decreases down a group and increases across a period. The inert gases can be ignored. The lower the electronegativity number, the more electropositive an element is said to be. The most electronegative element is in the upper right corner—F, fluorine. The most electropositive is in the lower left corner of the chart—Fr, francium.

Table 5. First Ionization Energies and Electronegativities

First Ionization Energy (kJ/mol of atoms)
Electronegativity*

1	2		13	14	15	16	17	18
1310 H 2.2								2372 He
523 Li 1.0	900 Be 1.5		799 B 2.0	1000 C 2.6	1406 N 3.1	1314 O 3.5	1682 F 4.0	2079 Ne
498 Na 0.9	736 Mg 1.2		577 Al 1.5	787 Si 1.9	1013 P 2.2	1000 S 2.6	1255 Cl 3.2	1519 Ar
418 K 0.8	590 Ca 1.0		577 Ga 1.6	761 Ge 1.9	946 As 2.0	941 Se 2.5	1142 Br 2.9	1351 Kr
401 Rb 0.8	549 Sr 1.0		556 In 1.7	707 Sn 1.8	833 Sb 2.1	870 Te 2.3	1008 I 2.7	1172 Xe
377 Cs 0.7	502 Ba 0.9		590 Tl 1.8	715 Pb 1.8	707 Bi 1.9	812 Po 2.0	At 2.2	1038 Rn
Fr 0.7	510 Ra 0.9							

*Arbitrary scale based on fluorine = 4.0

TIP

The most electronegative element is F, in the upper right corner.

TIP

The least electronegative element is Fr, in the lower left corner.

TIP

Know this definition of the first ionization energy.

Ionization Energy

Atoms hold their valence electrons, then, with different amounts of energy. If enough energy is supplied to one outer electron to remove it from its atom, this amount of energy is called the **first ionization energy**. With the first electron gone, the removal of succeeding electrons becomes more difficult because of the loss of repulsive effects that were present with a greater number of electrons. It should be noted that the lowest ionization energies are found with the least electronegative elements.

Ionization energies can be plotted against atomic numbers, as shown in the graph below. Follow this discussion on the graph to help you understand the peaks and valleys. Not surprisingly, the highest peaks on the graph occur for the ionization energy needed to remove the first electron from the outer energy level of the noble gases, He, Ne, Ar, Kr, Xe, and Rn, because of the stability of the filled *p* orbitals in the outer energy level. Notice that, even among these elements, the energy needed gradually declines. This can be explained by considering the distance of the involved energy level from the positively charged nucleus. With each succeeding noble gas, a more distant *p* orbital is involved, therefore making it easier to remove an electron from the positive attraction of the nucleus. Besides this consideration, as more energy levels are added to the atomic structure as the atomic number increases, the additional negative fields associated with the additional electrons screen out some of the positive attraction of the nucleus. Within a period such as that from Li to Ne, the ionization energy generally increases. The lowest occurs when a lone electron occupies the outer *s* orbital, as in Li. As the *s* orbital fills with two electrons at atomic number 4, Be, the added stability of a filled 2*s* orbital explains the small peak at 4. At atomic number 5, B, a lone electron occupies the 2*p* orbital. This electron can be removed with less energy, and therefore a dip occurs in the graph. With the 2*p* orbitals filling according to Hund's Rule (refer to Table 2, page 72), with only one electron in each orbital before pairing occurs, again a slightly more stable situation and, therefore, another small peak occurs at atomic number 7. After this peak, a dip and continual increases occur until the 2*p* orbitals are completely filled with paired electrons at the noble gas Ne. As you continue to associate the atomic number with the line in the chart, you find peaks occurring in the same general pattern. These peaks are always related to the state of filling of the orbitals involved and the distance of these orbitals from the nucleus.

TIP

Know how this trend relates to the chart below.

Can you explain the peaks?

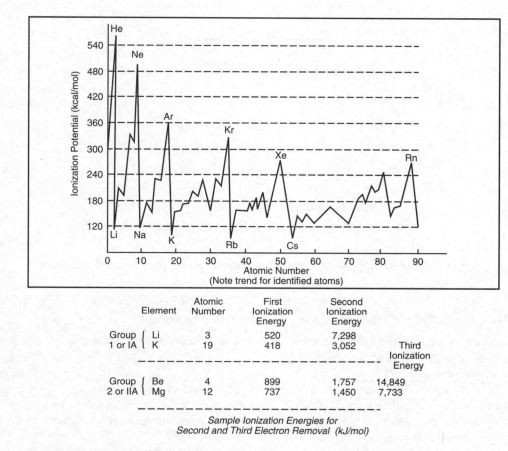

TIP

Know the reason for the peaks and valleys.

	Element	Atomic Number	First Ionization Energy	Second Ionization Energy	
Group 1 or IA {	Li	3	520	7,298	
	K	19	418	3,052	Third Ionization Energy
Group 2 or IIA {	Be	4	899	1,757	14,849
	Mg	12	737	1,450	7,733

Sample Ionization Energies for
Second and Third Electron Removal (kJ/mol)

NUCLEAR TRANSFORMATIONS AND STABILITY

At the same time advances in atomic theory were occurring, scientists were noticing phenomena associated with emissions from the nucleus of atoms in the form of "X-rays." While Roentgen announced the discovery of X-rays, Becquerel was exploring the phosphorescence of some materials. Becquerel's work received little attention until early in 1898, when Marie and Pierre Curie entered the picture.

Searching for the source of the intense radiation in uranium ore, Marie and Pierre Curie used tons of it to isolate very small quantities of two new elements, radium and polonium, both radioactive. Along with Becquerel, the Curies shared the Nobel Prize in Physics in 1903.

THE NATURE OF RADIOACTIVE EMISSIONS

While the early separation experiments were in progress, an understanding was slowly being gained of the nature of the spontaneous emission from the various radioactive elements. Becquerel thought at first that there were simply X-rays, but THREE different kinds of radioactive emission, now called **alpha particles**, **beta particles**, and **gamma rays**, were soon found. We now know that alpha particles are positively charged particles of helium nuclei, beta particles are streams of high-speed electrons, and gamma rays are high-energy radiations similar to X-rays. The emission of these three types of radiation is depicted below.

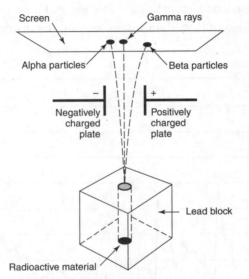

Deflection of Radioactive Emissions

TIP

Know the types of radiation and the characteristics of each.

The important characteristics of each type of radiation can be summarized as follows:

> ### *Alpha Particle*   (helium nucleus $^4_2$He) Positively charged, 2+
>
> 1. Ejection reduces the atomic number by 2, the atomic weight by 4 amu.
> 2. High energy, relative velocity.
> 3. Range: about 5 cm in air.
> 4. Shielding needed: stopped by the thickness of a sheet of paper, skin.
> 5. Interactions: produces about 100,000 ionizations per centimeter; repelled by the positively charged nucleus; attracts electrons, but does not capture them until its speed is much reduced.
> 6. An example: Thorium-230 has an unstable nucleus and undergoes radioactive decay through alpha emission. The nuclear equation that describes this reaction is:
>
> $$^{230}_{90}\text{Th} \rightarrow {}^{4}_{2}\text{He} + {}^{226}_{88}\text{Ra}$$
>
> In a decay reaction like this, the initial element (thorium-230) is called the parent nuclide and the resulting element (radium-226) is called the daughter nuclide.

> ### Beta Particle (fast electron) Negatively charged, 1–
>
> 1. Ejected when a *neutron* decays into a proton and an electron.
> 2. High velocity, low energy.
> 3. Range: about 12 m.
> 4. Shielding needed: stopped by 1 cm of aluminum or thickness of average book.
> 5. Interactions: weak because of high velocity, but produces about 100 ionizations per centimeter.
> 6. An example: Protactinium-234 is a radioactive nuclide that undergoes beta emission. The nuclear equation is:
>
> $$^{234}_{91}Pa \rightarrow {}^{234}_{92}U + {}^{0}_{-1}e$$

> ### Gamma Radiation (electromagnetic radiation identical with light; high energy) No charge
>
> 1. Beta particles and gamma rays are usually emitted together; after a beta is emitted, a gamma ray follows.
> 2. Arrangement in nucleus is unknown. Same velocity as visible light.
> 3. Range: no specific range.
> 4. Shielding needed: about 13 cm of lead.
> 5. Interactions: weak of itself; gives energy to electrons, which then perform the ionization.

METHODS OF DETECTION OF ALPHA, BETA, AND GAMMA RAYS

All methods of detection of these radiations rely on their ability to ionize. Three methods are in common use.

1. *Photographic plate.* The fogging of a photographic emulsion led to the discovery of radioactivity. If this emulsion is viewed under a high-power microscope, it is seen that beta and gamma rays cause the silver bromide grains to develop in a scattered fashion.
2. *Scintillation counter.* A fluorescent screen (e.g., ZnS) will show the presence of electrons and X-rays, as already mentioned. If the screen is viewed with a magnifying eyepiece, small flashes of light, called scintillations, will be observed. By observing the scintillations, one not only can detect the presence of alpha particles, but also can actually count them.
3. *Geiger counter.* This instrument is perhaps the most widely used at the present time for determining individual radiation. Any particle that will produce an ion gives rise to an avalanche of ions, so the type of particle cannot be identified. However, each individual particle can be detected.

DECAY SERIES, TRANSMUTATIONS, AND HALF-LIFE

The nuclei of uranium, radium, and other radioactive elements are continually disintegrating. It should be emphasized that spontaneous disintegration produces the gas known as radon. The time required for half of the atoms of a radioactive nuclide to decay is called its **half-life**.

TIP

Know how to use half-life to determine the age of a substance. See "Radioactive Dating" for more info.

For example, for radium, we know that, on the average, half of all the radium nuclei present will have disintegrated to radon in 1,590 years. In another 1,590 years, half of this remainder will decay, and so on. When a radium atom disintegrates, it loses an alpha particle, which eventually, upon gaining two electrons, becomes a neutral helium atom. The remainder of the atom becomes **radon**.

Such a conversion of an element to a new element (because of a change in the number of protons) is called a **transmutation**. This transmutation can be produced artificially by bombarding the nuclei of a substance with various particles from a particle accelerator, such as the cyclotron.

The following uranium-radium disintegration series shows how a radioactive atom may change when it loses each kind of particle. Note that an atomic number is shown by a subscript ($_{92}$U), and the isotopic mass by a superscript (^{238}U). The alpha particle is represented by the Greek symbol α, and the beta particle by β.

$$^{238}_{92}U \xrightarrow{-\alpha} {}^{234}_{90}Th \xrightarrow{-\beta} {}^{234}_{91}Pa \xrightarrow{-\beta} {}^{234}_{92}U \xrightarrow{-\alpha} {}^{230}_{90}Th \xrightarrow{-\alpha}$$

$$^{226}_{88}Ra \xrightarrow{-\alpha} {}^{222}_{86}Rn \xrightarrow{-\alpha} {}^{218}_{84}Po \xrightarrow{-\alpha} {}^{214}_{82}Pb \xrightarrow{-\beta} {}^{214}_{83}Bi \xrightarrow{-\beta}$$

$$^{214}_{84}Po \xrightarrow{-\alpha} {}^{210}_{82}Pb \xrightarrow{-\beta} {}^{210}_{83}Bi \xrightarrow{-\beta} {}^{210}_{84}Po \xrightarrow{-\alpha} {}^{206}_{82}Pb \text{ (stable)}$$

The changes that occur in radioactive reactions and the subatomic particles involved are summarized in the following charts.

Radioactive Decay and Nuclear Change

Type of Decay	Decay Particle	Particle Mass	Particle Charge	Change in Nucleon Number	Change in Atomic Number
Alpha decay	α	4	2+	Decreases by 4	Decreases by 2
Beta decay	β	0	1−	No change	Increases by 1
Gamma radiation	γ	0	0	No change	No change
Positron emission	β^+	0	1+	No change	Decreases by 1
Electron capture	e^-	0	1−	No change	Decreases by 1

Nuclear Symbols for Subatomic Particles

Particle	Symbols	Nuclear Symbols
Proton	p	^1_1p or ^1_1H
Neutron	n	^1_0n
Electron	e^- or β^-	$^0_{-1}e$ or $^0_{-1}\beta$
Positron	e^+ or β^+	$^0_{+1}e$ or $^0_{+1}\beta$
Alpha particle	α	^4_2He or $^4_{+2}\alpha$
Beta particle	β or β^-	$^0_{-1}e$ or $^0_{-1}\beta$
Gamma ray	γ	$^0_0\gamma$

This is shown graphically in the radioactive decay series below.

RADIOACTIVE DATING

A helpful application of radioactive decay is in the determination of the ages of substances such as rocks and relics that have bits of organic material trapped in them. Because carbon-14 has a half-life of about 5,700 years and occurs in the remains of organic materials, it has been useful in dating these materials. A small percentage of CO_2 in the atmosphere contains carbon-14. The stable isotope of carbon is carbon-12. Carbon-14 is a beta emitter and decays to form nitrogen-14:

$$^{14}_{6}C \rightarrow {}^{14}_{7}N + {}^{0}_{-1}e$$

In any living organism, the ratio of carbon-14 to carbon-12 is the same as in the atmosphere because of the constant interchange of materials between organism and surroundings. When an organism dies, this interaction stops, and the carbon-14 gradually decays to nitrogen. By comparing the relative amounts of carbon-14 and carbon-12 in the remains, the age of the organism can be established. Carbon-14 has a half-life of 5,700 years. If a sample of wood had originally contained 5 grams of carbon-14 and now had only half or 2.5 grams of carbon-14, its age would be 5,700 years. In other words, the old wood emits half as much beta radiation per gram of carbon as that emitted by living plant tissues. This method was used to determine the age of the Dead Sea Scrolls (about 1,900 years) and has been found to be in agreement with several other dating techniques.

NUCLEAR REACTIONS

Nuclear fission reactions have been in use since the 1940s. The first atomic bombs used in 1945 were nuclear fission bombs. Since that time, many countries, including our own, have put nuclear fission power plants into use to provide a new energy source for electrical energy. Basically, a nuclear fission reaction is the splitting of a heavy nucleus into two or more lighter nuclei.

➡ EXAMPLE:

U-235 is bombarded with slow neutrons to produce Ba-139, Kr-94, or other isotopes and also 3 fast-moving neutrons.

$$^{235}_{92}U + {}^{1}_{0}n \rightarrow {}^{139}_{56}Ba + {}^{94}_{36}Kr + 3\,{}^{1}_{0}n + \text{Energy}$$

A nuclear chain reaction is a reaction in which an initial step, such as the reaction above, leads to a succession of repeating steps that continues indefinitely. Nuclear chain reactions are used in nuclear reactors and nuclear bombs.

A **nuclear fusion reaction** is the combination of very light nuclei to make a heavier nucleus. Extremely high temperatures and pressures are required in order to overcome the repulsive forces of the two nuclei. Fusion has been achieved only in hydrogen bombs. Scientists are still trying to harness this reaction for domestic uses. The following examples show basically how the reactions occur.

➡ EXAMPLES: _____

Two deuterium atoms combining

$$_1^2H + _1^2H \rightarrow _2^4He + Energy$$

Tritium combining with hydrogen

$$_1^3H + _1^1H \rightarrow _2^4He + Energy$$

The energy released in a nuclear reaction (either fission or fusion) comes from the fractional amount of mass converted into energy. Nuclear changes convert matter into energy. Energy released during nuclear reactions is much greater than the energy released during chemical reactions.

CHAPTER SUMMARY

The following terms summarize all the concepts and ideas that were introduced in this chapter. You should be able to explain their meaning and how you would use them in chemistry. They appear in boldface type in this chapter to draw your attention to them. The boldface type also makes it easier for you to look them up if you need to. You could also use Internet search engines like _google.com_ on your computer to get a quick and expanded explanation of these terms, laws, and formulas.

alpha particle	half-life	oxidation number
atomic mass	Hund's Rule	Pauli Exclusion Principle
atomic number	inert atoms	periodic law
atomic radii	ionic radii	period or row
Aufbau Principle	ionization energy	phosphorescence
beta particle	isotopes	quantum numbers
Bohr model	Lewis structure	quantum theory
covalent radius	Mendeleev	radon
Dalton's atomic theory	metalloids	_s, p, d, f_ orbitals
electron	Moseley	transmutation
electronegativity	neutron	transition elements
gamma ray	nuclear fission	uncertainty principle
Geiger counter	nuclear fusion	valence electrons
group or family	nucleus	wave-mechanical model

Online content that reinforces the major concepts discussed in this chapter can be found at the following Internet addresses if they are still available. *Some may have been changed or deleted.*

Emission Spectroscopy

http://phys.educ.ksu.edu/vqm/html/emission.html

This site offers interactive computer visualizations of the quantum effects of emission spectra for a variety of substances.

Interactive Periodic Table

http://www.webelements.com

This is an interactive online periodic table available on the Web.

www.ptable.com/

This site shows properties, orbitals, and isotopes in chart form.

Radioactive Dating

http://chemistry.about.com/od/workedchemistryproblems/a/c14dating.htm

This Web page offers a tutorial on radioactive dating using carbon-14.

1. The two main regions of an atom are the

   (A) principal energy levels and energy sublevels
   (B) nucleus and kernel
   (C) nucleus and energy levels
   (D) planetary electrons and energy levels

2. The lowest principal quantum number that an electron can have is

   (A) 0
   (B) 1
   (C) 2
   (D) 3

3. The sublevel that has only one orbital is identified by the letter

   (A) s
   (B) p
   (C) d
   (D) f

4. The sublevel that can be occupied by a maximum of 10 electrons is identified by the letter

   (A) d
   (B) f
   (C) p
   (D) s

5. An orbital may never be occupied by

   (A) 1 electron
   (B) 2 electrons
   (C) 3 electrons
   (D) 0 electrons

6. An atom of beryllium consists of 4 protons, 5 neutrons, and 4 electrons. The mass number of this atom is

   (A) 13
   (B) 9
   (C) 8
   (D) 5

7. The number of orbitals in the second principal energy level, $n = 2$, of an atom is

   (A) 1
   (B) 9
   (C) 16
   (D) 4

8. Lewis structure consists of the symbol representing the element and an arrangement of dots that usually shows

   (A) the atomic number
   (B) the atomic mass
   (C) the number of neutrons
   (D) the electrons in the outermost energy level

9. Chlorine is represented by the Lewis structure $:\overset{\cdot}{Cl}:$. The atom that would be represented by an identical electron-dot arrangement has the atomic number

   (A) 7
   (B) 9
   (C) 15
   (D) 19

10. Radioactive changes differ from ordinary chemical changes because radioactive changes

    (A) involve changes in the nucleus
    (B) are explosive
    (C) absorb energy
    (D) release energy

11. Isotopes of uranium have different

    (A) atomic numbers
    (B) atomic masses
    (C) numbers of planetary electrons
    (D) numbers of protons

12. Atoms of ^{235}U and ^{238}U differ in structure by three

    (A) electrons
    (B) isotopes
    (C) neutrons
    (D) protons

13. The use of radioactive isotopes has produced promising results in the treatment of certain types of

    (A) cancer
    (B) heart disease
    (C) pneumonia
    (D) diabetes

14. The emission of a beta particle results in a new element with the atomic number

    (A) increased by 1
    (B) increased by 2
    (C) decreased by 1
    (D) decreased by 2

The following questions are in the format that is used on the **SAT Subject Test in Chemistry**. If you are not familiar with these types of questions, study pages xiii–xvi before doing the remainder of the review questions.

Questions 15–19

Use this abbreviated periodic table to answer the following questions.

Period ↓	Group 1	Group 2		Group 14	Group 15	Group 16	Group 17
2	(A) **Li**	**Be**		(D) **C**	**N**	**O**	(E) **F**
3	(B) **Na**	**Mg**					
4	(C) **K**						

15. Which neutral atom has an outer energy level configuration of $3s^1$?

16. An atom of which element shows the greatest affinity for an additional electron?

17. Which element is the most active metal of this group?

18. Which element has the lowest electronegativity of this group?

19. Which element has an outer orbital configuration of $2s^2 2p^2$?

	I		II

20. Si, with an atomic number of 14, will probably exhibit an oxidation number of +4 in a compound — BECAUSE — silicon is an element that is considered a metalloid.

21. Nonmetallic atoms have larger ionic radii than their atomic radii — BECAUSE — nonmetallic atoms generally gain electrons to form the ionic state and increase the size of the electron cloud.

22. Elements in the upper right corner of the Periodic Table form acid anhydrides — BECAUSE — nonmetallic oxides react with water to form acid solutions.

*Fill in oval CE only if II is a correct explanation of I.

	I	II	CE*
20.	T F	T F	◯
21.	T F	T F	◯
22.	T F	T F	◯

Answers and Explanations

1. **(C)** The two main parts of the atom are the nucleus and its energy levels.

2. **(B)** The principal quantum numbers start with the value of 1 to represent the first level.

3. **(A)** The letter s is used to represent the first orbital that can hold two electrons.

4. **(A)** The sublevel d has five orbitals that each can hold two electrons, totaling ten electrons.

5. **(C)** Each orbital can only hold two electrons.

6. **(B)** The mass number is the total of the number of protons and neutrons, which in this case is nine.

7. **(D)** The second principal energy level has an s orbital and three p orbitals, making a total of four.

8. **(D)** The Lewis electron dot notation shows the symbol and the outermost energy-level electrons, which are referred to as the valence electrons.

9. **(B)** Chlorine is a member of the halogen family found in group 17. The other element in the same family would be the element with the atomic number 9, fluorine.

10. **(A)** Radioactive changes differ because they involve changes in the nucleus.

11. **(B)** Isotopes differ in their atomic mass because of differences in the number of neutrons.

12. **(C)** These two isotopes, ^{235}U and ^{238}U, differ in their atomic mass by 3 neutrons.

13. **(A)** Radioactive isotopes have been successful in the treatment of certain cancers.

14. **(A)** A beta particle emission causes an increase of 1 in the atomic number.

15. **(B)** Because Na (sodium) has the position shown, it has the atomic number 11. The electron configuration is $1s^2 2s^2 2p^6 3s^1$.

16. **(E)** Fluorine has seven electrons in its outer energy level and needs only one more to complete its outer energy level octet. It therefore has the greatest affinity for one more electron.

17. **(C)** The most active metal of the group is found in the lower left-hand corner. Because its outer energy level is farthest from the nucleus and these are the most loosely held electrons, it is the most likely to lose an electron.

18. **(C)** The element that has the lowest electronegativity will be in the lower left corner. For this group, it is K.

19. **(D)** The element that has this configuration is carbon, which has six electrons. The first two are in the first level, and the next four are in the second level, as $2s^2 2p^2$.

20. **(T, T)** Both statements are true, but statement II does not explain statement I.

21. **(T, T, CE)** Because nonmetallic atoms gain electrons to form ions, the additional negative charge of the added electrons increases the size of the ion. This occurs because of the increased repulsion of the additional negative charge(s) and additional shielding from the positively charged nucleus.

22. **(T, T, CE)** Elements found in the upper right corner of the Periodic Table are nonmetals that form oxides and react with water to form acids. An example is sulfur that forms sulfur trioxide, and this reacts with water to form sulfuric acid.

Bonding

These are the skills that are usually tested on the SAT Subject Test in Chemistry. You should be able to...

→ Define ionic and covalent bonds, and explain how they form.

→ Identify the differences in the continuum that exists between ionic and covalent bonding.

→ Describe the implications of the type of bond on the structure of the compound.

→ Explain the implications of intermolecular forces and van der Waals forces.

→ Explain how VSEPR and hybridization solve the need to comply with known molecular shapes.

This chapter will review and strengthen these skills. Be sure to do the Practice Exercises at the end of the chapter.

Some elements show no tendency to combine with either like atoms or other kinds of elements. These elements are said to be monoatomic molecules; three examples are helium, neon, and argon. A **molecule** is defined as the smallest particle of an element or a compound that retains the characteristics of the original substance. Water is a triatomic molecule since two hydrogen atoms and one oxygen atom must combine to form the substance water with its characteristic properties. When atoms do combine to form molecules, there is a shifting of valence electrons, that is, the electrons in the outer energy level of each atom. Usually, this results in completion of the outer energy level of each atom. This more stable form may be achieved by the gain or loss of electrons or the sharing of pairs of electrons. The resulting attraction of the atoms involved is called a **chemical bond**. When a chemical bond forms, energy is released; when this bond is broken, energy is absorbed.

This relationship of bonding to the valence electrons of atoms can be further explained by studying the electron structures of the atoms involved. As already mentioned, the noble gases are monoatomic molecules. The reason can be seen in the electron distributions of these noble gases as shown in the following table.

Noble Gas	Electron Distribution	Electrons in Valence Energy Level
Helium	$1s^2$	2
Neon	$1s^2 2s^2 2p^6$	8
Argon	$1s^2 2s^2 2p^6 3s^2 3p^6$	8
Krypton	$1s^2 2s^2 2p^6 3s^2 3p^6 3d^{10} 4s^2 4p^6$	8
Xenon	$1s^2 2s^2 2p^6 3s^2 3p^6 3d^{10} 4s^2 4p^6 4d^{10} 5s^2 5p^6$	8
Radon	$1s^2 2s^2 2p^6 3s^2 3p^6 3d^{10} 4s^2 4p^6 4d^{10} 4f^{14} 5s^2 5p^6 5d^{10} 6s^2 6p^6$	8

TIP

Notice the recurrence of the octet (8) of electrons in noble gases.

The distinguishing factor in these very stable configurations is the arrangement of two *s* electrons and six *p* electrons in the valence energy level in five of the six atoms. (Note that helium, He, has only a single *s* valence energy level, which is filled with two electrons, making He a very stable atom.) This arrangement is called a **stable octet**. All other elements, other than the noble gases, have one to seven electrons in their outer energy levels. These elements are reactive to varying degrees. When they do react to form chemical bonds, usually the electrons shift in such a way that stable octets form. In other words, in bond formation, atoms usually attain the stable electron structure of one of the noble gases. The type of bond formed is directly related to whether this structure is achieved by gaining, losing, or sharing electrons.

TYPES OF BONDS
Ionic Bonds

TIP

A 1.7 or greater electronegativity difference between atoms will essentially form an ionic bond.

When the electronegativity values of two kinds of atoms differ by 1.7 or more (especially differences greater than 1.7), the more electronegative atom will borrow the electrons it needs to fill its energy level, and the other atom will lend electrons until it, too, has a complete energy level. Because of this exchange, the borrower becomes negatively charged and is called an anion; the lender becomes positively charged and is called a cation. They are now referred to as **ions**, and the bond or attraction between them is called an **ionic bond**. These ions do not retain the properties of the original atoms. An example can be seen in Figure 11.

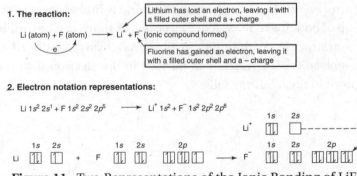

Figure 11. Two Representations of the Ionic Bonding of LiF

These ions do not form an individual molecule in the liquid or solid phase but are arranged into a crystal lattice or giant ion molecule containing many such ions. Ionic solids of this type tend to have high melting points and will not conduct a current of electricity until they are in the molten state.

Covalent Bonds

When the electronegativity difference between two or more atoms is 0 or very small (not greater than about 0.4), the atoms tend to share the valence electrons in their respective outer energy levels. This attraction is called a **nonpolar covalent bond**. Here is an example using electron-dot notation and orbital notation:

Covalent bonds involve a sharing of electrons between atoms. Their electronegativity difference is between 0 and 0.4.

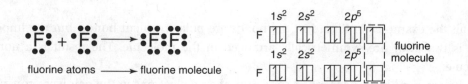

These covalent bonded molecules do not have electrostatic charges like those of ionic bonded substances. In general, covalent compounds are gases, liquids having fairly low boiling points, or solids that melt at relatively low temperatures. Unlike ionic compounds, they do not conduct electric currents.

When the electronegativity difference is between 0.4 and 1.6, there will not be an equal sharing of electrons between the atoms involved. The shared electrons will be more strongly attracted to the atom of greater electronegativity. As the difference in the electronegativities of the two elements increases above 0.4, the polarity or degree of ionic character increases. At a difference of 1.7 or more, the bond has more than 50% ionic character. However, when the difference is between 0.4 and 1.6, the bond is called a **polar covalent bond**. An example:

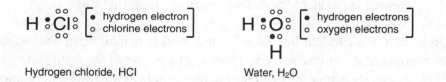

Polar covalent bonds have unequal sharing of electrons. Their electronegativity difference is between 0.4 and 1.6.

Notice that the electron pair in the bond is shown closer to the more electronegative atom. When these nonsymmetrical polar bonds are placed around a central atom, the overall molecule is polar. In the examples above, the chlorine (in HCl) and oxygen (in H_2O) are considered the central atoms. Both the bonds and the molecules could be described as **polar**. Polar molecules are also referred to as **dipoles** because the whole molecule itself has two distinct ends from a charge perspective. Because of this unequal sharing, the molecules shown are said to be polar molecules, or **dipoles**. However, polar covalent bonds exist in some nonpolar molecules. Examples are CO_2, CH_4, and CCl_4. (See Figure 12.)

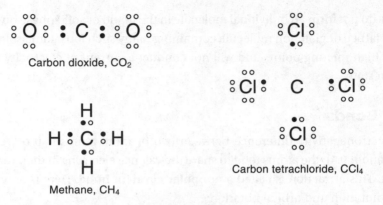

Carbon dioxide, CO_2

Methane, CH_4

Carbon tetrachloride, CCl_4

Figure 12. Polar Covalent Bonds in Nonpolar Molecules

In all the examples in Figure 12 the bonds are polar covalent bonds, but the important thing is that they are symmetrically arranged in the molecule. The result is a nonpolar molecule.

In the **covalent** bonds described so far, the shared electrons in the pair were contributed one each from the atoms bonded. In some cases, however, both electrons for the shared pair are supplied by only one of the atoms. Two examples are the bonds in NH_4^+ and H_2SO_4. (See Figure 13.)

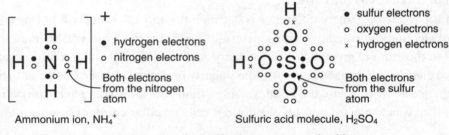

Ammonium ion, NH_4^+

• hydrogen electrons
○ nitrogen electrons

Both electrons from the nitrogen atom

Sulfuric acid molecule, H_2SO_4

• sulfur electrons
○ oxygen electrons
× hydrogen electrons

Both electrons from the sulfur atom

Figure 13. Covalent Bonds (both electrons supplied by one atom)

The formation of a covalent bond can be described in graphic form and related to the potential energies of the atoms involved. Using the formation of the hydrogen molecule as an example, we can show how the potential energy changes as the two atoms approach and form a covalent bond. In the illustration that follows, frames (1), (2), and (3) show the effect on potential energy as the atoms move closer to each other. In frame (3), the atoms have reached the condition of lowest potential energy, but the inertia of the atoms pulls them even closer, as shown in frame (4). The repulsion between them then forces the two nucleii to a stable position, as shown in frame (5).

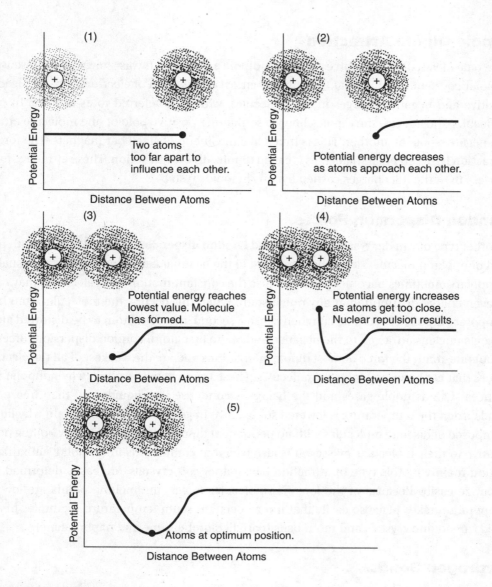

(1) Two atoms too far apart to influence each other.

(2) Potential energy decreases as atoms approach each other.

(3) Potential energy reaches lowest value. Molecule has formed.

(4) Potential energy increases as atoms get too close. Nuclear repulsion results.

(5) Atoms at optimum position.

Metallic Bonds

In most metals, one or more of the valence electrons become detached from the atom and migrate in a "sea" of free electrons among the positive metal ions. The attractive force strength varies with the nuclear positive charge of the metal atoms and the number of electrons in this electron sea. Both of these factors are reflected in the amount of heat required to vaporize the metal. The strong attraction between these differently charged particles forms a **metallic bond**. Because of this firm bonding, metals usually have high melting points, show great strength, and are good conductors of electricity.

TIP

Metallic bonds are like positive ions in a "sea" of electrons.

INTERMOLECULAR FORCES OF ATTRACTION

The term **intermolecular forces** refers to attractions *between* molecules. Although it is proper to refer to all intermolecular forces as **van der Waals forces**, named after Johannes van der Waals (Netherlands), this concept should be expanded for clarity.

Dipole-Dipole Attraction

One type of van der Waals forces is **dipole–dipole attraction**. It was shown in the discussion of polar covalent bonding that the unsymmetrical distribution of electronic charges leads to positive and negative charges in the molecules, which are referred to as **dipoles**. In polar molecular substances, the dipoles line up so that the positive pole of one molecule attracts the negative pole of another. This is much like the lineup of small bar magnets. The force of attraction between polar molecules is called **dipole–dipole attraction**. These attractive forces are less than the full charges carried by ions in ionic crystals.

London Dispersion Forces

TIP

Weakest of all, the London dispersion forces are one-tenth the force of most dipole attractions.

Another type of van der Waals forces is called **London dispersion forces**. Found in both polar and nonpolar molecules, it can be attributed to the fact that a molecule/atom that usually is nonpolar sometimes becomes polar because the constant motion of its electrons may cause uneven charge distribution at any one instant. When this occurs, the molecule/atom has a temporary dipole. This dipole can then cause a second, adjacent atom to be distorted and to have its nucleus attracted to the negative end of the first atom. London dispersion forces are about one-tenth the force of most dipole interactions and are the weakest of all the electrical forces that act between atoms or molecules. These forces help to explain why nonpolar substances such as noble gases and the halogens condense into liquids and then freeze into solids when the temperature is lowered sufficiently. In general, they also explain why liquids composed of discrete molecules with no permanent dipole attraction have low boiling points relative to their molecular masses. It is also true that compounds in the solid state that are bound mainly by this type of attraction have rather soft crystals, are easily deformed, and vaporize easily. Because of the low intermolecular forces, the melting points are low and evaporation takes place so easily that it may occur at room temperature. Examples of such solids are iodine crystals and moth balls (paradichlorobenzene and naphthalene).

Hydrogen Bonds

A proton or hydrogen nucleus has a high concentration of positive charge. When a hydrogen atom is bonded to a small, highly electronegative atom, its positive charge will have an enhanced attraction for neighboring electron pairs on the small, highly electronegative atom in neighboring molecules. This special kind of dipole–dipole attraction is called a **hydrogen bond**. The more strongly polar the molecule is, the more effective the hydrogen bonding is in binding the molecules into a larger unit. As a result, the boiling points of such molecules are higher than those of similar polar molecules not exhibiting hydrogen bonding. Hydrogen bonding typically occurs *between* molecules that *contain* hydrogen covalently bonded to nitrogen, oxygen or fluorine *within* the molecule. These three small but highly electronegative atoms create significantly polar bonds with the hydrogen atoms in the molecule, making the molecule very polar. The highly positive hydrogen end of the molecule, will be very attracted to the highly negative end of another molecule of the substance (where the nitrogen, oxygen, or fluorine atoms reside with their associated non-bonding electron pairs). This process creates an enhanced dipole–dipole bond.

Studying Figure 14 shows that in the series of compounds consisting of H_2O, H_2S, H_2Se, and H_2Te an unusual rise in the boiling point of H_2O occurs that is not in keeping with the typical slow increase of boiling point as molecular mass increases. Instead of the expected slope of the line between H_2O and H_2S, which is shown in Figure 14 as a dashed line, the actual boiling point of H_2O is quite a bit higher—100°C. The explanation is that hydrogen bonding occurs in H_2O but not to any significant degree in the other compounds.

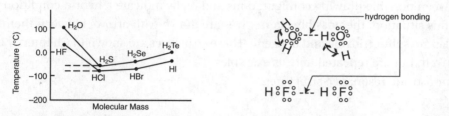

Figure 14. Boiling Points of Hydrogen Compounds with Similar Electron Dot Structures

This same phenomenon occurs with the hydrogen halides (HF, HCl, HBr, and HI). Note in Figure 14 that hydrogen fluoride, HF, which has strong hydrogen bonding, shows an unexpectedly high boiling point.

Hydrogen bonding also explains why some substances have unexpectedly low vapor pressures, high heats of vaporization, and high melting points. In order for vaporization or melting to take place, molecules must be separated. Energy must be expended to break hydrogen bonds and thus break down the larger clusters of molecules into separate molecules. As with the boiling point, the melting point of H_2O is abnormally high when compared with the melting points of the hydrogen compounds of the other elements having six valence electrons, which are chemically similar but which have no apparent hydrogen bonding. The hydrogen bonding effect in water is discussed on pages 189–190.

DOUBLE AND TRIPLE BONDS

To achieve the **octet** structure, which is an outer energy level resembling the noble gas configuration of eight electrons, it is necessary for some atoms to share two or even three pairs of electrons. Sharing two pairs of electrons produces a **double bond**. An example:

$$\overset{\times}{\underset{\times}{\text{O}}}\overset{\times}{\times}\ \overset{\circ}{\underset{\circ}{\text{C}}}\ \overset{\times}{\underset{\times}{\text{O}}}\overset{\times}{\times}, \text{ and by a line formula } O=C=O$$
carbon dioxide

In the line formula, only the shared pair of electrons is indicated by a bond (—). The sharing of three electron pairs results in a **triple bond**. An example:

$$H\overset{\times}{\circ}C\overset{\circ\circ}{\underset{\circ\circ}{}}C\overset{\times}{\circ}H \text{ , and by a line formula } H-C\equiv C-H$$
acetylene

It can be assumed from these structures that there is a greater electron density between the nuclei involved and hence a greater attractive force between the nuclei and the shared electrons. Experimental data verify that greater energy is required to break double bonds than single bonds, and triple bonds than double bonds. Also, since these stronger bonds tend to pull atoms closer together, the atoms joined by double and triple bonds have smaller interatomic distances and greater bond strengths, respectively.

RESONANCE STRUCTURES

TIP

Resonance structure is a hybrid of the possible drawings because no one Lewis structure can represent the situation.

It is not always possible to represent the bonding structure of a molecule by either the Lewis dot structure or the line drawing because data about the bonding distance and bond strength are between possible drawing configurations and really indicate a hybrid condition. To represent this situation, the possible alternatives are drawn with arrows between them. Classic examples are sulfur trioxide and benzene. These structures are shown in Chapters 13 and 14, respectively, but are repeated here as examples.

Sulfur trioxide resonance structures:

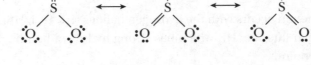

Benzene resonance structures:

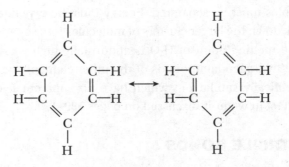

MOLECULAR GEOMETRY—VSEPR—AND HYBRIDIZATION
VSEPR—Electrostatic Repulsion

TIP

Two theories explain molecular structure: <u>VSEPR theory</u> uses valence shell electron pair repulsion. <u>Hybridization theory</u> uses changes in the orbitals of the valence electrons.

Properties of molecules depend not only on the bonding of atoms but also on the **molecular geometry**—the three-dimensional arrangement of the molecule's atoms in space. The combination of the polarity of the bonds and the geometry of the molecule determine the molecular polarity. This can be defined as the uneven distribution of the molecular charge. The chemical formula reveals little information about a molecule's geometry. It is only after doing many tests designed to reveal the shapes of the various molecules that chemists developed two different yet equally successful theories to explain certain aspects of their findings. One theory accounts structurally for molecular bond angles. The other is used to describe changes in the orbitals that contain the valence electrons of a molecule's atoms. The structural theory that deals with the bond angles is called the **VSEPR theory**, whereas the one that describes changes in the orbitals that contain the valence electrons is called the **hybridization theory**. (VSEPR represents Valence Shell Electron Pair Repulsion.)

VSEPR uses as its basis the fact that like charges will orient themselves in such a way as to diminish the repulsion between them.

1. Mutual repulsion of two electron clouds forces them to the opposite sides of a sphere. This is called a **linear arrangement**.

 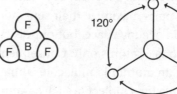

➡ **Example:** BeF₂, berylium fluoride _____

2. Minimum repulsion between three electron pairs occurs when the pairs are at the vertices of an equilateral triangle inscribed in a sphere. This arrangement is called a **trigonal-planar arrangement**.

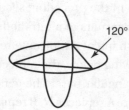 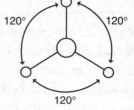

TIP

These basic arrangements are important to learn!

➡ **Example:** BF₃, boron trifluoride _____

3. Four electron pairs are farthest apart at the vertices of a tetrahedron inscribed in a sphere. This arrangement is called a **tetrahedral**-shaped distribution of electron pairs.

 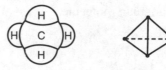

TIP

Configurations often appear as questions on the SAT test.

➡ **Example:** CH₄, methane _____

4. Mutual repulsion of six identical electron clouds directs them to the corners of an inscribed regular octahedron. This is said to have an **octahedral** arrangement.

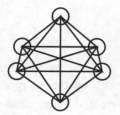 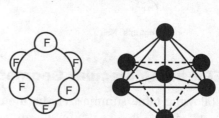

➡ **Example:** SF₆, sulfur hexafluoride _____

VSEPR and Unshared Electron Pairs

Ammonia, NH_3, and water, H_2O, are examples of molecules in which the central atom has both shared and unshared electron pairs. Here is how the VSEPR theory accounts for the geometries of these molecules.

The Lewis structure of ammonia shows that, in addition to the three electron pairs the central nitrogen atom shares with the three hydrogen atoms, it also has one unshared pair of electrons:

VSEPR theory postulates that the lone pair occupies space around the nitrogen atom just as the bonding pairs do. Thus, as in the methane molecule shown in the preceding section, the electron pairs maximize their separation by assuming the four corners of a tetrahedron. Lone pairs do occupy space, but our description of the observed shape of a molecule refers to the positions of atoms only. Consequently, as shown in the drawing below, the molecular geometry of an ammonia molecule is that of a pyramid with a triangular base. The general VSEPR formula for a molecule such as ammonia (NH_3) is AB_3E, where A replaces N, B replaces H, and E represents the unshared electron pair.

A water molecule has two unshared electron pairs and can be represented as an AB_2E_2 molecule. Here, the oxygen atom is at the center of a tetrahedron, with two corners occupied by hydrogen atoms and two by the unshared pairs, as shown below. Again, VSEPR theory states that the lone pairs occupy space around the central atom but that the actual shape of the molecule is determined only by the positions of the atoms. In the case of water, this results in a "bent," or angular, molecule.

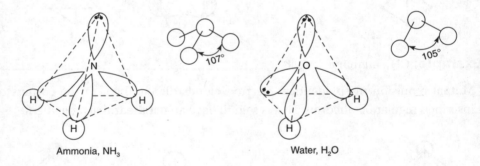

Ammonia, NH_3 Water, H_2O

VSEPR and Molecular Geometry

The following table summarizes the molecular shapes associated with particular types of molecules. Notice that, in VSEPR theory, double and triple bonds are treated in the same way as single bonds. It is helpful to use the Lewis structures and this table together to predict the shapes of molecules with double and triple bonds, as well as the shapes of polyatomic ions.

Summary of Molecular Shapes

Type of Molecule	Molecular Shape	Atoms Bonded to Central Atom	Lone Pairs of Electrons	Formula Example	Lewis Structure
Linear		2	0	BeF_2	$:\!\ddot{F}\!-\!Be\!-\!\ddot{F}\!:$
Bent		2	1	$SnCl_2$	Sn with :Cl and Cl:
Trigonal-planar		3	0	BF_3	B bonded to three F
Tetrahedral		4	0	CH_4	H—C—H with H above and H below
Trigonal-pyramidal		3	1	NH_3	N with H H H
Bent		2	2	H_2O	Ö with H H
Trigonal-bipyramidal	90° 120°	5	0	PCl_5	:Cl—P with Cl
Octahedral	90° 90°	6	0	SF_6	S bonded to six F

TIP

Know these molecular and Lewis structures.

Hybridization

The molecular configurations derived by VSEPR can also be arrived at through the concept of **hybridization**. Briefly stated, this means that chemists envision that two or more pure atomic orbitals (usually *s, p,* and *d*) can be mixed to form two or more new hybrid atomic orbitals that are identical and conform to the known shapes of molecules. Hybridization can be illustrated as follows:

1. *sp* Hybrid Orbitals

   Spectroscopic measurements of beryllium fluoride, BeF_2, reveal a bond angle of 180° and equal bond lengths.

The ground state of beryllium is:

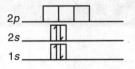

To accommodate the experimental data, we theorize that a 2s electron is excited to a 2p orbital; then the two orbitals hybridize to yield two identical orbitals called *sp* orbitals. Each contains one electron but is capable of holding two electrons.

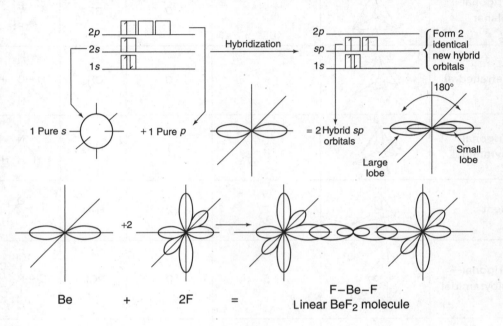

2. *sp²* Hybrid Orbitals

Boron trifluoride, BF_3, has bond angles of 120° of equal strength. To accommodate these data, the boron atom hybridizes from its ground state of $1s^2 2s^2 2p^1$ to:

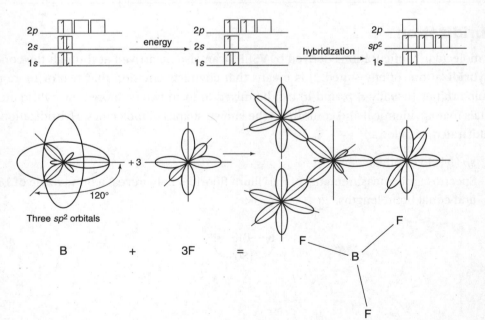

3. sp^3 Hybrid Orbitals

Methane, CH_4, can be used to illustrate this hybridization. Carbon has a ground state of $1s^2 2s^2 2p^2$. One $2s$ electron is excited to a $2p$ orbital, and the four involved orbitals then form four new identical sp^3 orbitals.

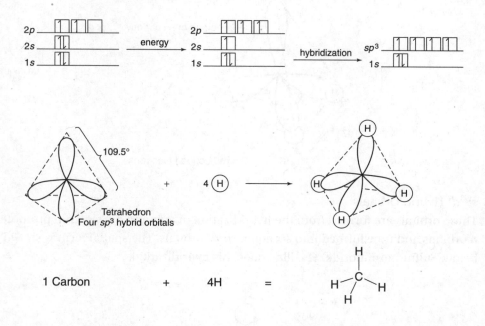

In some compounds where only certain sp^3 orbitals are involved in bonding, distortion in the bond angle occurs because of unbonded electron repulsion. Examples:

a. Water, H_2O

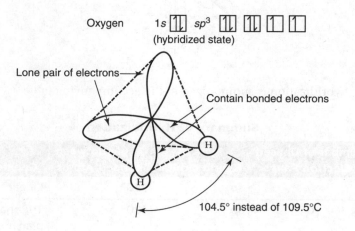

b. Ammonia, NH_3

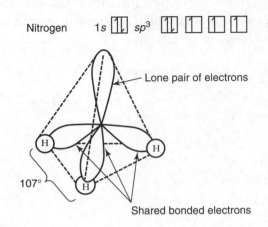

Nitrogen $1s$ sp^3

Lone pair of electrons

107°

Shared bonded electrons

4. sp^3d^2 Hybrid Orbitals

These orbitals are formed from the hybridization of an s and a p electron promoted to d orbitals and transformed into six equal sp^3d^2 orbitals. The spatial form is shown below. Sulfur hexafluoride, SF_6, illustrates this hybridization.

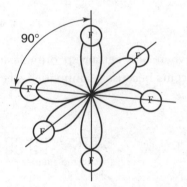

90°

The concept of hybridization is summarized in the accompanying table.

Summary of Hybridization

Number of Bonds	Number of Unused Electron Pairs	Type of Hybrid Orbital	Angle Between Bonded Atoms	Geometry	Example
2	0	sp	180°	Linear	BeF_2
3	0	sp^2	120°	Trigonal-planar	BF_3
4	0	sp^3	109.5°	Tetrahedral	CH_4
3	1	sp^3	90° to 109.5°	Pyramidal	NH_3
2	2	sp^3	90° to 109.5°	Angular	H_2O
6	0	sp^3d^2	90°	Octahedral	SF_6

TIP

Know these hybrid orbitals designations and their corresponding shapes.

SIGMA AND PI BONDS

When bonding occurs between *s* and *p* orbitals, each bond is identified by a special term. A *sigma bond* is a bond between *s* orbitals, or an *s* orbital and another orbital such as a *p* orbital. It includes bonding between hybrids of *s* orbitals such as sp, sp^2, and sp^3.

In the methane molecule, the sp^3 orbitals are each bonded to hydrogen atoms. These are sigma bonds.

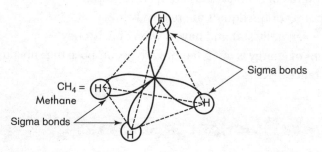

When two *p* orbitals share electrons in a covalent bond and the interaction is not symmetrical about a line between the two nuclei, the result is a **pi bond**. Here is an example:

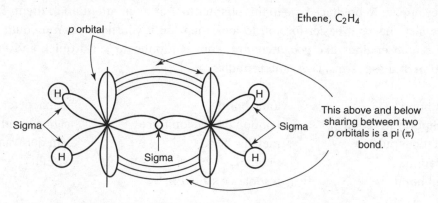

Chapter 14 gives more examples of sigma and pi bonding.

PROPERTIES OF IONIC SUBSTANCES

Laboratory experiments reveal that, in general, ionic substances are characterized by the following properties:

1. In the solid phase at room temperature they do not conduct appreciable electric current.
2. In the liquid phase they are relatively good conductors of electric current. The conductivity of ionic substances is much smaller than that of metallic substances.
3. They have relatively high melting and boiling points. There is a wide variation in the properties of different ionic compounds. For example, potassium iodide (KI) melts at 686°C and boils at 1,330°C, while magnesium oxide (MgO) melts at 2,800°C and boils at 3,600°C. Both KI and MgO are ionic compounds.
4. They have relatively low volatilities and low vapor pressures. In other words, they do not vaporize readily at room temperature.
5. They are brittle and easily broken when stress is exerted on them.
6. Those that are soluble in water form electrolytic solutions that are good conductors of electricity. There is, however, a wide range in the solubilities of ionic compounds. For example, at 25°C, 92 grams of sodium nitrate ($NaNO_3$) dissolves in 100 grams of water, while only 0.0002 grams of barium sulfate ($BaSO_4$) dissolves in the same mass of water.

PROPERTIES OF MOLECULAR CRYSTALS AND LIQUIDS

Experiments have shown that these substances have the following general properties:

1. Neither the liquids nor the solids conduct electric current appreciably.
2. Many exist as gases at room temperature and atmospheric pressure, and many solids and liquids are relatively volatile.
3. The melting points of solid crystals are relatively low.
4. The boiling points of the liquids are relatively low.
5. The solids are generally soft and have a waxy consistency.
6. A large amount of energy is often required to decompose the substance chemically into simpler substances.

CHAPTER SUMMARY

The following terms summarize all the concepts and ideas that were introduced in this chapter. You should be able to explain their meaning and how you would use them in chemistry. They appear in boldface type in this chapter to draw your attention to them. The boldface type also makes it easier for you to look them up if you need to. You could also use Internet search engines like *google.com* on your computer to get a quick and expanded explanation of these terms, laws, and formulas.

covalent bond	ionic bond	stable octet
dipole–dipole attraction	London dispersion forces	sigma bond
electrostatic repulsion	metallic bond	van der Waals forces
hybridization	pi bond	VSEPR
hydrogen bond	resonance structure	

INTERNET RESOURCES

Online content that reinforces major concepts discussed in this chapter can be found at the following Internet addresses if they are still available. *Some may have been changed or deleted.*

Chemical Bonding
www.chem.wisc.edu/deptfiles/genchem/sstutorial/Text7/tx71/tx71.html
This site offers a fairly complete look at the concept of chemical bonding for the beginning chemistry student.

VSEPR (Valence Shell Electron Pair Repulsion) Theory
winter.group.shef.ac.uk/vsepr/
This site is a good tutorial on molecular shapes utilizing the Valence Shell Electron Pair Repulsion Theory.

The following questions are in the format that is used on the **SAT Subject Area Test in Chemistry**. If you are not familiar with these types of questions, study pages xiii–xvi before doing these review questions.

> **Directions:** The following set of lettered choices refers to the numbered questions immediately below it. For each numbered item, choose the one lettered choice that fits it best. Then fill in the corresponding oval on the answer sheet. Each choice in the set may be used once, more than once, or not at all.

Questions 1–7

    (A) ionic

    (B) covalent

    (C) polar covalent

    (D) metallic

    (E) hydrogen bonding

1. When the electronegativity difference between two atoms is 2, what type of bond can be predicted?

2. If two atoms are bonded in such a way that both members of the pair equally share one electron with the other, what is the bond called?

3. Which of the five choices is considered the weakest bond in the group?

4. Which of the above bonds explains water's abnormally high boiling point?

5. If the sharing of an electron pair is unequal and the atoms have an electronegativity difference of 1.4 to 1.6, what is this type of sharing called?

6. If an electron is lost by one atom and completely captured by another, what is this type of bond called?

7. If one or more valence electrons become detached from the atoms and migrate in a "sea" of free electrons among the positive metal ions, what is this type of bonding called?

I		II
8. Maximum repulsion between two electron pairs in a molecular compound will result in a linear structure	BECAUSE	the VSEPR model says that like charges will orient themselves so as to diminish the repulsion between them.
9. Sodium chloride is an example of ionic bonding	BECAUSE	sodium and chlorine have the same electronegativity.
10. Ammonia has a trigonal pyramidal molecular structure	BECAUSE	ammonia has a tetrahedral electron pair geometry with three atoms bonded to the central atom.

*Fill in oval CE only if II is a correct explanation of I.

	I	II	CE*
8.	T F	T F	◯
9.	T F	T F	◯
10.	T F	T F	◯

Answers and Explanations

1. **(A)** When the electronegativity difference between the two atoms is greater than 1.7, the bond between them is considered more than 50% ionic.

2. **(B)** When two atoms equally share a pair of electrons, the bond is considered covalent. With an electronegativity difference of 0 to 0.3, the bond is still considered covalent.

3. **(E)** Hydrogen bonding is the weakest because it is the weak attraction of the hydrogen end of a polar molecule to the partial negative charge of an adjacent molecule. Especially strong in water, it is responsible for many of water's properties.

4. **(E)** Because of the strong hydrogen bonding in water, it takes much more energy to cause the molecules to break away from each other in the liquid state and change to steam.

5. **(C)** When the electronegativity difference between the two bonding atoms is between 0.3 and 1.7, the electrons are not equally shared and result in a polar covalent bond.

6. **(A)** This is the definition of a pure ionic bond.

7. **(D)** Metallic bonds are defined as a "sea" of free electrons that migrate through the metal.

8. **(T, T, CE)** The mutual repulsion of two electron clouds forces them to the opposite sides of a sphere. An example is BeF_2, which forms a linear molecular structure like this, F –Be – F.

9. **(T, F)** The bond between sodium and chlorine in sodium chloride is ionic, because the difference between their electronegativity is greater than 1.7.

10. **(T, T, CE)** The molecular structure of NH_3 is a trigonal pyramidal structure like this.

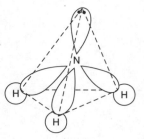

Ammonia, NH_3

Chemical Formulas

4

These skills are usually tested on the SAT Subject Test in Chemistry. You should be able to...

→ Recall and use the basic rules about ionic charges to write correct formulas for ionic compounds. This includes writing formulas with polyatomic ions.

→ Recall and use the basic rules about writing compounds for covalent (molecular) compounds.

→ Name compounds (acids, bases, and salts) using the Stock system and the prefix system, and write their formulas.

→ Calculate the formula mass of a compound and the percent composition of each element.

→ Calculate the empirical formula when given the percent composition of each element. When given the formula mass, you should be able to find the true formula.

→ Write a simple balanced equation, indicating the phase (or state) of the reactants and products.

This chapter will review and strengthen these skills. Be sure to do the Practice Exercises at the end of the chapter.

NAMING AND WRITING CHEMICAL FORMULAS

With the knowledge you have about atomic structure, the significance of each element's placement in the periodic table, and the bonding of atoms in ionic and covalent arrangements, you can now use this information to write appropriate formulas and name the resulting products. Obviously, many compounds can result. Some system of writing the names and formulas of these many combinations was needed. The system explained in this text is an organized way of accomplishing this. It uses three categories for those compounds containing only two elements:

CATEGORY I—Binary ionic compounds where the metal present forms only a single type of positively charged ion (cation)

CATEGORY II—Binary ionic compounds where the metal forms more than one type of ionic compound with a given negatively charged ion (anion)

CATEGORY III—Binary covalent compounds formed between two nonmetals

Table 6 is a list of ions that are often encountered in a first-year chemistry course. **You should know them**. Although using the periodic table can help you write the symbol and apparent charge of cations and anions, knowing these common ions can help you write formulas and equations.

Table 6. Table of Common Ions Used in a First-Year Course

Metals Cations (+)

Monovalent I		Bivalent II		Trivalent III		Tetravalent IV		V	
Hydrogen	H	Barium	Ba	Aluminum	Al	Carbon	C	Arsenic(V)*	As
Potassium	K	Calcium	Ca	Gold(III)*	Au	Silicon	Si	Phosphorus(V)*	P
Sodium	Na	Cobalt	Co	Arsenic (III)*	As	Manganese(IV)*	Mn	Antimony(V)*	Sb
Silver	Ag	Magnesium	Mg	Chromium	Cr	Tin(IV)*	Sn	Bismuth(V)*	Bi
Mercury(I)*	Hg	Lead (II)*	Pb	Iron(III)*	Fe	Platinum	Pt		
Copper(I)*	Cu	Zinc	Zn	Phosphorus(III)*	P	Sulfur	S		
Gold(I)*	Au	Mercury(II)*	Hg	Antimony(III)*	Sb	Lead (IV)*	Pb		
Ammonium†	(NH_4)	Copper(II)*	Cu	Bismuth(III)*	Bi				
		Iron(II)*	Fe						
		Manganese(II)*	Mn						
		Tin(II)*	Sn						

*Use of the Roman numeral, instead of the suffix, is now preferred; for example, iron(II) oxide instead of ferrous oxide.

†Polyatomic ion

Nonmetals‡ Anions (−)

Monovalent I		Bivalent II		Trivalent III		Tetravalent IV	
Fluorine	F	Oxygen	O	Nitrogen	N	Carbon	C
Chlorine	Cl	Sulfur	S	Phosphorus	P		
Bromine	Br						
Iodine	I						

‡Last syllable of nonmetal name is changed to -ide in binary compound.

Polyatomic Ions (−)

Monovalent I		Bivalent II		Trivalent III		Tetravalent IV	
Hydroxide	(OH)	Carbonate	(CO_3)	Borate	(BO_3)	Ferrocyanide	$[Fe(CN)_6]$
Hydrogen carbonate (or Bicarbonate)	(HCO_3)	Sulfite	(SO_3)	Phosphate	(PO_4)		
Nitrite	(NO_2)	Sulfate	(SO_4)	Phosphite	(PO_3)		
Nitrate	(NO_3)	Tetraborate	(B_4O_7)	Ferricyanide	$[Fe(CN)_6]$		
Hypochlorite	(ClO)	Silicate	(SiO_3)				
Chlorate	(ClO_3)	Chromate	(CrO_4)				
Chlorite	(ClO_2)	Oxalate	(C_2O_4)				
Perchlorate	(ClO_4)						
Acetate	$(C_2H_3O_2)$						
Permanganate	(MnO_4)						
Hydrogen sulfate (or Bisulfate)	(HSO_4)						

Category I—Binary Ionic Compounds

Category I binary ionic compounds contain metallic ions from groups 1 and 2 of the Periodic Table. These metallic ions have only one type of charge. The binary ionic compounds formed are composed of a positive ion (cation) that is written first and a negative ion (anion). The following rules show how to name and write the formulas for binary ionic compounds. $CaCl_2$ is used as an example.

1. Name the cation first and then the anion.
2. The monoatomic (one-atom) cation takes its name from the name of the element. Therefore the calcium ion, Ca^{2+}, is called calcium and its chemical symbol appears first.
3. The monoatomic anion with which the cation combines is named by taking the root of the element's name and adding –ide. You must know this rule. The anion's name comes second. Therefore, the chlorine ion, Cl^-, is called chloride.
4. The name of this compound is calcium chloride.

A quick way to determine the formula of a binary ionic compound is to use the **crisscross rule.**

➡ Example 1 _____

To determine the formula for calcium chloride, first write the ionic forms with their associated charges.

Next move the numerical value of the metal ion's superscript (without the charge) to the subscript of the nonmetal's symbol. Then take the numerical value of the nonmetal's superscript and make it the subscript of the metal as shown above.

Note that the numerical value 1 is not shown in the final formula.

You now have the chlorine's 1 as the subscript of the calcium and the calcium's 2 as the subscript of the chloride. As a result, you have $CaCl_2$ as the final formula for calcium chloride.

➡ Example 2 _____

Write the name and formula for the product formed when aluminum reacts with oxygen. First write the name.

1. Name the cation first and then the anion.
2. The monoatomic (one-atom) cation takes its name from the name of the element. Therefore the aluminum ion, Al^{3+}, is called aluminum and its chemical symbol appears first.
3. The monoatomic anion with which the cation combines is named by taking the root of the element's name and adding –ide. You must know this rule. The anion's name comes second. Therefore, the oxygen ion, O^{2-}, is called oxide.
4. The name of this compound is aluminum oxide.

To determine the formula for aluminum oxide, first write the ionic forms with their associated charges.

Next move the numerical value of the Al's superscript (without the charge) to the subscript of the O symbol. Do the same with the 2 of the O. In other words, crisscross the values. You now have the 2 as the subscript of the aluminum and the 3 as a subscript of the oxygen. You now have Al_2O_3 as the final formula for aluminum oxide.

This crisscross rule generally works very well. In one situation, though, you have to be careful. Suppose you want to write the compound formed when magnesium reacts with oxygen. Magnesium, an alkaline earth metal in group 2 forms a 2^+ cation, and oxygen forms a 2^- anion. You would predict its formula be Mg_2O_2, but this is incorrect. After you do the crisscrossing (unless you know that the compound actually exists, like H_2O_2), you need to reduce all the subscripts by a common factor. In this example, you can divide all the subscripts by 2 to get the correct formula for magnesium oxide, MgO.

When you attempt to write a formula, you should know whether the substance actually exists. For example, you could easily write the formula for carbon nitrate, but no chemist has ever prepared this compound.

> **REMEMBER**
>
> **Reduce all subscripts by a common factor unless you are sure the compound exists, like H_2O_2.**

Examples of Category I Binary Ionic Compounds

Ions Present*	Formula	Name
K^+, Cl^-	KCl	Potassium chloride
Na^+, I^-	NaI	Sodium iodide
Ca^{2+}, S^{2-}	CaS	Calcium sulfide
Al^{3+}, F^-	AlF_3	Aluminum fluoride
Li^+, N^{3-}	Li_3N	Lithium nitride

Ionic charges are shown as numerical exponents followed by the charge.

Category II—Binary Ionic Compounds

In category II binary ionic compounds, the metals form more than one ion, each with a different charge. The metallic ions (cation) ionically bind with a negatively charged ion (anion).

The following chart lists many of the metals that form more than one type of ionic cation and therefore more than one binary ionic compound with a given anion.

Common Category II Cations (Multivalent Metals)

Ion	Systematic Name	Ion	Systematic Name
Fe^{3+}	Iron(III)	Sn^{4+}	Tin(IV)
Fe^{2+}	Iron(II)	Sn^{2+}	Tin(II)
Cu^{1+}	Copper(I)	Pb^{4+}	Lead(IV)
Cu^{2+}	Copper(II)	Pb^{2+}	Lead(II)
Hg_2^{2+}	Mercury(I)*	Hg^{2+}	Mercury(II)

This form of mercury(I) ions always occurs bonded together as a Hg_2^{2+} ion.

Although the following metals are "transition" metals, they form only one type of cation. So a Roman numeral is not used when naming their compounds.

Ag^{1+} Silver
Cd^{2+} Cadmium
Zn^{2+} Zinc

➡ Example

The compound containing the Fe^{2+} ion and the compound containing the Fe^{3+} ion both combine with the chloride ion to form two different compounds. Using the crisscross system, you get the formula $FeCl_2$ for iron(II) chloride.

The compound formed using the Fe^{3+} ion and the chloride ion is $FeCl_3$, which is iron(III) chloride.

The names iron(II) chloride and iron(III) chloride are arrived at by using the Roman numerals in parentheses to indicate the charge of the metallic ion used as the cation. Using Roman numerals this way—to indicate the charge on the ion—is called the Stock system.

Another, older system of naming category II binary ionic compounds is still seen in some books. Simply stated, for metals that form only two ions, the ion with the higher charge has a name ending in –ic and the ion with the lower charge has a name ending in –ous. In this system, Fe^{3+} is called the ferric ion and Fe^{2+} is called the ferrous ion. The names for $FeCl_2$ and $FeCl_3$ are then ferric chloride and ferrous chloride, respectively.

Examples of Category II Binary Ionic Compounds

Formula	Name
CuCl	Copper(I) chloride
HgO*	Mercury(II) oxide
FeO*	Iron(II) oxide
MnO_2†	Manganese(IV) oxide
$PbCl_2$	Lead(II) chloride

*The subscripts are reduced and are not written because subscripts of 1 are understood.
†The subscripts are reduced.

The modified periodic chart that follows shows the location of the common category I and category II ions. Also shown in this chart are the common nonmetallic monoatomic ions as anions.

GROUPS→																
1	2	3	4	5	6	7	8	9	10	11	12	13	14	15	16	17
Li^+														N^{3-}	O^{2-}	F^-
Na^+	Mg^{2+}											Al^{3+}			S^{2-}	Cl^-
K^+	Ca^{2+}				Cr^{2+} Cr^{3+}	Mn^{2+} Mn^{3+}	Fe^{2+} Fe^{3+}	Co^{2+} Co^{3+}		Cu^+ Cu^{2+}	Zn^{2+}					Br^-
Rb^+	Sr^{2+}									Ag^+	Cd^{2+}		Sn^{2+} Sn^{4+}			I^-
Cs^+	Ba^{2+}									Hg_2^{2+} Hg^{2+}			Pb^{2+} Pb^{4+}			

■ Metallic + ions
Uses only one charge
Category I

▨ Metallic + ions, Ionic
Uses > 1 type of charge
Category II

▧ Nonmetallic – ions
Uses only one charge

Category I and II Ionic Compounds Formed with Polyatomic Ions

Another group of ionic compounds contains polyatomic ions. A **polyatomic ion** is a group of elements that *act like a single ion* when forming a compound. The bonds within these polyatomic ions are predominantly covalent. However, the group as a whole has an excess charge, which is usually negative, because of an excess of electrons. If the compounds formed with the polyatomic ions consist of three elements, they are called **ternary** compounds.

Polyatomic ions have special names and formulas that you must memorize. Table 6 contains the names and ionic charges of the common polyatomic ions encountered in a first-year chemistry course. Note that only one commonly used positively charged polyatomic ion is in Table 6, the ammonium ion, NH_4^+.

Also notice in Table 6 that several of the polyatomic anions contain an atom of a given element and a different number of oxygen atoms, such as NO_2 and NO_3. When there are two members of such a series, the name of the one with fewer oxygen atoms ends in -ite and the name of the one with more oxygen atoms ends in –ate. The following table shows examples of polyatomic ions of sulfur.

TIP

You should memorize all the polyatomic ions in Table 6 to help you use them in formulas and equations.

Polyatomic Ions Containing the *-ite* and *-ate* Forms of Sulfur

Ionic Formula	Name of the Ion	Sample Formula	Name of Compound
SO_3^{2-}	Sulfite	Na_2SO_3	Sodium sulfite
SO_4^{2-}	Sulfate	Na_3SO_4	Sodium sulfate

Sometimes an element combines with oxygen to form more than just two polyatomic ions, such as ClO^-, ClO_2^-, ClO_3^-, and ClO_4^-. When this occurs, the prefix *hypo-* is used to name the polyatomic ion with the fewest oxygen ions and the prefix *per-* to name the polyatomic ion with the most oxygen ions.

Polyatomic Ions Containing Chlorine and Oxygen

Ionic Formula	Name of the Ion
ClO^-	Hypochlorite
ClO_2^-	Chlorite
ClO_3^-	Chlorate
ClO_4^-	Perchlorate

Writing Formulas for Compounds with Polyatomic Ions

When writing formulas using polyatomic anions, the rules do not change. Simply treat the polyatomic ion as if it were a single anion. If the cation is from category I, follow the rules for category I. If the cation is from category II, follow the rules for category II. The crisscross method does not change, either.

➡ Example 1 _____

Use the crisscross method to write the formula for calcium sulfate, a category I cation and a polyatomic ion.

Calcium sulfate

$$Ca^{2+} \quad (SO_4)^{2-}$$
$$Ca_2(SO_4)_2$$

The final formula is $CaSO_4$. Notice that the subscripts "2" are omitted.

➡ Example 2 _____

Use the crisscross method to write the formula for iron(III) sulfate, a category II cation and a polyatomic ion.

Iron(III) sulfate

$$Fe^{3+} \quad (SO_4)^{2-}$$
$$Fe_2(SO_4)_3$$

The final formula is $Fe_2(SO_4)_3$.

Examples of Ionic Compounds with Polyatomic Ions and Either Category I or II Cations

Name	Formula	Comment
Sodium sulfate	Na_2SO_4	Category I—the Na^+ always is 1^+
Potassium dihydrogen phosphate	KH_2PO_4	The $H_2PO_4^-$ ion has a 1^- charge and the K^+, from category I, is 1^+
Iron(III) nitrate	$Fe(NO_3)_3$	Category II—transition metal, must contain a Roman numeral
Cesium perchlorate	$CsClO_4$	The *per-* prefix is used because the polyatomic ion has 1 more oxygen than the chlorate ion
Manganese(II) hydroxide	$Mn(OH)_2$	Category II—transition metal, must contain a Roman numeral

Category III—Binary Covalent Compounds

Binary covalent compounds are formed between two nonmetals. Although these compounds do not contain ions, they are named very similarly to binary ionic compounds. To name binary covalent compounds, use these steps.

1. The first element in the formula is named first, using the full elemental name.
2. The second element is named as if it were an anion and uses its elemental name.
3. Prefixes are used to denote the number of the second element present. These prefixes are shown in the table below.
4. The prefix *mono-* is never used for naming the first element. For example, CO is called carbon monoxide, not monocarbon monoxide.

Prefixes Used to Indicate Numbers in Covalent Compounds

Prefix	Number	Prefix	Number
mono-	1	hexa-	6
di-	2	hepta-	7
tri-	3	octa-	8
tetra-	4	nona-	9
penta-	5	deca-	10

The following are examples of covalent compounds formed from the nonmetals nitrogen and oxygen, using the rules above.

Compound	Systematic Name	Common Name
N_2O	Dinitrogen monoxide*	Nitrous oxide
NO	Nitrogen monoxide*	Nitric oxide
NO_2	Nitrogen dioxide	
N_2O_3	Dinitrogen trioxide	
N_2O_4	Dinitrogen tetroxide*	
N_2O_5	Dinitrogen pentoxide*	

*Notice that for ease of pronunciation, the final "a" or "o" of the prefix is dropped if the element begins with a vowel.

To write the formula for binary covalent compounds, use the same steps as when writing the formula of ionic compounds.

1. The symbol of the first element in the formula is written first, followed by the second element.
2. Use the prefix(es) denoted in the name for the number of each element present in the formula.

The following show some examples of binary covalent compounds.

Name	Formula
Sulfur hexafluoride	SF_6
Phosphorus trichloride	PCl_3

NAMES AND FORMULAS OF COMMON ACIDS AND BASES

The definition of an acid and a base is expanded later in a first-year chemistry course. For now, common acids are aqueous solutions of hydrogen compounds that contain hydrogen ions, H^+. Common bases are aqueous solutions containing hydroxide ions, OH^-.

A binary acid is named by placing the prefix *hydro-* in front of the stem or full name of the nonmetallic element, and adding the ending *-ic*. Examples are *hydro*chlor*ic* acid (HCl) and *hydro*sulfur*ic* acid (H_2S).

A **ternary compound** consists of three elements, usually an element and a polyatomic ion. To name the compound, you merely name each component in the order of positive first and negative second.

Ternary acids usually contain hydrogen, a nonmetal, and oxygen. Because the amount of oxygen often varies, the name of the most common form of the acid in the series consists of merely the stem of the nonmetal with the ending *-ic*. The acid containing one less atom of oxygen than the most common acid is designated by the ending *-ous*. The name of the acid containing one more atom of oxygen than the most common acid has the prefix *per-* and the ending *-ic*; that of the acid containing one less atom of oxygen than the *-ous* acid has the prefix *hypo-* and the ending *-ous*. This is evident in Table 7 with the acids containing H, Cl, and O.

You can remember the names of the common acids and their salts by learning the following simple rules:

TIP

Learn these rules.

Rule	Example
-ic acids form *-ate* salts.	Sulfuric acid forms sulfate salts.
-ous acids form *-ite* salts.	Sulfurous acid forms sulfite salts.
hydro-(stem)-*ic* acids form *-ide* salts.	Hydrochloric acid forms chloride salts.

When the name of the ternary acid has the prefix *hypo-* or *per-*, that prefix is retained in the name of the salt (hypochlorous acid = sodium hypochlorite).

The names and formulas of some comon acids and bases are listed in Table 7.

Table 7. Formulas of Common Acids and Bases

ACIDS, BINARY		ACIDS, TERNARY	
Name	*Formula*	*Name*	*Formula*
Hydrofluoric	HF	Nitric	HNO_3
Hydrochloric	HCl	Nitrous	HNO_2
Hydrobromic	HBr	Hypochlorous	HClO
Hydriodic	HI	Chlorous	$HClO_2$
Hydrosulfuric	H_2S	Chloric	$HClO_3$
		Perchloric	$HClO_4$
BASES		Sulfuric	H_2SO_4
Sodium hydroxide	NaOH	Sulfurous	H_2SO_3
Potassium "	KOH	Phosphoric	H_3PO_4
Ammonium "	NH_4OH	Phosphorous	H_3PO_3
Calcium "	$Ca(OH)_2$	Carbonic	H_2CO_3
Magnesium "	$Mg(OH)_2$	Acetic	$HC_2H_3O_2$
Barium "	$Ba(OH)_2$	Oxalic	$H_2C_2O_4$
Aluminum "	$Al(OH)_3$	Boric	H_3BO_3
Iron(II) "	$Fe(OH)_2$		
Iron(III) "	$Fe(OH)_3$		
Zinc "	$Zn(OH)_2$		
Lithium "	LiOH		

CHEMICAL FORMULAS: THEIR MEANING AND USE

As you have seen, a chemical formula is an indication of the makeup of a compound in terms of the kinds of atoms and their relative numbers. It also has some quantitative applications. By using the atomic masses assigned to the elements, we can find the **formula mass** of a compound. If we are sure that the formula represents the actual makeup of one molecule of the substance, the term **molecular mass** may be used as well. In some cases the formula represents an ionic lattice and no discrete molecule exists, as in the case of table salt, NaCl, or the formula merely represents the simplest ratio of the combined substances and not a molecule of the substance. For example, CH_2 is the simplest ratio of carbon and hydrogen united to form the actual compound ethylene, C_2H_4. This simplest ratio formula is called the **empirical formula**, and the actual formula is the **true formula**. The formula mass is determined by multiplying the atomic mass units (as a whole number) by the subscript for that element and then adding these values for all the elements in the formula. For example:

$Ca(OH)_2$ (one calcium amu + two hydrogen and two oxygen amu = formula mass).

$$1Ca\ (amu = 40) = 40$$
$$2O\ \ (amu = 16) = 32$$
$$2H\ \ (amu = \ 1) = 2.0$$

Formula mass $Ca(OH)_2 = 74$ amu (or μ)

In Chapter 6, the concept of a mole is introduced. If you have 6.02×10^{23} atoms of an element, then the atomic mass units can be expressed in grams, and then the formula mass can be called the **molar mass**. This is further explained on page 163. Another example is Fe_2O_3.

$$2Fe\,(amu = 56) = 112$$
$$3O\;(amu = 16) = 48.0$$
$$\text{Formula mass } Fe_2O_3 = 160.\,amu$$

It is sometimes useful to know what percent of the total weight of a compound is made up of a particular element. This is called finding the **percentage composition**. The simple formula for this is:

$$\frac{\text{Total amu of the element in the compound}}{\text{Total formula amu}} \times 100\% = \text{Percentage composition of that element}$$

TIP

Know how to compute the percentage composition of an element in a compound.

To find the percent composition of calcium in calcium hydroxide in the example above, we set the formula up as follows:

$$\frac{Ca = 40.\,amu}{\text{Formula mass} = 74\,amu} \times 100\% = 54\%\,\text{Calcium}$$

To find the percent composition of oxygen in calcium hydroxide:

$$\frac{O = 32\,amu}{\text{Formula mass} = 74\,amu} \times 100\% = 43\%\,\text{Oxygen}$$

To find the percent composition of hydrogen in calcium hydroxide:

$$\frac{H = 2.0\,amu}{\text{Formula mass} = 74\,amu} \times 100\% = 2.7\%\,\text{Hydrogen}$$

➥ Another Example

Find the percent compositions of Cu and H_2O in the compound $CuSO_4 \cdot 5H_2O$ (the dot is read "with").

First, we calculate the formula mass:

$$
\begin{aligned}
1\,Cu &= 64\,amu \\
1\,S &= 32\,amu \\
4\,O &= 64\,(4 \times 16)\,amu \\
\underline{5\,H_2O} &= \underline{90.\,(5 \times 18)\,amu} \\
&\;\;\;250\,amu
\end{aligned}
$$

TIP

This type of question *always* appears on the test.

and then find the percentages:

Percentage Cu:

$$\frac{Cu = 64\,amu}{\text{Formula mass} = 250\,amu} \times 100\% = 26\%$$

Percentage H_2O:

$$\frac{5.0H_2O = 90.\,amu}{\text{Formula mass} = 250\,amu} \times 100\% = 36\%$$

When you are given the percentage of each element in a compound, you can find the empirical formula as shown with the following example:

Given that a compound is composed of 60.0% Mg and 40.0% O, find the empirical formula of the compound.

1. It is easiest to think of 100 mass units of this compound. In this case, the 100 mass units are composed of 60. amu of Mg and 40. amu of O. Because you know that 1 unit of Mg is 24 amu (from its atomic mass) and, likewise, 1 unit of O is 16, you can divide 60 by 24 to find the number of units of Mg in the compound and divide 40. by 16 to find the number of units of O in the compound.

$$\text{Mg} \qquad\qquad \text{O}$$
$$\frac{60.}{24} = 2.5 \text{ units Mg} \qquad \frac{40.}{16} = 2.5 \text{ units O}$$

2. Now, because we know formulas are made up of whole-number units of the elements, which are expressed as subscripts, we must manipulate these numbers to get whole numbers. This is usually accomplished by dividing these numbers by the smallest quotient. In this case they are equal, so we divide by 2.5.

$$\text{Mg} \qquad\qquad \text{O}$$
$$\frac{2.5}{2.5} = 1 \qquad\qquad \frac{2.5}{2.5} = 1$$

3. So the empirical formula is MgO.

➥ Another Example

Given: Ba = 58.81%, S = 13.73%, and O = 27.46%.
Find the empirical formula.

1. Divide each percent by the amu of the element.

$$\text{Ba} \qquad\qquad \text{S} \qquad\qquad \text{O}$$
$$\frac{58.8}{137} = 0.43 \qquad \frac{13.7}{32} = 0.43 \qquad \frac{27.5}{16} = 1.72$$

2. Manipulate numbers to get small whole numbers. Try dividing them all by the smallest first. In this case, divide each result by 0.43, as shown below.

$$\text{Ba} \qquad\qquad \text{S} \qquad\qquad \text{O}$$
$$\frac{0.43}{0.43} = 1 \qquad \frac{0.43}{0.43} = 1 \qquad \frac{1.72}{0.43} = 4$$

3. The formula is $BaSO_4$.

In some cases you may be given the true formula mass of the compound. To check if your empirical formula is correct, add up the formula mass of the empirical formula and compare it with the given formula mass. If it is *not* the same, multiply the empirical formula by the small whole number that gives you the correct formula mass. For example, if your empirical formula is CH_2 (which has a formula mass of 14) and the true formula mass is given as 28, you can see that you must double the empirical formula by doubling all the subscripts. The true formula is C_2H_4.

LAWS OF DEFINITE COMPOSITION AND MULTIPLE PROPORTIONS

In the problems involving percent composition, we have depended on two things: each unit of an element has the same atomic mass, and every time the particular compound forms, it forms in the same percent composition. That this latter statement is true no matter the source of the compound is the **Law of Definite Composition**. There are some compounds formed by the same two elements in which the mass of one element is constant, but the mass of the other varies. In every case, however, the mass of the other element is present in a small-whole-number ratio to the weight of the first element. This is called the **Law of Multiple Proportions**. An example is H_2O and H_2O_2.

In H_2O the proportion of $H:O = 2:16$ or $1:8$

In H_2O_2 the proportion of $H:O = 2:32$ or $1:16$

The ratio of the mass of oxygen in each is $8:16$ or $1:2$ (a small-whole-number ratio).

WRITING AND BALANCING SIMPLE EQUATIONS

An equation is a simplified way of recording a chemical change. Instead of words, chemical symbols and formulas are used to represent the **reactants** and the **products**. Here is an example of how this can be done. The following is the word equation of the reaction of burning hydrogen with oxygen:

Hydrogen + oxygen yields water

Replacing the words with the chemical formulas, we have

$$H_2 + O_2 \rightarrow H_2O$$

We replaced hydrogen and oxygen with the formulas for their diatomic molecular states and wrote the appropriate formula for water based on the respective oxidation (valence) numbers for hydrogen and oxygen. Note that the word **yields** was replaced with the arrow.

Although the chemical statement tells what happened, it is not an equation because the two sides are not equal. While the left side has two atoms of oxygen, the right side has only one. Knowing that the Law of Conservation of Matter dictates that matter cannot easily be created or destroyed, we must get the number of atoms of each element represented on the left side to equal the number on the right. To do this, we can only use numbers, called **coefficients**, in front of the formulas. It is important to note that in attempting to balance equations THE SUBSCRIPTS IN THE FORMULAS MAY NOT BE CHANGED.

Looking again at the skeleton equation, we notice that if 2 is placed in front of H_2O the numbers of oxygen atoms represented on the two sides of the equation are equal. However, there are now four hydrogens on the right side with only two on the left. This can be corrected by using a coefficient of 2 in front of H_2. Now we have a balanced equation:

$$2H_2 + O_2 \rightarrow 2H_2O$$

This equation tells us more than merely that hydrogen reacts with oxygen to form water. It has quantitative meaning as well. It tells us that two molecular masses of hydrogen react with one molecular mass of oxygen to form two molecular masses of water. Because molecular masses are indirectly related to grams, we may also relate the masses of reactants and products in grams.

TIP

You cannot change subscripts of formulas to attempt to balance an equation!

$$2H_2 \quad\quad + \quad O_2 \quad\quad\quad\quad \rightarrow \quad 2H_2O$$

2(2)   32   2(18)

4 units   +   32 units   =   36 units

4 grams of H_2 +   32 grams of O_2 =   36 grams of water

This aspect will be important in solving problems related to the masses of substances in a chemical equation.

Here is another, more difficult example: Write the balanced equation for the burning of butane (C_4H_{10}) in oxygen. First, we write the skeleton equation:

$$C_4H_{10} + O_2 \text{ yields } CO_2 + H_2O.$$

TIP

First deal with multiatomic reactants.

Looking at the oxygens, we see that there are an even number on the left but an odd number on the right. This is a good place to start. If we use a coefficient of 2 for H_2O, that will even out the oxygens but introduce four hydrogens on the right while there are ten on the left. A coefficient of 5 will give us the right number of hydrogens but introduces an odd number of oxygens. Therefore, we have to go to the next even multiple of 5, which is 10. Ten gives us 20 hydrogen atoms on the right. By placing another coefficient of 2 in front of C_4H_{10}, we also have 20 hydrogen atoms on the left. Now the carbons need to be balanced. By placing an 8 in front of CO_2, we have eight carbons on both sides. The remaining step is to balance the oxygens. We have 26 on the right side, so we need a coefficient of 13 in front of the O_2 on the left to give us 26 oxygens on both sides. Our balanced equation is:

TIP

Be sure you have included all sources of a particular element since it may occur in two or more compounds.

$$2C_4H_{10} + 13O_2 \rightarrow 8CO_2 + 10H_2O$$

SHOWING PHASES IN CHEMICAL EQUATIONS

Once an equation is balanced, you may choose to give additional information in the equation. This can be done by indicating the phases of substances, telling whether each substance is in the liquid phase (ℓ), the gaseous phase (g), or the solid phase (s). Since many solids will not react to any appreciable extent unless they are dissolved in water, the notation (aq) is used to indicate that the substance exists in a water (aqueous) solution. Information concerning phase is given in parentheses following the formula for each substance. Several illustrations of this notation are given below:

TIP

(g) = gaseous state
(ℓ) = liquid state
(s) = solid state

Formula with Phase Notation	Meaning
$Cl_2(g)$	Chlorine gas
$H_2O(\ell)$	Water in the liquid state as opposed to ice or steam
$NaCl(s)$	Sodium chloride as a solid
$NaCl(aq)$	A water solution of dissolved sodium chloride

An example of phase notation in an equation:

$$2HCl(aq) + Zn(s) \rightarrow ZnCl_2(aq) + H_2(g)$$

In words, this says that a water solution of hydrogen chloride (called hydrochloric acid) reacts with solid zinc to produce zinc chloride dissolved in water plus hydrogen gas.

WRITING IONIC EQUATIONS

At times, chemists choose to show only the substances that react in the chemical action. These equations are called **ionic** equations because they stress the reaction and production of ions. If we look at the preceding equation, we see the complete cast of "actors":

Reactants

$2HCl(aq)$ releases $\rightarrow$ $2H^+(aq) + 2Cl^-(aq)$ in solution

$Zn(s)$ stay as $\rightarrow$ $Zn(s)$ particles

Products

$ZnCl_2(aq)$ releases $\rightarrow$ $Zn^{2+}(aq) + 2Cl^-(aq)$ in solution

$H_2(g)$ stay as $\rightarrow$ $H_2(g)$ molecules

Writing the complete reaction using these results, we have:

$$2H^+(aq) + 2Cl^-(aq) + Zn(s) \rightarrow Zn^{2+}(aq) + 2Cl^-(aq) + H_2(g)$$

Notice that nothing happened to the chloride ion. It appears the same on both sides of the equation. It is referred to as a spectator ion. In writing the **net ionic equation**, spectator ions are omitted, so the net ionic equation is:

$$2H^+(aq) + Zn(s) \rightarrow Zn^{2+}(aq) + H_2(g)$$

TIP

In net ionic equations, do not show "spectator" ions that do not change.

CHAPTER SUMMARY

The following terms summarize all the concepts and ideas that were introduced in this chapter. You should be able to explain their meaning and how you would use them in chemistry. They appear in boldface type in this chapter to draw your attention to them. The boldface type also makes it easier for you to look them up if you need to. You could also use Internet search engines like *google.com* on your computer to get a quick and expanded explanation of these terms, laws, and formulas.

binary compound	net ionic equation	reactants
coefficient	percentage composition	Stock system
empirical formula	polyatomic ion	ternary compound
formula mass	products	true formula
molecular mass		

Law of Definite Composition
Law of Multiple Proportions

Online content that reinforces major concepts discussed in this chapter can be found at the following Internet addresses if they are still available. *Some may have been changed or deleted.*

Interactive Practice on Naming Ionic Compounds

http://www.chemistrywithmsdana.org/wp-content/uploads/2012/07/ionic.html

This site offers an interactive practice on naming ionic compounds.

Chemical Formula Writing

http://chemsite.lsrhs.net/FlashMedia/html/compoundsAll.html

This site offers an interactive quiz on writing chemical formulas.

Naming Chemical Compounds

http://www.fernbank.edu/Chemistry/nomen.html

This site offers a Shockwave Flash exercise on chemical nomenclature.

Balancing Chemical Reaction Equations

http://www.wfu.edu/~ylwong/balanceeq/balanceq.html

This site offers a tutorial on balancing simple reaction equations with helpful step-by-step hints.

Write the formula or name in questions 1 through 10:

1. AgCl _____

2. $CaSO_4$ _____

3. $Al_2(SO_4)_3$ _____

4. NH_4NO_3 _____

5. $FeSO_4$ _____

6. Potassium chromate _____

7. Sodium fluoride _____

8. Magnesium sulfite _____

9. Copper(II) sulfate _____

10. Iron(III) chloride _____

Directions: The following set of lettered choices refers to the numbered questions immediately below it. For each numbered item, choose the one lettered choice that fits it best. Then fill in the corresponding oval on the answer sheet. Each choice in the set may be used once, more than once, or not at all.

Questions 11–15

Use the following choices to indicate the charge on the multivalent metal whose symbol is underlined in the given formulas.

(A) 1+
(B) 2+
(C) 3+
(D) 4+
(E) 5+

11. $\underline{Fe}_2S_3$

12. $\underline{Pb}O$

13. $\underline{Pb}O_2$

14. $\underline{Cu}Cl_2$

15. $\underline{Sn}_3N_4$

_____

16. Find the percentage of sulfur in H_2SO_4.

17. What are the empirical formula and the true formula of a compound composed of 85.7% C and 14.3% H with a true formula mass of 42?

	I		II
18.	The compound HF dissolved in water is called hydrofluoric acid	BECAUSE	the prefix *hydro-* indicates that a hydrogen compound is dissolved in water.
19.	Net ionic equations must include spectator ions	BECAUSE	net ionic equations must be balanced.
20.	Balanced equations have the same number of reactant atoms as the product atoms	BECAUSE	the conservation of matter must apply in all regular chemical equations.

*Fill in oval CE only if II is a correct explanation of I.

	I	II	CE*
18.	Ⓣ Ⓕ	Ⓣ Ⓕ	◯
19.	Ⓣ Ⓕ	Ⓣ Ⓕ	◯
20.	Ⓣ Ⓕ	Ⓣ Ⓕ	◯

21. Write the complete ionic equation for this reaction. Then write the net ionic equation.

$$3NaOH(aq) + Fe(NO_3)_3(aq) \rightarrow 3NaNO_3(aq) + Fe(OH)_3(s)$$

Complete ionic equation:

_____ → _____

Net ionic equation:

_____ → _____

Answers and Explanations

1. Silver chloride

2. Calcium sulfate

3. Aluminum sulfate

4. Ammonium nitrate

5. Iron(II) sulfate or ferrous sulfate

6. K_2CrO_4

7. NaF

8. $MgSO_3$

9. $CuSO_4$

10. $FeCl_3$

11. **(C)** 3+ because the sulfur has a 2– charge

12. **(B)** 2+ because the oxygen has a 2– charge

13. **(D)** 4+ because the oxygen has a 2– charge

14. **(B)** 2+ because the chlorine has a 1– charge

15. **(D)** 4+ because the nitrogen has a 3– charge

16. H_2SO_4 is composed of:

$$2H = 2.0$$
$$1S = 32$$
$$4O = 64$$
$$Total = 98$$

Percentage of S:

$$\frac{S = 32}{Total = 98} \times 100\% = \underline{\underline{32.65}} \text{ or } \underline{\underline{33\%}}$$

17. **(C)** = 12) 85.7% C   H = 1) 14.3% H
       7.14   14.3

To find the lowest ratio of the whole numbers:

$$7.14)\underline{7.14} \qquad 7.14)\underline{14.3}$$
$$1.0 \qquad\qquad 2.0$$

The empirical formula is $\underline{\underline{CH_2}}$.

Since the formula mass is given as 42, the empirical formula CH_2, which represents a mass of 14, divides into 42 three times. Therefore, the true formula is $\underline{\underline{C_3H_6}}$.

18. **(T, F)** The compound hydrogen fluoride, HF, dissolved in water is called hydrofluoric acid, but the prefix *hydro-* refers to the fact that it is a binary acid that does not contain oxygen. Although the name hydrofluoric acid does refer to the compound hydrogen fluoride dissolved in water, the prefix *hydro-* is not what indicates this. For first year chemistry students, most substances referred to as acids are water solutions.

19. **(F, T)** Net ionic equations *eliminate* spectator ions. The elimination of spectator ions does not preclude the equation from being balanced. All properly written equations should be balanced.

20. **(T, T, CE)** The Law of Conservation of Matter does require that there are equal numbers of respective atoms on both sides of a regular (not atomic) equation. Because the formula of each substance is determined by oxidation numbers, the equation is balanced by using coefficients.

21. From this formula equation:

$$3NaOH(aq) + Fe(NO_3)_3(aq) \rightarrow 3NaNO_3(aq) + Fe(OH)_3(s)$$

The complete ionic equation is:

$$3Na^+(aq) + 3OH^-(aq) + Fe^{3+}(aq) + 3NO_3^-(aq) \rightarrow 3Na^+(aq) + 3NO_3^-(aq) + Fe(OH)_3(s)$$

The net ionic equation is:

$$Fe^{3+}(aq) + 3OH^-(aq) \rightarrow Fe(OH)_3(s)$$

Gases and the Gas Laws 5

These skills are usually tested on the SAT Subject Test in Chemistry. You should be able to...

→ Describe the physical and chemical properties of oxygen and hydrogen and the electronic makeup of their diatomic molecules.

→ Explain how atmospheric pressure is measured, how to read the pressure in a manometer, and the units used to measure pressure.

→ Read and explain a graphic distribution of the number of molecules versus kinetic energy at different temperatures.

→ Know and use the following laws to solve gas problems: Graham's, Charles's, Boyle's, Dalton's, the Combined Gas Law, and the Ideal Gas Law.

This chapter will review and strengthen these skills. Be sure to do the Practice Exercises at the end of the chapter.

When we discuss gases today, the most pressing concern is the gases in our atmosphere. These are the gases that are held against Earth by the gravitational field. The principal constituents of the atmosphere of Earth today are nitrogen (78%) and oxygen (21%). The gases in the remaining 1% are argon (0.9%), carbon dioxide (0.03%), varying amounts of water vapor, and trace amounts of hydrogen, ozone, methane, carbon monoxide, helium, neon, krypton, and xenon. Oxides and other pollutants added to the atmosphere by factories and automobiles have become a major concern because of their damaging effects in the form of acid rain. In addition, a strong possibility exists that the steady increase in atmospheric carbon dioxide, mainly attributed to fossil fuel combustion over the past century, may affect Earth's climate by causing a **greenhouse effect**, resulting in a steady rise in temperatures worldwide.

Studies of air samples show that up to 55 miles above sea level the composition of the atmosphere is substantially the same as at ground level; continuous stirring produced by atmospheric currents counteracts the tendency of the heavier gases to settle below the lighter ones. In the lower atmosphere, ozone is normally present in extremely low concentrations. The atmospheric layer 12 to 30 miles up contains more ozone that is produced by the action of ultraviolet radiation from the Sun. In this layer, however, the percentage of ozone is only 0.001 by volume. Human activity adds to the ozone concentration in the lower atmosphere where it can be a harmful pollutant.

The ozone layer became a subject of concern in the early 1970s when it was found that chemicals known as fluorocarbons, or chlorofluoromethanes, were rising into the atmosphere in large quantities because of their use as refrigerants and as propellants in aerosol dispensers. The concern centered on the possibility that these compounds, through the action of sunlight, could chemically attack and destroy stratospheric ozone, which protects

TIP

The major components of Earth's atmosphere: 78% nitrogen 21% oxygen

Earth's surface from excessive ultraviolet radiation. As a result, U.S. industries and the Environmental Protection Agency phased out the use of certain chlorocarbons and fluorocarbons as of the year 2000. There is still ongoing concern about both these environmental problems: the greenhouse effect and the deterioration of the ozone layer as it relates to possible global warming.

SOME REPRESENTATIVE GASES
Oxygen

Of the gases that occur in the atmosphere, the most important one to us is oxygen. Although it makes up only approximately 21% of the atmosphere, by volume, the oxygen found on Earth is equal in weight to all the other elements combined. About 50% of Earth's crust (including the waters on Earth, and the air surrounding it) is oxygen. (Note Figure 15.)

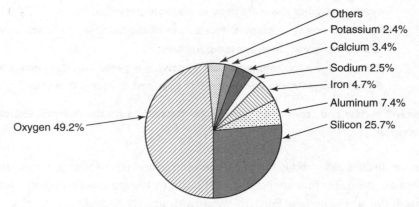

Figure 15. Composition of Earth's Crust

The composition of air varies slightly from place to place because air is a mixture of gases. The composition by volume is approximately as follows: nitrogen, 78%; oxygen, 21%; argon, 1%. There are also small amounts of carbon dioxide, water vapor, and trace gases.

PREPARATION OF OXYGEN. In 1774, an English scientist named Joseph Priestley discovered oxygen by heating mercuric oxide in an enclosed container with a magnifying glass. That mercuric oxide decomposes into oxygen and mercury can be expressed in an equation: $2HgO \rightarrow 2Hg + O_2$. After his discovery, Priestley visited one of the greatest of all scientists, Antoine Lavoisier, in Paris. As early as 1773 Lavoisier had carried on experiments concerning burning, and they had caused him to doubt the phlogiston theory (that a substance called phlogiston was released when a substance burned; the theory went through several modifications before it was finally abandoned). By 1775, Lavoisier had demonstrated the true nature of burning and called the resulting gas "oxygen."

Today oxygen is usually prepared in the lab by heating an easily decomposed oxygen compound such as potassium chlorate ($KClO_3$). The equation for this reaction is:

$$2KClO_3 + MnO_2 \rightarrow 2KCl + 3O_2(g) + MnO_2$$

A possible laboratory setup is shown in Figure 16.

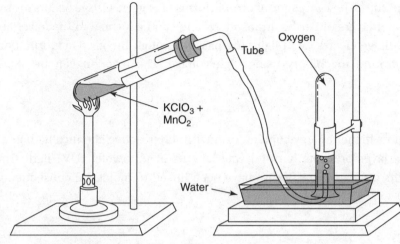

Figure 16. A Possible Laboratory Preparation of Oxygen

TIP

A catalyst speeds up the rate of reaction by lowering the activation energy needed for the reaction. A catalyst is not consumed.

In this preparation manganese dioxide (MnO_2) is often used. This compound is not used up in the reaction and can be shown to have the same composition as it had before the reaction occurred. The only effect it has is that it lowers the temperature needed to decompose the $KClO_3$, and thus speeds up the reaction. Substances that behave in this manner are referred to as **catalysts**. The mechanism by which a catalyst acts is not completely understood in all cases, but it is known that in some reactions the catalyst does change its structure temporarily. Its effect is shown graphically in the reaction graphs in Figure 17.

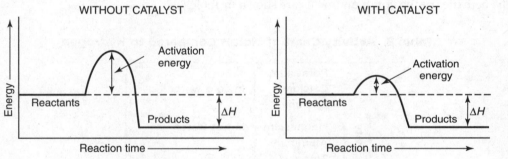

Figure 17. Effect of Catalyst on Reaction

TIP

Graphic representation of how a catalyst lowers the required activation energy

PROPERTIES OF OXYGEN. Oxygen is a gas under ordinary conditions of temperature and pressure, and it is a gas that is colorless, odorless, tasteless, and slightly heavier than air; all these physical properties are characteristic of this element. Oxygen is only slightly soluble in water, thus making it possible to collect the gas over water, as shown in Figure 16.

Although oxygen will support combustion, it will not burn. This is one of its chemical properties. The usual test for oxygen is to lower a glowing splint into the gas and see if the oxidation increases in its rate to reignite the splint. (Note: This is not the only gas that does this. N_2O reacts the same.)

OZONE. Ozone is another form of oxygen and contains three atoms in its molecular structure (O_3). Since ordinary oxygen and ozone differ in energy content and form, they have slightly different properties. They are called allotropic forms of oxygen. Ozone occurs in small quantities in the upper layers of Earth's atmosphere, and can be formed in the lower atmosphere, where high-voltage electricity in lightning passes through the air. This formation of ozone also occurs around machinery using high voltage. The reaction can be shown by this equation:

$$3O_2 + \text{elec.} \rightarrow 2O_3$$

Because of its higher energy content, ozone is more reactive chemically than oxygen.

The ozone layer prevents harmful wavelengths of ultraviolet (UV) light from passing through Earth's atmosphere. UV rays have been linked to biological consequences such as skin cancer.

Hydrogen

PREPARATION OF HYDROGEN. Although there is evidence of the preparation of hydrogen before 1766, Henry Cavendish was the first person to recognize this gas as a separate substance. He observed that, whenever it burned, it produced water. Lavoisier named it **hydrogen**, which means "water former."

Electrolysis of water, which is the process of passing an electric current through water to cause it to decompose, is one method of obtaining hydrogen. This is a widely used commercial method, as well as a laboratory method.

Another method of producing hydrogen is to displace it from the water molecule by using a metal. To choose the metal you must be familiar with its activity with respect to hydrogen. The activities of the common metals are shown in Table 8.

Table 8. Activity Chart of Metals Compared to Hydrogen

Increasing Activity		
Potassium, Calcium, Sodium	In cold water
Magnesium	In hot water
Aluminum, Zinc, Iron, Tin	In most dilute acids
HYDROGEN	
Copper, Mercury, Silver, Gold	Not active enough

As noted in Table 8, any of the first three metals will react with cold water; the reaction is as follows:

Very active metal + Water = Hydrogen + Metal hydroxide

Using sodium as an example:

$$2Na + 2HOH \rightarrow H_2(g) + 2NaOH$$

With the metals that react more slowly, a dilute acid reaction is needed to produce hydrogen in sufficient quantities to collect in the laboratory. This general equation is:

Active metal + Dilute acid → Hydrogen + Salt of the acid

An example:

$$Zn + dil. \ H_2SO_4 \rightarrow H_2(g) + ZnSO_4$$

This equation shows the usual laboratory method of preparing hydrogen. Mossy zinc is used in a setup as shown in Figure 18. The acid is introduced down the thistle tube after the zinc is placed in the reacting bottle. In this sort of setup, you would not begin collecting the gas that bubbles out of the delivery tube for a few minutes so that the air in the system has a chance to be expelled and you can collect a rather pure volume of the gas generated.

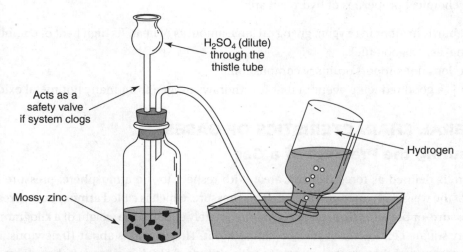

Figure 18. Preparation of an Insoluble Gas by the Addition of Liquid to Other Reactant

In industry, hydrogen is produced by (1) the electrolysis of water, (2) passing steam over red-hot iron or through hot coke, or (3) by decomposing natural gas (mostly methane, CH_4) with heat ($CH_4 + H_2O \rightarrow CO + 3H_2$).

PROPERTIES OF HYDROGEN. Hydrogen has the following important physical properties:

1. It is ordinarily a gas; colorless, odorless, tasteless when pure.

2. It weighs 0.9 gram per liter at 0°C and 1 atmosphere pressure. This is $\frac{1}{14}$ as dense as air.

3. It is slightly soluble in water.

4. It becomes a liquid at a temperature of −240°C and a pressure of 13 atmospheres.

5. It diffuses (moves from place to place in gases) more rapidly than any other gas. This property can be demonstrated as shown in Figure 19.

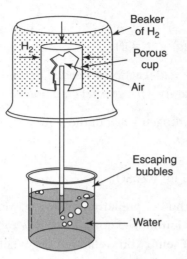

Here the H_2 in the beaker that is placed over the porous cup diffuses faster through the cup than the air can diffuse out. Consequently, there is a pressure buildup in the cup, which pushes the gas out through the water in the lower beaker.

Figure 19. Diffusion of Hydrogen

The chemical properties of hydrogen are:

1. It burns in air or in oxygen, giving off large amounts of heat. Its high heat of combustion makes it a good fuel.
2. It does not support ordinary combustion.
3. It is a good reducing agent in that it withdraws oxygen from many hot metal oxides.

GENERAL CHARACTERISTICS OF GASES
Measuring the Pressure of a Gas

Pressure is defined as force per unit area. With respect to the atmosphere, pressure is the result of the weight of a mixture of gases. This pressure, which is called **atmospheric pressure**, **air pressure**, or **barometric pressure**, is approximately equal to the weight of a kilogram mass on every square centimeter of surface exposed to it. This weight is about 10 newtons.

The pressure of the atmosphere varies with altitude. At higher altitudes, the weight of the overlying atmosphere is less, so the pressure is less. Air pressure also varies somewhat with weather conditions as low- and high-pressure areas move with weather fronts. On the average, however, the air pressure at sea level can support a column of mercury 760 millimeters in height. This average sea-level air pressure is known as **normal atmospheric pressure**, also called **standard pressure**.

The instrument most commonly used for measuring air pressure is the **mercury barometer**. The diagram below shows how it operates. Atmospheric pressure is exerted on the mercury in the dish, and this in turn holds the column of mercury up in the tube. This column at standard pressure will measure 760 millimeters above the level of the mercury in the dish below.

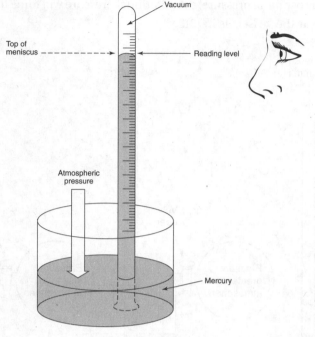

Vacuum

Top of
meniscus

Reading level

Atmospheric
pressure

Mercury

Mercury Barometer

TIP

**Read the top of
the meniscus for
mercury but the
bottom of the
meniscus for
water.**

In gas-law problems pressure may be expressed in various units. One standard atmosphere (1 atm) is equal to 760 millimeters of mercury (760 mm Hg) or 760 **torr**, a unit named for Evangelista Torricelli. In the SI system, the unit of pressure is the **pascal** (Pa), named in honor of the scientist of the same name, and standard pressure is 101,325 pascals or 101.325 kilopascals (kPa). One pascal (Pa) is defined as the pressure exerted by the force of one newton (1 N) acting on an area of one square meter. In many cases, as in atmospheric pressure, it is more convenient to express pressure in kilopascals (kPa).

Summary of Units of Pressure

Unit	Abbreviation	Unit Equivalent to 1 atm
Atmosphere	atm	1 atm
Millimeters of Hg	mm Hg	760 mm Hg
Torr	torr	760 torr
Pascal	Pa	101,325 Pa
Kilopascal	kPa	101.325 kPa

A device similar to the barometer can be used to measure the pressure of a gas in a confined container. This apparatus, called a **manometer**, is illustrated below. A manometer is basically a U-tube containing mercury or some other liquid. When both ends are open to the air, as in (1) in the diagram, the level of the liquid will be the same on both sides since the same pressure is being exerted on both ends of the tube. In (2) and (3), a vessel is connected to one end of the U-tube. Now the height of the mercury column serves as a means of reading the pressure inside the vessel if the atmospheric pressure is known. When the pressure inside the vessel is the same as the atmospheric pressure outside, the levels of liquid are the same. When the pressure inside is greater than outside, the column of liquid will be higher on the side that is exposed to the air, as in (2). Conversely, when the pressure inside the vessel is less than the outside atmospheric pressure, the additional pressure will force the liquid to a higher level on the side near the vessel, as in (3).

TIP

Know how to calculate the pressure in a closed vessel like in the manometer shown.

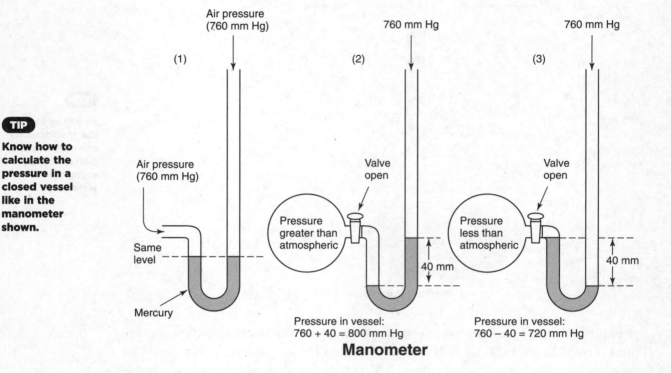

Manometer

Kinetic-Molecular Theory

By indirect observations, the Kinetic-Molecular Theory has been arrived at to explain the forces between molecules and the energy the molecules possess. There are three basic assumptions to the Kinetic-Molecular Theory:

TIP

Know these basic assumptions of the Kinetic-Molecular Theory.

1. Matter in all its forms (solid, liquid, and gas) is composed of extremely small particles. In many cases these are called molecules. The space occupied by the gas particles themselves is ignored in comparison with the volume of the space in which they are contained.

2. The particles of matter are in constant motion. In solids, this motion is restricted to a small space. In liquids, the particles have a more random pattern but still are restricted to a kind of rolling over one another. In a gas, the particles are in continuous, random, straight-line motion.

3. When these particles collide with each other or with the walls of the container, there is no loss of energy.

Some Particular Properties of Gases

As the temperature of a gas is increased, its kinetic energy is increased, thereby increasing the random motion. At a particular temperature not all the particles have the same kinetic energy, but the temperature is a measure of the average kinetic energy of the particles. A graph of the various kinetic energies resembles a normal bell-shaped curve with the average found at the peak of the curve (see Figure 20).

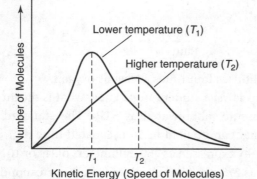

Figure 20. Molecular Speed Distribution in a Gas at Different Temperatures

TIP

When you read the temperature of a substance, you are measuring its average kinetic energy.

When the temperature is lowered, the gas reaches a point at which the kinetic energy can no longer overcome the attractive forces between the particles (or molecules) and the gas condenses to a liquid. The temperature at which this condensation occurs is related to the type of substance the gas is composed of and the type of bonding in the molecules themselves. This relationship of bond type to condensation point (or boiling point) is pointed out in Chapter 3, "Bonding."

The random motion of gases in moving from one position to another is referred to as **diffusion**. You know that, if a bottle of perfume is opened in one corner of a room, the perfume, that is, its molecules, will move or diffuse to all parts of the room in time. The rate of diffusion is the rate of the mixing of gases.

TIP

Diffusion means spreading out.

Effusion is the term used to describe the passage of a gas through a tiny orifice into an evacuated chamber. The rate of effusion measures the speed at which the gas is transferred into the chamber.

TIP

Effusion means passing of a gas through an orifice (like through the neck of a balloon).

GAS LAWS AND RELATED PROBLEMS
Graham's Law of Effusion (Diffusion)

This law relates the rate at which a gas diffuses (or effuses) to the type of molecule in the gas. It can be expressed as follows:

> The Rate of Effusion of a Gas Is Inversely Proportional to the Square Root of Its Molecular Mass.

Hydrogen, with the lowest molecular mass, can diffuse more rapidly than other gases under similar conditions.

➡ **Example** _____

Compare the rate of diffusion of hydrogen to that of oxygen under similar conditions.

The formula is

$$\frac{\text{Rate A}}{\text{Rate B}} = \frac{\sqrt{\text{Molecular mass of B}}}{\sqrt{\text{Molecular mass of A}}}$$

Let A be H_2 and B be O_2.

$$\frac{\text{Rate } H_2}{\text{Rate } O_2} = \frac{\sqrt{32}}{\sqrt{2}} = \frac{\sqrt{16}}{\sqrt{1}} = \frac{4}{1}$$

Therefore hydrogen diffuses four times as fast as oxygen.

In dealing with the gas laws, a student must know what is meant by standard conditions of temperature and pressure (abbreviated as STP). The standard pressure is defined as the height of mercury that can be held in an evacuated tube by 1 atmosphere of pressure (14.7 lb/in.^{2}). This is usually expressed as 760 millimeters of Hg or 101.3 kilopascals. **Standard temperature** is defined as 273 Kelvin or absolute (which corresponds to 0° Celsius).

Charles's Law $\left(\dfrac{V}{T} = k\right)$

Jacques Charles, a French chemist of the early nineteenth century, discovered that, when a gas under constant pressure is heated from 0°C to 1°C, it expands 1/273 of its volume. It contracts this amount when the temperature is dropped 1 degree to –1°C. Charles reasoned that, if a gas at 0°C was cooled to –273°C (actually found to be –273.15°C), its volume would be zero. Actually, all gases are converted into liquids before this temperature is reached. By using the Kelvin scale to rid the problem of negative numbers, we can state Charles's Law as follows:

> **If the Pressure Remains Constant, the Volume of a Gas Varies Directly as the Absolute Temperature. Then**
>
> Initial $\dfrac{V_1}{T_1}$ = Final $\dfrac{V_2}{T_2}$ at constant pressure or $\dfrac{V}{T} = k$

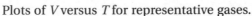

Graphic relationship—Charles's Law. The dashed lines represent extrapolation of the data into regions where the gas would become liquid or solid. Extrapolation shows that each gas, if it remained gaseous, would reach zero volume at 0 K or –273°C.

Plots of V versus T for representative gases.

➥ Example 1 _____

Assume dry gases unless otherwise stated.

The volume of a gas at 20°C is 500. mL. Find its volume at standard temperature if pressure is held constant.

Convert temperatures:

$$20°C = 20° + 273 = 293 \text{ K}$$
$$0°C = 0° + 273 = 273 \text{ K}$$

If you know that cooling a gas decreases its volume, then you know that 500. mL will have to be multiplied by a fraction (made up of the Kelvin temperatures) that has a smaller numerator than the denominator. So

$$500. \text{ mL} \times \frac{273}{293} = 465 \text{ mL}$$

TIP

STP = standard temperature of 273 K standard pressure of 760 mm Hg or 1 atmosphere (atm) or 101.3 kilopascals.

Or you can use the formula and substitute known values:

$$\frac{V_1}{T_1} = \frac{V_2}{T_2}$$
$$\frac{500. \text{ mL}}{293} = \frac{x \text{ mL}}{273}$$
$$x \text{ mL} = \textbf{465 mL}$$

➥ Example 2 _____

A sample of gas occupies 24 L at 175.0 K. What volume would the gas occupy at 400.0 K?

The temperature of the gas is increased. Charles's Law predicts that the gas volume will also increase. So

$$\frac{V_1}{T_1} = \frac{V_2}{T_2}$$
$$V_2 = \frac{V_1 T_2}{T_1}$$
$$V_2 = 24 \text{ L} \times \frac{400.0 \text{ K}}{175.0 \text{ K}}$$
$$V_2 = 55 \text{ L}$$

The final volume has increased as predicted.

Boyle's Law (*PV = k*)

Robert Boyle, a seventeenth-century English scientist, found that the volume of a gas decreases when the pressure on it is increased, and vice versa, when the temperature is held constant. Boyle's Law can be stated as follows:

TIP

Boyle's Law

$P_1V_1 = P_2V_2$

temperature is held constant

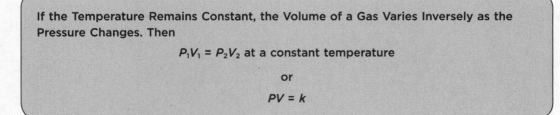

If the Temperature Remains Constant, the Volume of a Gas Varies Inversely as the Pressure Changes. Then

$P_1V_1 = P_2V_2$ at a constant temperature

or

$PV = k$

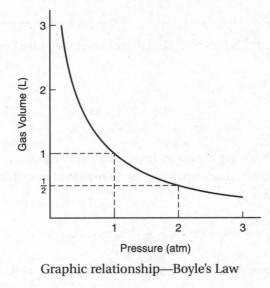

Graphic relationship—Boyle's Law

➥ **Example 1**

Given the volume of a gas as 200. mL at 1.05 atm pressure, calculate the volume of the same gas at 1.01 atm. Temperature is held constant.

If you know that this decrease in pressure will cause an increase in the volume, then you know 200. mL must be multiplied by a fraction (made up of the two pressures) that has a larger numerator than the denominator. So

$$200. \, \text{mL} \times \frac{1.05 \, \text{atm}}{1.01 \, \text{atm}} = 208 \, \text{mL}$$

Or you can use the formula:

$$P_1 V_1 = P_2 V_2$$

$$V_2 = V_1 \times \frac{P_1}{P_2}$$

$$= 200. \, \text{mL} \times \frac{1.05 \, \text{atm}}{1.01 \, \text{atm}} = 208 \, \text{mL}$$

➥ **Example 2**

The gas in a balloon has a volume of 7.5 L at 100. kPa. The balloon is released into the atmosphere, and the gas in it expands to a volume of 11. L. Assuming a constant temperature, what is the pressure on the balloon at the new volume?

The volume of the gas has increased. Boyle's Law predicts that the gas pressure will decrease. So

$$P_1 V_1 = P_2 V_2$$

$$P_2 = \frac{P_1 V_1}{V_2}$$

$$\text{Or} \quad P_2 = 100. \, \text{kPa} \times \frac{7.5 \, \text{L}}{11. \, \text{L}}$$

$$P_2 = 100. \, \text{kPa} \times \frac{7.5 \, \text{L}}{11. \, \text{L}} = 68. \, \text{kPa}$$

The final pressure has decreased as predicted.

Combined Gas Law

This is a combination of the two preceding gas laws. The formula is as follows:

$$\frac{P_1V_1}{T_1} = \frac{P_2V_2}{T_2}$$

In the combined gas law, all subscripts on the left side of the formula are 1's and on the right side are 2's.

➡ **Example 1** _____

The volume of a gas at 780. mm pressure and 30.°C is 500. mL. What volume would the gas occupy at STP?

You again can use reasoning to determine the kind of fractions the temperatures and pressures must be to arrive at your answer. Since the pressure is going from 780. mm to 760. mm, the volume should increase. The fraction must then be $\frac{780.}{760.}$. Also, since the temperature is going from 30°C (303 K) to 0°C (273 K), the volume should decrease; this fraction must be $\frac{273}{303}$. So

$$500.\,\text{mL} \times \frac{780.}{760.} \times \frac{273}{303} = 462\,\text{mL}$$

Or you can use the formula:

$$\frac{P_1V_1}{T_1} = \frac{P_2V_2}{T_2}$$

Solve for $V_2 = V_1 \times \dfrac{P_1}{P_2} \times \dfrac{T_2}{T_1}$

$$V_2 = 500.\,\text{mL} \times \frac{780.\,\text{mm Hg}}{760.\,\text{mm Hg}} \times \frac{273\,\text{K}}{303\,\text{K}} = 462\,\text{mL}$$

➡ **Example 2** _____

The volume of a gas is 27.5 mL at 22.0°C and 0.974 atm. What will the volume be at 15.0°C and 0.993 atm?

Using the combined gas law, $\dfrac{P_1V_1}{T_1} = \dfrac{P_2V_2}{T_2}$

and solving for $V_2 = V_1 \times \dfrac{P_1 T_2}{P_2 T_1}$

$$V_2 = 27.5\,\text{mL} \times \frac{0.974\,\text{atm}}{0.993\,\text{atm}} \times \frac{(15.0°C + 273 = 288\,\text{K})}{(22.0°C + 273 = 295\,\text{K})}$$

$$V_2 = 26.3\,\text{mL}$$

Pressure Versus Temperature (Gay-Lussac's Law)

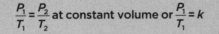

At Constant Volume, the Pressure of a Given Mass of Gas Varies Directly with the Absolute Temperature. Then

$$\frac{P_1}{T_1} = \frac{P_2}{T_2} \text{ at constant volume or } \frac{P_1}{T_1} = k$$

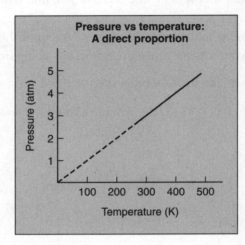

Pressure vs temperature: A direct proportion

➥ Example 1

A steel tank contains a gas at 27°C and a pressure of 12.0 atms. Determine the gas pressure when the tank is heated to 100.°C.

Reasoning that an increase in temperature will cause an increase in pressure at constant volume, you know the pressure must be multiplied by a fraction that has a larger numerator than denominator. The fraction must be $\frac{373\,\text{K}}{300.\,\text{K}}$. So

$$12.0\,\text{atm} \times \frac{373\,\cancel{K}}{300.\,\cancel{K}} = 14.9\,\text{atm or }15.0\,\text{atm}$$

Or you can use the formula:

$$\frac{P_1}{T_1} = \frac{P_2}{T_2}$$

$$P_2 = P_1 \times \frac{T_2}{T_1}$$

$$P_2 = 12.0\,\text{atm} \times \frac{373\,\text{K}}{300.\,\text{K}} = 14.9\,\text{atm}$$

➥ Example 2

At 120.°C, the pressure of a sample of nitrogen is 1.07 atm. What will the pressure be at 205°C, assuming constant volume?

Using the relationship of: $\dfrac{P_1}{T_1} = \dfrac{P_2}{T_2}$

and solving for $P_2 = \dfrac{P_1 T_2}{T_1}$

So $P_2 = \dfrac{1.07\,\text{atm} \times (205° + 273° = 478\ K)}{(120.° + 273° = 393\ K)}$

$P_2 = 1.30\,\text{atm}$

Dalton's Law of Partial Pressures

> **When a Gas Is Made Up of a Mixture of Different Gases, the Total Pressure of the Mixture Is Equal to the Sum of the Partial Pressures of the Components; That Is, the Partial Pressure of the Gas Would Be the Pressure of the Individual Gas if It Alone Occupied the Volume. The Formula Is:**
>
> $$P_{\text{total}} = P_{\text{gas 1}} + P_{\text{gas 2}} + P_{\text{gas 3}} + \ldots$$

Know Dalton's Law of Partial Pressures.

➥ Example

A mixture of gases at 760. mm Hg pressure contains 65.0% nitrogen, 15.0% oxygen, and 20.0% carbon dioxide by volume. What is the partial pressure of each gas?

$$0.650 \times 760. = 494 \text{ mm pressure } (N_2)$$
$$0.150 \times 760. = 114 \text{ mm pressure } (O_2)$$
$$0.200 \times 760. = 152 \text{ mm pressure } (CO_2)$$

If the pressure was given as 1.0 atm, you would substitute 1.0 atm for 760. mm Hg. The answers would be:

$$0.650 \times 1.0 \text{ atm} = 0.650 \text{ atm } (N_2)$$
$$0.150 \times 1.0 \text{ atm} = 0.150 \text{ atm } (O_2)$$
$$0.200 \times 1.0 \text{ atm} = 0.200 \text{ atm } (CO_2)$$

Corrections of Pressure

CORRECTION OF PRESSURE WHEN A GAS IS COLLECTED OVER WATER. When a gas is collected over a volatile liquid, such as water, some of the water vapor is present in the gas and contributes to the total pressure. Assuming that the gas is saturated with water vapor at the given temperature, you can find the partial pressure due to the water vapor in a table of such water vapor values. This vapor pressure, which depends only on the temperature, must be subtracted from the total pressure to find the partial pressure of the gas being measured.

CORRECTION OF DIFFERENCE IN THE HEIGHT OF THE FLUID. When gases are collected in eudiometers (glass tubes closed at one end), it is not always possible to get the level of the liquid inside the tube to equal the level on the outside. This deviation of levels must be taken into

When a gas is collected over water, subtract the water vapor pressure at the given temperature from the atmospheric pressure to find the partial pressure of the gas.

$P_{\text{gas}} = P_{\text{atm}} - P_{H_2O}$

account when determining the pressure of the enclosed gas. There are then two possibilities: (1) When the level inside is higher than the level outside the tube, the pressure on the inside is less, by the height of fluid in excess, than the outside pressure. If the fluid is mercury, you simply subtract the difference from the outside pressure reading (also in height of mercury and in the same units) to get the corrected pressure of the gas. If the fluid is water, you must first convert the difference to an equivalent height of mercury by dividing the difference by 13.6 (since mercury is 13.6 times as heavy as water, the height expressed in terms of Hg will be 1/13.6 the height of water). This is shown pictorially in Figure 21. Again, care must be taken that this equivalent height of mercury is in the same units as the expression for the outside pressure before it is subtracted to obtain the corrected pressure for the gas in the eudiometer. (2) When the level inside is lower than the level outside the tube, a correction must be added to the outside pressure. If the difference in height between the inside and the outside is expressed in terms of water, you must take 1/13.6 of this quantity to correct it to millimeters of mercury. This quantity is then added to the expression of the outside pressure, which must also be in millimeters of mercury. If the tube contains mercury, then the difference between the inside and outside levels is merely added to the outside pressure to get the corrected pressure for the enclosed gas.

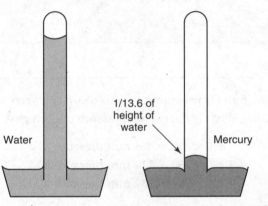

Figure 21. Same Pressure Exerted on Both Liquids

➥ Example

Hydrogen gas was collected in a eudiometer tube over water. It was impossible to level the outside water with that in the tube, so the water level inside the tube was 40.8 mm higher than that outside. The barometric pressure was 730. mm of Hg. The water vapor pressure at the room temperature of 29°C was found in a handbook to be 30.0 mm. What is the pressure of the dry hydrogen?

STEP 1 To find the true pressure of the gas, we must first subtract the water-level difference expressed in mm of Hg:

$$\frac{40.8}{13.6} = 3.00 \text{ mm of Hg}$$

Then 730. mm − 3.00 mm = 727 mm total pressure of gases in the eudiometer

STEP 2 Correcting for the partial pressure due to water vapor in the hydrogen, we subtract the vapor pressure (30.0 mm) from 727 mm and get our answer: 697 mm.

Ideal Gas Law

The preceding laws do not include the relationship of number of moles of a gas to the pressure, volume, and temperature of the gas. A law derived from the Kinetic-Molecular Theory relates these variables. It is called the Ideal Gas Law and is expressed as

$$PV = nRT$$

Know how to use the Ideal Gas Law:

$PV = nRT$

P, V, and T retain their usual meanings, but n stands for the number of moles of the gas and R represents the ideal gas constant.

Boyle's Law and Charles's Law are actually derived from the ideal gas law. Boyle's Law applies when the number of moles and the temperature of the gas are constant. Then in $PV = nRT$, the number of moles, n, is constant; the gas constant (R) remains the same; and by definition T is constant. Therefore, $PV = k$. At the initial set of conditions of a problem, $P_1 V_1$ = a constant (k). At the second set of conditions, the terms on the right side of the equation are equal to the same constant, so $P_1 V_1 = P_2 V_2$. This matches the Boyle's Law equation introduced earlier.

The same can also be done with Charles's Law, because $PV = nRT$ can be expressed with the variables on the left and the constants on the right:

$$\frac{V}{T} = \frac{nR}{P}$$

In Charles's Law the number of moles and the pressure are constant. Substituting k for the constant term, $\frac{nR}{P}$, we have

$$\frac{V}{T} = k$$

The expression relating two sets of conditions can be written as

$$\frac{V_1}{T_1} = \frac{V_2}{T_2}$$

To use the ideal gas law in the form $PV = nRT$, the gas constant, R, must be determined. This can be done mathematically as shown in the following example.

One mole of oxygen gas was collected in the laboratory at a temperature of 24.0°C and a pressure of exactly 1 atmosphere. The volume was 24.38 liters. Find the value of R.

Know the ideal gas constant:

$R = 0.0821 \dfrac{L \cdot atm}{mol \cdot K}$

$$PV = nRT$$

Rearranging the equation to solve for R gives

$$R = \frac{PV}{nT}$$

Substituting the known values on the right, we have

$$R = \frac{1\ atm \times 24.38\ L}{1\ mol \times 297\ K}$$

Calculating R, we get

$$R = 0.0821 \frac{L \cdot atm}{mol \cdot K}$$

Once R is known, the Ideal Gas Law can be used to find any of the variables, given the other three.

For example, calculate the pressure, at 16.0°C, of 1.00 gram of hydrogen gas occupying 2.54 liters.

Rearranging the equation to solve for P, we get

$$P = \frac{nRT}{V}$$

Remember to use appropriate units:

moles (mol)

liters (L)

atmosphere (atm)

The molar mass of hydrogen is 2.00 g/mol, so the number of moles in this problem would be

$$\frac{1.00\,g}{2.00\,g/mol} = 0.500\,mol$$

Substituting the known values, we have

$$P = \frac{(0.500\,\cancel{mol})\left(0.0821\frac{\cancel{L}\cdot atm}{\cancel{mol}\cdot\cancel{K}}\right)289\,\cancel{K}}{2.54\,\cancel{L}}$$

Calculating the value, we get

$$P = 4.66\,atm$$

Another use of the ideal gas law is to find the number of moles of a gas when P, T, and V are known.

For example, how many moles of nitrogen gas are in 0.38 liter of gas at 0°C and 0.50 atm pressure?

Rearranging the equation to solve for n gives

$$n = \frac{PV}{RT}$$

Changing temperature to kelvins and pressure to atmospheres gives

$$T = 0° + 273 = 273\,K$$

$$P = \frac{0.50\,atm}{1.00\,atm} = 0.50\,atm$$

Substituting in the equation, we have

$$n = \frac{(0.50\,\cancel{atm})(0.38\,\cancel{L})}{\left(0.0821\frac{\cancel{L}\cdot\cancel{atm}}{mol\cdot\cancel{K}}\right)(273\,\cancel{K})} = 0.0085\,mol$$

$$= 0.0085\,mol\ of\ nitrogen\ gas$$

TIP

Least deviations occur at low pressures and high temperatures.

High deviations occur at high pressures and very low temperatures.

Ideal Gas Deviations

In the use of the gas laws, we have assumed that the gases involved were "ideal" gases. This means that the molecules of the gas were not taking up space in the gas volume and that no intermolecular forces of attraction were serving to pull the molecules closer together. You will find that a gas behaves like an ideal gas at low pressures and high temperatures, which move the molecules as far as possible from conditions that would cause condensation. In general, pressures below a few atmospheres will cause most gases to exhibit sufficiently ideal properties for the application of the gas laws with a reliability of a few percent or better.

If, however, high pressures are used, the molecules will be forced into closer proximity with each other as the volume decreases until the attractive force between molecules becomes a factor. This factor decreases the volume, and therefore the PV values at high pressure conditions will be less than those predicted by the ideal gas law, where PV remains a constant.

Examining what occurs at very low temperatures creates a similar situation. Again, the molecules, because they have slowed down at low temperatures, come into closer proximity with each other and begin to feel the attractive force between them. This tends to make the gas volume smaller and, therefore, causes the PV to be lower than that expected in the ideal gas situation. Thus, under conditions of very high pressures and low temperatures, deviations from the expected results of the ideal gas law will occur.

CHAPTER SUMMARY

The following terms summarize all the concepts and ideas that were introduced in this chapter. You should be able to explain their meaning and how you would use them in chemistry. They appear in boldface type in this chapter to draw your attention to them. The boldface type also makes it easier for you to look them up if you need to. You could also use Internet search engines like *google.com* on your computer to get a quick and expanded explanation of these terms, laws, and formulas.

atmosphere	mercury barometer	standard pressure
atmospheric pressure	ozone	standard temperature
greenhouse effect	pascal	torr
manometer		

Boyle's Law $PV = k$ (constant T and n)

Charles's Law $\dfrac{V}{T} = k$ (constant P and n)

combined gas law $\dfrac{P_1 V_1}{T_1} = \dfrac{P_2 V_2}{T_2}$ (constant n)

Dalton's Law of Partial Pressures $P_{total} = P_A + P_B + P_C \ldots$

Graham's Law
ideal gas law
Kinetic-Molecular Theory

Online content that reinforces major concepts discussed in this chapter can be found at the following Internet address if they are still available. *Some may have been changed or deleted.*

Kinetic-Molecular Theory

http://www.chem.tamu.edu/class/majors/tutorialnotefiles/kinetic.htm

This site reviews the basic tenets of Kinetic-Molecular Theory.

The Ideal Gas Law

http://jersey.uoregon.edu/vlab/Piston/index.html

This site offers an experimental exercise to interact virtually with the behavior of gases.

Gas Law Problems

http://www.sciencegeek.net/Chemistry/taters/Unit7GasLaws.htm

This site gives you an opportunity to practice solving gas law problems.

1. The most abundant element in Earth's crust is

   (A) sodium
   (B) oxygen
   (C) silicon
   (D) aluminum

2. A compound that can be decomposed to produce oxygen gas in the lab is

   (A) MnO_2
   (B) NaOH
   (C) CO_2
   (D) $KClO_3$

3. In the usual laboratory preparation equation for the reaction in question 2, what is the coefficient of O_2?

   (A) 1
   (B) 2
   (C) 3
   (D) 4

4. In the graphic representation of the energy contents of the reactants and the resulting products in an exothermic reaction, the energy content would be

   (A) higher for the reactants
   (B) higher for the products
   (C) the same for both
   (D) impossible to determine

5. The process of separating components of a mixture by making use of the difference in their boiling points is called

   (A) destructive distillation
   (B) displacement
   (C) fractional distillation
   (D) filtration

6. When oxygen combines with an element to form a compound, the resulting compound is called

   (A) a salt
   (B) an oxide
   (C) oxidation
   (D) an oxalate

7. According to the activity chart of metals, which metal would react most vigorously in a dilute acid solution?

   (A) zinc
   (B) iron
   (C) aluminum
   (D) magnesium

8. Graham's Law refers to

   (A) boiling points of gases
   (B) gaseous diffusion
   (C) gas compression problems
   (D) volume changes of gases when the temperature changes

9. When 200 milliliters of a gas at constant pressure is heated, its volume

   (A) increases
   (B) decreases
   (C) remains unchanged

10. When 200 milliliters of a gas at constant pressure is heated from 0°C to 100°C, the volume must be multiplied by

    (A) 0/100
    (B) 100/200
    (C) 273/373
    (D) 373/273

11. If you wish to find the corrected volume of a gas that was at 20°C and 1 atmosphere pressure and conditions were changed to 0°C and 0.92 atmosphere pressure, by what fractions would you multiply the original volume?

(A) $\dfrac{293}{273} \times \dfrac{1}{0.92}$

(B) $\dfrac{273}{293} \times \dfrac{0.92}{1}$

(C) $\dfrac{273}{293} \times \dfrac{1}{0.92}$

(D) $\dfrac{293}{273} \times \dfrac{0.92}{1}$

12. When the level of mercury inside a gas tube is higher than the level in the reservoir, you find the correct pressure inside the tube by taking the outside pressure reading and __?__ the difference in the height of mercury.

(A) subtracting
(B) adding
(C) dividing by 13.6
(D) doing both (C) and (A)

13. If water is the liquid in question 12 instead of mercury, you can change the height difference to an equivalent mercury expression by

(A) dividing by 13.6
(B) multiplying by 13.6
(C) adding 13.6
(D) subtracting 13.6

14. Standard conditions are

(A) 0°C and 14.7 mm
(B) 273 K and 760 mm Hg
(C) 273°C and 760 mm Hg
(D) 4°C and 7.6 mm Hg

15. When a gas is collected over water, the pressure is corrected by

(A) adding the vapor pressure of water
(B) multiplying by the vapor pressure of water
(C) subtracting the vapor pressure of water at that temperature
(D) subtracting the temperature of the water from the vapor pressure

16. At 5.00 atmospheres pressure and 70°C how many moles are present in 1.50 liters of O_2 gas?

(A) 0.036
(B) 0.103
(C) 0.266
(D) 0.536

Directions: Every set of the given lettered choices below refers to the numbered questions immediately below it. For each numbered item, choose the one lettered choice that fits it best. Every choice in a set may be used once, more than once, or not at all.

<u>Questions 17–20</u> refer to the following graphs, assuming that the axes on each are appropriately labeled:

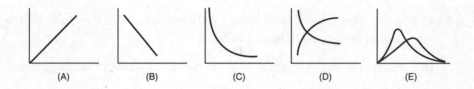

(A) (B) (C) (D) (E)

17. Which is a graphic depiction of Boyle's Law?

18. Which is a graphic depiction of Charles's Law?

19. Which is a graphic depiction of the relationship of the pressure of a given volume of gas with the absolute temperature?

20. Which graph shows the distribution of molecules with respect to their kinetic energy at different temperatures?

Answers and Explanations

1. **(B)** Oxygen is the most abundant element in Earth's crust, with 50% of its composition.

2. **(D)** $KClO_3$ can be decomposed with heat to form KCl and O_2.

3. **(C)** The equation is $2KClO_3(s) \rightarrow 2KCl(s) + 3O_2(g)$

4. **(A)** Since energy is given off in an exothermic reaction, the heat content of the reactants would be higher than the products.

5. **(C)** Fractional distillation can separate liquids of different boiling points.

6. **(B)** Oxygen compounds formed in combination reactions are called oxides.

7. **(D)** Magnesium, which is the most active of the given metals, will react to release hydrogen.

8. **(B)** Graham's Law refers to gaseous diffusion or effusion.

9. **(A)** When confined gases at a constant pressure are heated, they expand.

10. **(D)** The temperature must be changed to absolute temperature by adding 273 to the Celsius readings.

11. **(C)** Since the temperature is decreasing, the volume must decrease with the temperature fraction, and, because the pressure is decreasing, the volume must increase. The correct answer does this.

12. **(A)** The level is higher because the pressure inside the tube is less than outside. You must subtract the height inside the tube from the outside pressure.

13. **(A)** If water is the liquid, you must divide by 13.6 to change the height to the equivalent height of mercury.

14. **(B)** The only correct indication of standard conditions is (B). It is usually stated as 273 K and 760 mm Hg. (Note: "mm" and "torr" are interchangeable.)

15. **(C)** Because the vapor pressure of water is a part of the pressure reading, it must be subtracted to get the atmospheric pressure.

16. **(C)** Using the general gas law, $PV = nRT$, convert 70°C to K by adding 273 = 343 K. Solving for n, you get

$$n = \frac{PV}{RT} = \frac{5.00 \ \cancel{atm} \times 1.50 \ \cancel{L}}{0.0821 \dfrac{\cancel{L} \cdot \cancel{atm}}{mol \cdot \cancel{K}} 343 \ \cancel{K}} = 0.266 \ mol$$

17. **(C)** Boyle's Law is an inverse relationship. As the pressure increases, the volume decreases, as shown by the graph.

18. **(A)** Charles's Law is a direct relationship. As the temperature increases, the volume increases.

19. **(A)** The relationship of pressure to absolute temperature while the volume is held constant is a direct one. As the temperature on a given volume increases, the pressure will increase.

20. **(E)** The graphs in (E) show the distributions of the kinetic energy of molecules at two different temperatures.

Stoichiometry (Chemical Calculations) and the Mole Concept

6

These skills are usually tested on the SAT Subject Test in Chemistry. You should be able to...

→ Use the mole concept to find the molar mass of formulas and of monoatomic and diatomic molecules, and how gas volumes are related to molar mass.

→ Solve stoichiometric problems involving Gay-Lussac's Law, density, mass–volume relation, mass–mass problems, volume–volume problems, problems with a limiting reactant (excess of one reactant), and finding the percent yield.

This chapter will review and strengthen these skills. Be sure to do the Practice Exercises at the end of the chapter.

This chapter deals with the solving of a variety of quantitative chemistry problems, which is often referred to as **stoichiometry**. Solving problems should be done in an organized manner, and it would be to your advantage to go back to the Introduction to this book and review the section called, "How Can You Improve Your Problem Solving Skills?" It describes a well-planned method for attacking the process of solving problems that you will find helpful in this chapter.

Although there are many specific methods for solving the types of problems in this chapter, we will focus primarily on the technique called **dimensional analysis**. Dimensional analysis provides a very clear problem-solving pathway for stoichiometry problems. It emphasizes not only the numerical values involved in the calculations but also the units describing the quantities in question. Dimensional analysis was introduced in Chapter 1 to make unit conversions. It will be used here in the same manner but specifically to relate quantities of reactants and products in a chemical reaction.

THE MOLE CONCEPT

Providing a name for a quantity of things taken as a whole is common in everyday life. Some examples are a dozen, a gross, and a ream. Each of these represents a specific number of items and is not dependent on the commodity. A dozen eggs, oranges, or bananas will always represent 12 items. In chemistry we have a unit that decribes a quantity of particles. It is called the **mole** (sometimes abbreviated as **mol**). A mole is 6.02×10^{23} particles. Technically, that's the number of carbon atoms found in exactly 12 grams of carbon-12. Since the atomic masses of all the elements' atoms are related to the mass of carbon-12, a mole is also the number of atoms found in the atomic mass of *any* element if it is *expressed in grams*. Keep in mind that the masses found on the Periodic Table for any element are actually weighted averages of all the isotopes that exist for that element (based on their relative natural abundances). Those masses, if expressed in atomic mass units (amu), represent just one average atom for that element. If, however, the value for mass was expressed in grams, that sample of the element

TIP

Just as a dozen is 12 units of an item, a mole is 6.02×10^{23} units of an item. It is called *Avogadro's number*.

would contain 6.02×10^{23} atoms of that element. This value is also known as **Avogadro's number** in honor of the Italian scientist whose hypothesis concerning the volumes of gases led to its determination. More on Avogadro's hypothesis will be discussed in an upcoming section on gas volumes and molar mass. You should recognize that Avogadro's number is very large because the items being counted (atoms) are very small. So 6.02×10^{23} atoms of most elements represent samples of atoms that are conveniently sized for working in the laboratory.

MOLAR MASS AND MOLES

The mass of a mole of particles is referred to as its **molar mass**. For moles of atoms, the atomic mass found on the Periodic Table for that element expressed in grams is the molar mass for that element. Some elements naturally exist as molecules, however. The molar mass of those elements takes into account the number of atoms in the molecule *in an additive manner*. Most elements are considered in a monatomic way (one atom). However, a few (hydrogen, nitrogen, oxygen, fluorine, chlorine, bromine, and iodine) are typically considered in a diatomic manner (two atoms) based on the way they are generally found to exist. This is not to say that you could not count moles of hydrogen atoms (H) as opposed to hydrogen molecules (H_2). You should, though, always be cognizant of the type of particle involved in any mole calculation.

➡ **Example 1** _____

Determine the molar mass of silicon, nitrogen, and iron using the Periodic Table.

The atomic mass of silicon is 28.1 amu as found on the Periodic Table. Therefore, the molar mass of silicon is 28.1 g and represents 6.02×10^{23} *atoms* of silicon or 1 mole of silicon atoms.

The atomic mass of nitrogen is 14.0 amu as found on the Periodic Table. Nitrogen (N_2) is a diatomic element, however. It has a molar mass of 28.0 g, which represents 6.02×10^{23} *molecules* of nitrogen or 1 mole of nitrogen molecules. A mole of nitrogen (N) atoms would have a mass of 14.0 g if they were the appropriate particle to be considered in a given circumstance.

The atomic mass of iron is 55.8 amu. So iron has a molar mass of 55.8 g. This represents a sample of 6.02×10^{23} atoms of iron or 1 mole of iron atoms. Iron is typically not found in nature as a diatomic molecule.

Elements are just one type of substance for which the molar mass can be found. The molar masses of compounds can be found in a way similar to that of diatomic elements. Just add up the molar masses of the individual elements found in the compound based on the compound's formula.

➡ **Example 2** _____

Find the molar mass of NH_3 (this is a molecular compound known as ammonia).

The molar mass of nitrogen is 14.0 g, and the molar mass of hydrogen is 1.0 g. Since there are 3 hydrogen atoms per molecule of ammonia, the molar mass of ammonia is 17.0 g and represents 6.02×10^{23} molecules of ammonia or 1 mole of ammonia molecules.

➥ Example 3

Find the molar mass of $CaCO_3$ (this is an ionic compound known as calcium carbonate).

The molar masses of calcium, carbon, and oxygen are 40.1 g, 12.0 g, and 16.0 g, respectively. Since there are 3 oxygen particles per formula unit of calcium carbonate in addition to the single particles of calcium and carbon, the molar mass of calcium carbonate is 40.1 g + 12.0 g + (3)16.0 g or 100.1 g. This represents 6.02×10^{23} formula units (the particle for an ionic compound) or 1 mole of calcium carbonate.

➥ Example 4

Find the molar mass of $CuSO_4 \cdot 5\,H_2O$ (this is a hydrated ionic compound known as copper(II) sulfate pentahydrate).

The molar masses of copper, sulfur, oxygen, and water are 63.5 g, 32.1 g, 16.0 g, and 18.0 g, respectively. Since there are 4 oxygen particles and 5 water molecules per formula unit in addition to the single particles of copper and sulfur, the molar mass of copper(II) sulfate pentahydrate is 63.5 g + 32.1 g + (4)16.0 g + (5)18.0 g or 249.6 g. This represents 6.02×10^{23} formula units or 1 mole of copper(II) sulfate pentahydrate.

MOLAR MASS AND GAS VOLUMES

In 1811, Amedeo Avogadro made a far-reaching scientific assumption that also bears his name. **Avogadro's Hypothesis** states that equal volumes of different gases contain equal numbers of particles at the same temperature and pressure. It means that under the same conditions, the number of molecules of hydrogen in a 1-liter container is exactly the same as the number of molecules of carbon dioxide (or of any other gas) in a 1-liter container even though the individual molecules of the different gases have different masses and sizes. Because of the substantiation of this hypothesis by much data since its inception, it is often referred to as **Avogadro's Law** and can be added to the list of gas laws discussed in Chapter 5. Avogadro's Law shows the relationship between the volume and the number of particles of a gas sample when the temperature and pressure are constant:

$$\frac{V}{n} = k$$

In other words, volume and the number of gas particles are directly related.

Because the volume of a gas may vary depending on the temperature and pressure, a standard is set for comparing gases. As stated in Chapter 5, the standard conditions of temperature and pressure (abbreviated STP) are 273 K and 1 atmosphere. Because the relationship between volume and number of particles of a gas is direct when the temperature and pressure are constant, the molar mass of a gas (which represents a *set* number of particles, namely 1 mole) occupies a *set* volume. The volume of 22.4 L is recognized as the molar volume of any gas at STP.

USING MOLAR MASS AND MOLAR VOLUME

Molar mass and molar volume are typically used as conversion factors to change quantities of reactants expressed as masses or as volumes to moles for use in stoichiometry problems via dimensional analysis.

➡ Example 1

Find the number of moles of silicon present in 4.30 g of silicon.

Recall that silicon is an element that is not diatomic. It has a molar mass of 28.1 g. So 28.1 g of silicon contains 1.00 mol of silicon atoms if significant figures are kept in mind.

Use dimensional analysis:

$$\frac{4.30\,\text{g Si}}{1} \times \frac{1.00\,\text{mol Si}}{28.1\,\text{g Si}} = 0.0153\,\text{mol Si}$$

Note that when using dimensional analysis, the given quantity is simply multiplied by a factor that is equal to the value 1 since the numerator and denominator in that factor are equal to each other. Using this method causes the magnitude of the given quantity not to change. What does change is the units in which the quantity is expressed. The given units cancel out, leaving only the unit in the numerator of the factor to describe the quantity.

➡ Example 2

Find the number of moles of calcium carbonate in 0.750 g of calcium carbonate.

Recall that calcium carbonate is an ionic compound. It has a molar mass of 100.1 g. So 100.1 g of calcium carbonate contains 1.00 mol of calcium carbonate.

Use dimensional analysis:

$$\frac{0.750\,\text{g CaCO}_3}{1} \times \frac{1.00\,\text{mol CaCO}_3}{100.1\,\text{g CaCO}_3} = 0.00749\,\text{mol CaCO}_3$$

➡ Example 3

Find the number of moles of ammonia in 0.300 L of ammonia at STP.

Recall that ammonia is a gas at STP. Since the given quantity is supplied as a volume and not as mass, molar volume (not molar mass) should be used in the dimensional analysis equation:

$$\frac{0.300\,\text{L NH}_3}{1} \times \frac{1.00\,\text{mol NH}_3}{22.4\,\text{L NH}_3} = 0.0134\,\text{mol NH}_3$$

DENSITY AND MOLAR MASS

Since the density of a gas is usually given in grams/liter of gas at STP, we can use the molar volume at STP to solve the following types of problems.

➡ Example 1

Find the molar mass of a gas when the density is given as 1.25 grams/liter.

Because it is known that 1 mole of a gas occupies 22.4 liters at STP, we can solve this problem by multiplying the mass of 1 liter by 22.4 liters/mole using dimensional analysis:

$$\frac{1.25\,\text{g}}{\cancel{L}} \times \frac{22.4\,\cancel{L}}{1\,\text{mol}} = 28.0\,\text{g/mol}$$

Even if the mass given is not for 1 liter, the same setup can be used.

> **REMEMBER**
>
> For a dry gas at STP, the mass of 1 liter × 22.4 = the molar mass of the gas.

➡ Example 2 _____

If 3.00 liters of a gas weighs 2.00 grams, find the molar mass.
   Use dimensional analysis:

$$\frac{2.00\,\text{g}}{3.00\,\cancel{L}} \times \frac{22.4\,\cancel{L}}{1\,\text{mol}} = 14.9\,\text{g/mol}$$

   You can also find the density of a gas if you know the molar mass. Since the molar mass occupies 22.4 liters at STP, dividing the molar mass by 22.4 liters will give you the mass per liter, or the density.

➡ Example 3 _____

Find the density of oxygen at STP.
   Oxygen exists naturally as a diatomic molecule. The molar mass of O_2 is 32.0 g/mol.
   Use dimensional analysis:

$$\frac{32.0\,\text{g}}{1\,\text{mol}} \times \frac{1\,\text{mol}}{22.4\,\text{L}} = 1.43\,\text{g/L}$$

Example 3 shows that you can find the density of any gas at STP by dividing its molar mass by 22.4 L.

For a dry gas at STP:

density

$$= \frac{\text{molar mass}}{22.4\,\text{L}}$$

STOICHIOMETRY: MOLE-MOLE PROBLEMS

The types of mole problems investigated so far have been ones involving only one substance. Chemical calculations often involve more than one substance and take into consideration information found in **balanced reaction equations** discussed in Chapter 4. Recall that coefficients from a balanced equation can be used to describe the numbers of atoms, ions, or molecules involved in the chemical process. Also recall that the quantities of each must balance on both sides of the equation to satisfy the Law of Conservation of Matter. Those coefficients can also represent larger numbers of particles, namely moles of those substances, that are reacting or being produced. Therefore, the balanced chemical reaction

$$2\text{NaClO(s)} \rightarrow 2\text{NaCl(s)} + O_2\text{(g)}$$

can be interpreted in two ways.

   1. Decomposing 2 **formula units** of sodium hypochlorite produces 2 **formula units** of sodium chloride and 1 **molecule** of oxygen.
   2. Decomposing 2 **moles** of sodium hypochlorite produces 2 **moles** of sodium chloride and 1 **mole** of oxygen molecules.

Coefficients in balanced chemical reaction equations therefore provide **mole ratios** for reacting substances and substances produced.

Example 1

How many moles of sodium chloride can be produced from 0.0253 moles of sodium hypochlorite?

Use dimensional analysis:

$$\frac{0.0253 \text{ mol NaClO}}{1} \times \frac{2 \text{ mol NaCl}}{2 \text{ mol NaClO}} = 0.0253 \text{ mol NaCl}$$

Example 2

How many moles of sodium hypochlorite are needed to produce 0.750 moles of oxygen?

Use dimensional analysis:

$$\frac{0.750 \text{ mol O}_2}{1} \times \frac{2 \text{ mol NaClO}}{1 \text{ mol O}_2} = 1.50 \text{ mol NaClO}$$

STOICHIOMETRY: MASS-MASS PROBLEMS

In order to work with substances in the laboratory, chemists must work with quantities they can easily measure. A mole is an amount impractical to actually count out in the lab. That's why the concept of molar mass, which relates moles to mass, was discussed in a previous section of this chapter. Using molar mass allows problems to be based on mass, which is a measurable quantity. **Mass–mass** problems often involve determining the masses of other substances needed to react with a given mass of a substance or the mass of other substances that can be produced from that given mass.

Example

Consider the following balanced chemical reaction equation:

$$CaCO_3(s) + 2HCl(aq) \rightarrow CaCl_2(aq) + H_2O(l) + CO_2(g)$$

If 2.59 g of $CaCO_3$ was reacted with enough HCl to use up all of the $CaCO_3$, what mass of HCl would be needed and what mass of CO_2 would be produced?

Use dimensional analysis:

$$\frac{2.59 \text{ g CaCO}_3}{1} \times \underset{\uparrow \text{Factor \#1}}{\frac{1 \text{ mol CaCO}_3}{100.1 \text{ g CaCO}_3}} \times \underset{\uparrow \text{Factor \#2}}{\frac{2 \text{ mol HCl}}{1 \text{ mol CaCO}_3}} \times \underset{\uparrow \text{Factor \#3}}{\frac{36.5 \text{ g HCl}}{1 \text{ mol HCl}}} = 1.89 \text{ g HCl}$$

The first factor in the dimensional analysis equation converts the mass of the $CaCO_3$ to moles by using information concerning molar mass as found from the Periodic Table. The second factor converts moles of $CaCO_3$ to moles of HCl needed to react with the $CaCO_3$. The mole ratio was found from the coefficients in front of each substance in the balanced reaction. The third factor converts moles of HCl to grams of HCl, once again using molar mass as found from the Periodic Table.

To find the mass of CO_2 produced, set up a similar dimensional analysis equation. First convert the mass of $CaCO_3$ to moles. This time, however, use the mole ratio for CO_2 and $CaCO_3$ for the second factor. Finally, use the molar mass of CO_2 to convert moles of CO_2 to grams.

$$\frac{2.59 \text{ g CaCO}_3}{1} \times \underset{\uparrow \text{Factor \#1}}{\frac{1 \text{ mol CaCO}_3}{100.1 \text{ g CaCO}_3}} \times \underset{\uparrow \text{Factor \#2}}{\frac{1 \text{ mol CO}_2}{1 \text{ mol CaCO}_3}} \times \underset{\uparrow \text{Factor \#3}}{\frac{44.0 \text{ g CO}_2}{1 \text{ mol CO}_2}} = 1.14 \text{ g CO}_2$$

STOICHIOMETRY: VOLUME–VOLUME PROBLEMS

In the lab, gases are often used and their volumes are easily measured. Recall that the molar volume of any gas at STP is 22.4 L. So the volume of a gas at standard temperature and pressure can be converted to moles of that gas via dimensional analysis. Combining those conversions with the mole ratios found in balanced equations, for reactions involving gases, allows for **volume–volume** stoichiometry problems to be calculated. In volume–volume problems, you are given the volume of one gas at STP and asked to determine the volume(s) of other gases involved in the reaction.

➡ Example

Consider the following balanced chemical reaction equation:

$$N_2(g) + 3H_2(g) \rightarrow 2NH_3(g)$$

To produce 0.400 L of NH_3 at STP, what volumes of nitrogen and hydrogen, also at STP, would be required?

Use dimensional analysis:

$$\frac{0.400 \text{ L } NH_3}{1} \times \underbrace{\frac{1 \text{ mol } NH_3}{22.4 \text{ L } NH_3}}_{\uparrow Factor\,\#1} \times \underbrace{\frac{1 \text{ mol } N_2}{2 \text{ mol } NH_3}}_{\uparrow Factor\,\#2} \times \underbrace{\frac{22.4 \text{ L } N_2}{1 \text{ mol } N_2}}_{\uparrow Factor\,\#3} = 0.200 \text{ L } N_2$$

The first factor in the dimensional analysis equation converts the volume of NH_3 to moles. The second factor uses the mole ratio from the balanced equation to convert moles of NH_3 to moles of N_2. Finally, the third factor converts moles of N_2 to the volume of N_2. Mathematically, factors #1 and #3 simply undo each other. They display a relationship that was suggested in Chapter 5 by the Ideal Gas Law as well as by Avogadro's Hypothesis (discussed in this chapter). Namely, there is a direct relationship between the volumes of gases, at the same temperature and pressure, and their numbers of particles (measured in moles). In other words, mole ratios can be construed as volume ratios between gases existing at the same temperature and pressure. The only factor needed to solve the previous problem mathematically was factor #2. To express the dimensional analysis equation properly, though, requires using the mole ratios as volume ratios as is done here:

$$\frac{0.400 \text{ L } NH_3}{1} \times \frac{1 \text{ L } N_2}{2 \text{ L } NH_3} = 0.200 \text{ L } N_2$$

To find the amount of H_2 required to produce the 0.400 L of NH_3 requires a similar equation but with a different ratio between the gases:

$$\frac{0.400 \text{ L } NH_3}{1} \times \frac{3 \text{ L } H_2}{2 \text{ L } NH_3} = 0.600 \text{ L } H_2$$

Using mole ratios, i.e., the coefficients from the balanced equation, as volume ratios saves time. It also makes volume-volume problems less cumbersome to solve. The relationship between the volumes of reacting gases was first noted by the French scientist Joseph Louis Gay-Lussac and is sometimes called **Gay-Lussac's Law of Combining Gases**. This law states that when only gases are involved in a chemical reaction, the volumes of the reacting gases and the volumes of the gaseous products are in small whole-number ratios with each other. Those small whole numbers are the coefficients in the balanced reaction equation.

Often reactions between gases do not occur at STP. However, Gay-Lussac's Law still applies. The reason is fundamentally due to Avogadro's Law. That law shows that the only requirement for the volumes of gases to be related to the number of particles of those gases is that the temperature and pressure of the gases be the same. Whether or not the temperature and pressure of the gases are at STP is inconsequential. Using the coefficients from the balanced reaction equation as volume ratios between reacting gases in dimensional analysis/stoichiometry problems is encouraged.

STOICHIOMETRY: MASS-VOLUME OR VOLUME-MASS PROBLEMS

Reactions often involve gases and other phases of matter. In those reactions, it is common to know the mass of one substance involved in the chemical process and the need to determine the volume of a different substance, such as a gas. Likewise, it is not unusual to know the volume of a gas taking part in a reaction and the need to determine the mass of another substance, often a solid or liquid. Even if the reaction is taking place at STP, Gay-Lussac's Law cannot be taken advantage of here since both of the substances are not gases and the information desired is not restricted to just volumes. In other words, Gay-Lussac's Law applies only when all the substances being considered are gases.

➡ **Example 1** _____

In the reaction below, what mass of magnesium is required to produce 0.250 L of H_2 at STP?

$$Mg(s) + 2HCl(aq) \rightarrow MgCl_2(aq) + H_2(g)$$

The solution to this problem uses both molar mass and molar volume. Use dimensional analysis:

$$\frac{0.250 \text{ L } H_2}{1} \times \frac{1 \text{ mol } H_2}{22.4 \text{ L } H_2} \times \frac{1 \text{ mol Mg}}{1 \text{ mol } H_2} \times \frac{24.3 \text{ g Mg}}{1 \text{ mol Mg}} = 0.271 \text{ g Mg}$$

Reactions involving gases and other phases of matter NOT at STP are very common. Obviously, Gay-Lussac's Law cannot be used to solve such problems since both of the substances in question are not gases. Additionally, the relationship 22.4 L = 1 mole of the gas cannot be used since the reaction is not taking place at STP. To solve problems such as these, the Ideal Gas Law, $PV = nRT$, must be considered. Recall from Chapter 5 that the Ideal Gas Law can be manipulated to solve for the moles of gas at any temperature and pressure as long as the volume is supplied ($n = \frac{PV}{RT}$). Likewise, the volume of a gas can be determined if the number of moles of the gas is known along with its temperature and pressure ($V = \frac{nRT}{P}$).

➡ Example 2 _____

Based on the reaction below, what volume of oxygen is produced from the decomposition of 5.00 g of hydrogen peroxide if the oxygen produced was collected at 70.0°C and 1.25 atm pressure?

$$2H_2O_2(aq) \rightarrow 2H_2O(l) + O_2(g)$$

To solve this problem, dimensional analysis should be used to determine the **number of moles** of O_2 that would be produced:

$$\frac{5.00\,g\,H_2O_2}{1} \times \frac{1\,mol\,H_2O_2}{34.0\,g\,H_2O_2} \times \frac{1\,mol\,O_2}{2\,mol\,H_2O_2} = 0.0735\,mol\,O_2$$

The number of moles of O_2 produced can now be plugged into the Ideal Gas Law to find the volume of that amount of O_2 at the temperature and pressure outlined in the problem:

$$V = \frac{(0.0735\,mol)\left(0.0821\,\frac{L \cdot atm}{mol \cdot K}\right)(343\,K)}{1.25\,atm}$$

$$V = 1.66\,L$$

PROBLEMS WITH AN EXCESS OF ONE REACTANT OR A LIMITING REACTANT

It will not always be true that the amounts given in a particular problem are exactly in the proportion required for the reaction to use up all of the reactants. In other words, at times some of one reactant will be left over after the other has been used up. This is similar to the situation in which two eggs are required to mix with one cup of flour in a particular recipe, and you have four eggs and four cups of flour.

Since two eggs require only one cup of flour, four eggs can use only two cups of flour and two cups of flour will be left over.

A chemical equation is very much like a recipe.

TIP

The paragraph describes a practical example of a "limiting reactant."

TIP

Remember this recipe analogy to solve limiting-reactant questions!

➡ Example _____

Consider the following reaction:

$$CH_4(g) + 2O_2(g) \rightarrow CO_2(g) + 2H_2O(g)$$

If you are given 15.0 grams of methane (CH_4) and 15.0 grams of oxygen (O_2), how many grams of carbon dioxide gas can be produced? Which reactant will be left over? How much of this reactant will not be used?

Once again, dimensional analysis will be used to solve this problem. However, two equations will be required as it will be necessary to determine how much carbon dioxide can be produced from each reactant:

$$\frac{15.0\,g\,CH_4}{1} \times \frac{1\,mol\,CH_4}{16.0\,g\,CH_4} \times \frac{1\,mol\,CO_2}{1\,mol\,CH_4} \times \frac{44.0\,g\,CO_2}{1\,mol\,CO_2} = 41.3\,g\,CO_2$$

$$\frac{15.0\,g\,O_2}{1} \times \frac{1\,mol\,O_2}{32.0\,g\,O_2} \times \frac{1\,mol\,CO_2}{2\,mol\,O_2} \times \frac{44.0\,g\,CO_2}{1\,mol\,CO_2} = 10.3\,g\,CO_2$$

Since the oxygen can produce only 10.3 g CO_2 (the lesser of the quantities of CO_2 shown above), it is referred to as the **limiting reactant**. There simply is not enough of oxygen available to make what the methane has the potential to produce. That smaller amount of CO_2 is referred to as the **theoretical yield**. A portion of the CH_4 (the **reactant in excess**) will be used, however, to produce the 10.3 g of CO_2. To determine that amount, another dimensional analysis is needed.

$$\frac{15.0\,g\,O_2}{1} \times \frac{1\,mol\,O_2}{32.0\,g\,O_2} \times \frac{1\,mol\,CH_4}{2\,mol\,O_2} \times \frac{16.0\,g\,CH_4}{1\,mol\,CH_4} = 3.75\,g\,CH_4\,(needed)$$

Since only 3.75 g of CH_4 is needed, the amount of CH_4 in excess can be determined by subtraction:

$$\overset{(have)}{15.0\,g\,CH_4} - \overset{(need)}{3.75\,g\,CH_4} = \overset{(left\ over)}{11.25\,g\,CH_4} \quad (or\ 11.3\,g\ when\ considering\ significant\ figures)$$

PERCENT YIELD OF A PRODUCT

In most stoichiometric problems, we assume that the results are exactly what we would theoretically expect. In reality, the resulting theoretical yield is rarely the actual yield. Why the actual yield of a reaction may be less than the theoretical yield occurs for many reasons. Some of the product is often lost during the purification or collection process.

Chemists are usually interested in the efficiency of a reaction. The efficiency is expressed by comparing the actual and the theoretical yields.

The **percent yield** is the ratio of the actual yield to the theoretical yield, multiplied by 100%.

$$percent\ yield = \frac{actual\ yield}{theoretical\ yield} \times 100$$

➡ **Example**

Aluminum is commonly produced by the smelting of aluminum oxide into aluminum metal by the reaction below:

$$2Al_2O_3\ (dissolved) + 3C(s) \rightarrow 4Al(\ell) + 3CO_2(g)$$

If 3.89 kg of aluminum oxide is smelted and the actual yield of Al is 1.95 kg, what is the percent yield associated with the process?

The *theoretical yield* of aluminum from that amount of aluminum oxide can be found using dimensional analysis:

$$\frac{3,890\,g\,Al_2O_3}{1} \times \frac{1\,mol\,Al_2O_3}{102\,g\,Al_2O_3} \times \frac{4\,mol\,Al}{2\,mol\,Al_2O_3} \times \frac{27.0\,g\,Al}{1\,mol\,Al} = 2,060\,g\,Al$$

The *percent yield* can then be found:

$$percent\ yield = \frac{1,950\,g\,Al\,(actual\ yield)}{2,060\,g\,Al\,(theoretical\ yield)} \times 100\% = 94.7\%$$

The following terms summarize all the concepts and ideas that were introduced in this chapter. You should be able to explain their meaning and how you would use them in chemistry. They appear in boldface type in this chapter to draw your attention to them. The boldface type also makes it easier for you to look them up if you need to. You could also use Internet search engines like *google.com* on your computer to get a quick and expanded explanation of these terms, laws, and formulas.

dimensional analysis	mole	Avogadro's Law
molar mass	STP	Avogadro's number = (6.02×10^{23})
molar volume		Gay-Lussac's Law
		Density of gases

$$\text{percent yield} = \frac{\text{actual yield}}{\text{theoretical yield}} \times 100$$

INTERNET RESOURCES

Online content that reinforces major concepts discussed in this chapter can be found at the following Internet addresses if they are still available. *Some may have been changed or deleted.*

The Mole Concept
http://web.clark.edu/nfattaleh/classes/100/LectureNotes/Ch6.pdf
This site gives a general review of the mole concept and some sample problems to make understanding of the mole concept easier.

Stoichiometry
http://www.chemtutor.com/mols.htm
This site gives a good review of solving typical stoichiometry problems.

1. What is the molar mass of carbon monoxide?

   (A) 44.0 amu
   (B) 44.0 g
   (C) 28.0 amu
   (D) 28.0 g

2. Describe the particles of argon in 79.8 grams of elemental argon by both number and type.

   (A) 2 mol of argon atoms
   (B) 2 mol of argon molecules
   (C) 6.02×10^{23} argon molecules
   (D) 6.02×10^{23} argon atoms

3. A sample of oxygen gas with a volume of 11.2 L at STP would contain how many particles of that gas?

   (A) 6.02×10^{23} molecules
   (B) 3.01×10^{23} molecules
   (C) 1.12×10^{23} atoms
   (D) 3.01×10^{23} atoms

4. The density of gases in g/L

   (A) is independent of the identity of the gas
   (B) can be found by dividing its molar mass by 22.4 as long as it's at STP
   (C) can be found by multiplying its molar mass by 22.4 as long as it's at STP
   (D) can be found by dividing its molar mass by 22.4 as long as it's not at STP

5. In the reaction between magnesium and oxygen to produce magnesium oxide, the number of moles of oxygen needed to react with 4 moles of magnesium is

   (A) 1 mole
   (B) 2 moles
   (C) 3 moles
   (D) 4 moles

6. In the reaction

   $$H_2(g) + Cl_2(g) \rightarrow 2HCl(g)$$

   what volume of HCl could be produced when 3 L of H_2 reacts with 2 L of Cl_2?

   (A) 2 L
   (B) 3 L
   (C) 4 L
   (D) 5 L

7. Consider the *unbalanced* reaction equation describing the combustion of butane.

   $$C_4H_{10}(g) + O_2(g) \rightarrow CO_2(g) + H_2O(g)$$

   How many moles of carbon dioxide could be produced if 29.0 grams of butane reacted with an excess of O_2?

   (A) 1.00 mol
   (B) 2.00 mol
   (C) 4.00 mol
   (D) 8.00 mol

8. What volume of hydrogen would be produced in the following reaction

   $$2K(s) + 2H_2O(\ell) \rightarrow 2KOH(aq) + H_2(g)$$

   if 39.1 g of potassium were completely reacted and the hydrogen gas was collected and then stored at STP?

   (A) 11.2 L
   (B) 22.4 L
   (C) 44.8 L
   (D) 60.2 L

9. After a reaction between a silver nitrate solution and copper in the lab, 45.0 grams of silver was retrieved. Stoichiometry concerning the reaction shows that 50.0 grams was produced. The percent yield for this reaction was

   (A) 100.%
   (B) 95.0%
   (C) 90.0%
   (D) 45.0%

Answers and Explanations

1. **(D)** The molar mass of a compound is found by adding the masses of the individual elements in the compound based on the subscripts associated with that element in the formula. In this case, the formula for carbon monoxide is CO. The molar mass of carbon is 12.0 grams and that of oxygen is 16.0 g. Therefore, the molar mass of carbon monoxide is 28.0 g. The masses in the answers expressed in amu would be appropriate for 1 molecule of a substance, not for 1 mole.

2. **(A)** The molar mass of argon is 39.9 grams. Dimensional analysis shows that 79.8 grams of argon represents 2.00 mol.

$$\frac{79.8\,\text{g Ar}}{1} \times \frac{1\,\text{mol Ar}}{39.9\,\text{g Ar}} = 2.00\,\text{mol Ar}$$

Since argon is a monatomic element, the 2.00 mol of Ar refers to moles of atoms.

3. **(B)** Since 1 mole of any gas at STP has a volume of 22.4 L, then 11.2 L of oxygen contains 0.500 moles of oxygen or 3.01×10^{23} molecules of the gas. This can be shown by using dimensional analysis.

$$\frac{11.2\,\text{L O}_2}{1} \times \frac{1\,\text{mol O}_2}{22.4\,\text{L O}_2} \times \frac{6.02 \times 10^{23}\,\text{mol O}_2}{1\,\text{mol O}_2} = 3.01 \times 10^{23}\,\text{molecules O}_2$$

4. **(B)** The molar mass of a gas divided by 22.4 gives the density of any gas at STP because the molar volume of a gas at STP is 22.4 L. Use dimensional analysis:

$$\frac{\text{grams}}{\text{mole}} \times \frac{1\,\text{mole}}{22.4\,\text{liters}} = \frac{\text{grams}}{\text{liter}}$$

5. **(B)** The balanced reaction between magnesium and oxygen to produce magnesium oxide is

$$2\text{Mg(s)} + \text{O}_2\text{(g)} \rightarrow 2\text{MgO(s)}$$

Dimensional analysis shows

$$\frac{4\,\text{mol Mg}}{1} \times \frac{1\,\text{mol O}_2}{2\,\text{mol Mg}} = 2\,\text{mol O}_2$$

6. **(C)** Since the volumes for both reactants were given, this is a limiting-reactant problem. Use dimensional analysis to find the theoretical yield by choosing the smaller volume of HCl that could be made by each. The coefficients in the equation can also be construed as volume ratios since both reactants are gases (Gay-Lussac's Law).

$$\frac{3\,\text{L H}_2}{1} \times \frac{2\,\text{L HCl}}{1\,\text{L H}_2} = 6\,\text{L HCl}$$

$$\frac{2\,\text{L Cl}_2}{1} \times \frac{2\,\text{L HCl}}{1\,\text{L Cl}_2} = 4\,\text{L HCl (theoretical yield)}$$

7. **(B)** The balanced reaction equation for the combustion of butane is

$$2C_4H_{10}(g) + 13O_2(g) \rightarrow 8CO_2(g) + 10H_2O(g)$$

Use dimensional analysis:

$$\frac{29.0 \text{ g } C_4H_{10}}{1} \times \frac{1 \text{ mol } C_4H_{10}}{58.0 \text{ g } C_4H_{10}} \times \frac{8 \text{ mol } CO_2}{2 \text{ mol } C_4H_{10}} = 2.00 \text{ mol } CO_2$$

(Please note: you will not have a calculator when you solve problems like this while taking the test. By using mental math, you should be able to evaluate the first part of the problem as 0.500 moles of butane. Since 8 is 4 times 2, the next factor shows that you need to multiply 0.500 by 4 to get the final answer, 2.00 moles CO_2.)

8. **(A)** Use dimensional analysis and mental math:

$$\frac{39.1 \text{ g } K}{1} \times \frac{1 \text{ mol } K}{39.1 \text{ g } K} \times \frac{1 \text{ mol } H_2}{2 \text{ mol } K} \times \frac{22.4 \text{ L } H_2}{1 \text{ mol } H_2} = 11.2 \text{ L } H_2$$

9. **(C)** 45.0 g is the actual yield of silver, while 50.0 g is the theoretical yield of silver. Percent yield can be found by

$$\text{percent yield} = \frac{45.0 \text{ g } Ag}{50.0 \text{ g } Ag} \times 100\% = 90.0\%$$

Liquids, Solids, and Phase Changes

7

These skills are usually tested on the SAT Subject Test in Chemistry. You should be able to...

→ Describe the states of matter (solids, liquids, and gases).
→ Describe and compare the important properties of solids, liquids, and gases.
→ Describe the various types of phase change.
→ Analyze a phase diagram.
→ Explain the use of liquid water as a common solvent.
→ Solve solubility and concentration problems.
→ Solve colligative property problems in a qualitative manner.
→ Describe the continuum of water mixtures, including solutions, colloids, and suspensions.

This chapter will review and strengthen these skills. Be sure to do the Practice Exercises at the end of the chapter.

Liquids and solids are considered **condensed phases of matter**. By itself, this is a significant differentiation from the phase of matter called a gas, which was discussed extensively in Chapter 5. But there are many other important characteristics that distinguish solids and liquids from gases as well as those that distinguish solids from liquids. Macroscopically (i.e., in a form visible to the naked eye), there are several ways to differentiate solids, liquids, and gases. But in order to understand the states of matter well, and in addition how they change from one to another, it is important to be able to envision these phases from a mind's-eye perspective that takes into account what they are like on an atomic/molecular level.

GENERAL CHARACTERISTICS OF SOLIDS, LIQUIDS, AND GASES

Macroscopically, solids are seen as rigid materials that maintain their shape independent of the container they are in. This rigidity is what differentiates them from liquids and gases, which can both be described as **fluid** because they can flow and take on the shape of their containers. Although liquids and solids are different with regard to their visible shapes, they are similar with regard to their measurable volumes. Liquids, unlike gases, do not assume the volume of their container. Gases spread out to take up the volume of the container in which they find themselves. Solids act like liquids: their volume is not dependent upon or determined by the container they occupy. The table below summarizes these points.

Phase:	Solid	Liquid	Gas
Shape:	Definite	Variable	Variable
Volume:	Definite	Definite	Variable

In order to explain these observable differences and similarities between the phases of matter, an understanding of the states of matter from a particle perspective is needed. Recall that liquids and solids were described above as *condensed* phases of matter. This is so because, on a particulate level, the atoms and/or molecules in a liquid or solid are in relatively close proximity. In the gas phase, on the other hand, the particles are relatively far apart. This is true whether different states of the *same* substance at *different* temperatures or different states of *different* substances at the *same* temperature are being compared. Notice how important it is that the temperature and the substance being evaluated be taken into consideration. The temperature of the material is critical because it is a measure of the average kinetic energy possessed by the substance. Recall from Chapter 1 that kinetic energy can be described as the energy associated with the motion of a substance's particles. The substance itself is important because the attractive forces between the pertinent particles making up the substance dictate the degree to which the particles will want to cling to each other and make the phase of the substance condensed or not. All the types of attractive forces (i.e., ionic and covalent bonding, intermolecular attractions, and metallic bonding) have already been highlighted in Chapter 3, but they will be emphasized and used in this chapter to explain the properties of matter dealing with phase as well.

In order to encourage a substance to behave as a solid, it makes sense that its temperature (i.e., its average kinetic energy) be relatively low. This way the motion of the particles will not inhibit the particles from interacting and thereby cause them to cling to each other. Likewise, substances are encouraged to be in the solid state when the attractive force strength between the particles is relatively high. This way the particles will be able to overcome their independent motion and be made to cling to each other once again. Similarly, gases may be said to exist when the temperature is relatively high and the attractive force strength between the particles is relatively low. In these circumstances the particles can spread out without interacting with each other. The liquid state is generally present when the temperature and attractive force strength for a particular substance are in a range between that for solids and gases. Since the liquid phase is *between* the gas and solid phases, liquids are, under certain conditions, more like solids (i.e., condensed), while under other conditions, more like gases (i.e., fluid). The tables below summarize general occurrences of the above points for most substances.*

States of Matter for the *Same* Substance at *Different* Temperatures

Phase:	Solid	Liquid	Gas
Particle Proximity:	Very close	Close	Far apart
Average Kinetic Energy:	Low	Moderate	High
Attractive Force Strength:	------------------ All the same** ------------------		

*Water constitutes an exception to the rule of relative particle proximity. That is, the molecules of water are closer in the liquid state than they are in the solid state. Ice floats on liquid water because ice is less dense. See page 189 for more information.

**The attractive force strength is the same for all the phases because with the substance held constant, the particle interactions are unchanging.

States of Matter for *Different* Substances at the *Same* Temperature

Phase:	Solid	Liquid	Gas
Particle Proximity	Very close	Close	Far apart
Average Kinetic Energy:	------------------ All the same*** ------------------		
Attractive Force Strength:	High	Moderate to high	Low

***The average kinetic energy is the same for all phases because the temperature is unchanging.

It is noteworthy that at the same temperature, different substances have the same average kinetic energy despite being in different phases. If a substance has a temperature higher than absolute zero (or 0 K)—which all substances must have according to the Third Law of Thermodynamics, discussed in Chapter 9—all phases contain particles that are moving. The types of motion available to the particles of matter include vibration (shaking and stretching), rotation (spinning), and translation (moving from one point to another in 3-D space). Since solids generally contain particles that are very close and held in fixed positions by forces strong enough to make these substances rigid, for them the most significant type of motion realized is vibration. On the other hand, gases, which generally contain particles that are far apart and are little attracted to each other, exhibit all three modes of motion—particularly translation. These characteristics account for their ability to expand to the shape and volume of their container. Liquid particles exhibit all three kinds of motion; though translation is limited, there is enough to allow for the mixing of liquids left alone over time.

It is also worth emphasizing that temperature is directly related *not* to the *total* kinetic energy possessed by a sample of a substance but *only* to its *average* kinetic energy. Because of the condensed nature of solids and liquids, many more of their particles than of gases exist in a given amount of space. On the average, less than 1 percent of the space of a gas sample is occupied by gaseous particles, whereas the space taken up by particles in liquids and solids measures in the 70 percent range. This explains why gases are described as compressible in contrast to solids and liquids, which are generally incompressible and also exhibit a greater total kinetic energy at a given temperature. Since solids and liquids have many more particles to which the *average* kinetic energy can be assigned, their *total* kinetic energy is consequently higher.

Also noteworthy is that not all particles of a given substance possess the same kinetic energy; instead, particles exhibit a distribution of kinetic energies. Some particles in a sample at a given temperature possess a relatively low amount of kinetic energy, yet other particles exhibit a relatively high amount of kinetic energy. Most particles, however, possess an amount near the average kinetic energy. This is shown in see Figure 22, which graphs the number of particles versus kinetic energy. This type of graph is often referred to as a Maxwell-Boltzmann distribution. Note that the distribution of kinetic energies skews to the right and spreads out a bit as the temperature increases. It is important to recognize that though the average kinetic energy of the particles in the sample increases at the higher temperature, the actual number of particles *with* the average kinetic energy decreases. Nevertheless, the sample still possesses particles of lower and higher kinetic energy than the average.

TIP

An increase in temperature increases the average kinetic energy of the molecules.

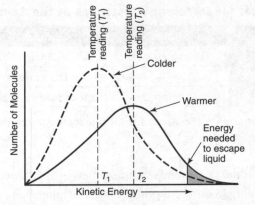

Figure 22. Distribution of the Kinetic Energy of Molecules

A final but very important aspect in the differentiation between the phases of matter is the orderliness found in the arrangement of the particles in a particular phase. As has been stated, gases are composed of particles exhibiting all types of motion (particularly translation) with considerable space between the particles. Consequently, the word used to describe gases is **random**, a word that implies that a lack of order is associated with gases. The condensed solid and liquid phases are different in that they often display order, although to different degrees. The orderliness in liquids is commonly described as *short-range*, while the order in solids is commonly described as *long-range*. Liquids can be viewed as containing **clusters** of particles in which order is found within the cluster but not from cluster to cluster. Generally, clusters contain about 100 to 1,000 particles, with the constituents of clusters in constant change. The attractive forces are strong enough to generally maintain the clustery nature of the liquid but not strong enough to keep the particles in the fixed positions found in the solid state. Although there are many examples of solid substances that do not display sufficient long-range order to maintain definite, fixed positions (and so they are referred to as being **amorphous**), solids most often display significant orderliness. Such solids are described as being **crystalline**. There are times when the three-dimensional arrangement involves such a large number of particles that this orderliness can be observed macroscopically and the geometry of the arrangement can be seen with the naked eye. Most other times, the solids are composed of an aggregation of smaller crystals, with these ordered domains arranged randomly; these solids are referred to as **polycrystalline**. The table below summarizes the points made in the last few paragraphs.

Phase:	Solid	Liquid	Gas
Particle Motion:	Vibration	Vibration, rotation, translation	Vibration, rotation, translation
Particle Orderliness:	Long-range (crystalline)	Short-range (clustery)	None (random)

IMPORTANT PROPERTIES OF LIQUIDS

Evidence of attractive forces between particles of matter is easily found by investigating the properties of liquids. Although each property is different, the magnitude of the measured properties discussed below is fundamentally related to the strength of the interactions between the particles and provides direct evidence of their existence.

Surface Tension

Surface tension can be defined as the resistance of a liquid to create new surfaces as the result of an imbalance of attractive forces. The particles on the surface of a liquid lack an attraction upward and as such experience a net force inward; that is, they are drawn to the particles below. In other words, particles on the surface of a liquid have a "desire" to be interior particles. Liquids, therefore, have an impetus to possess the least amount of surface possible and often "bead up." The effect essentially creates a "skin" for a liquid that can support other materials that may, for reasons of higher density, normally want to sink. This sinking would expose the interior particles of the liquid and turn them into surface particles. Resistance to this occurrence is what constitutes surface tension.

Generally, the higher the attractive force strength between the particles, the higher the surface tension. This is so because a stronger net force inward will create a greater resistance to creating new surface particles. The specific intermolecular attractions (a topic discussed in Chapter 3) that need to be evaluated for typical liquids at common conditions include **hydrogen bonding** (generally the strongest) followed by **permanent dipole–dipole** and then **temporary dipole–dipole** interactions (generally the weakest). Recall that temporary dipole–dipole interactions are also known as **London dispersion forces**.

➡ Example

Of the room-temperature liquids hexane (C_6H_{14}), water (H_2O), and acetone (CH_3COCH_3), which would exhibit the greatest surface tension?

Water would show the greatest surface tension; it contains molecules between which the strongest intermolecular bonding is to be found. In a pertinent manner, water exhibits hydrogen bonding, while acetone displays permanent dipole–dipole interactions and hexane only London dispersion forces. These are best recognized by drawing Lewis diagrams of the molecules. (See page 190 for a complete analysis of water in this regard, or see Chapter 3 to review intermolecular bonding in general.)

Viscosity

Viscosity can be defined as the resistance of a liquid to flow. The ability of a liquid to flow is again fundamentally related to the strength of the intermolecular attractions between the molecules making up the substance. The stronger the attractions between the molecules, the less likely a liquid's ability to flow and the greater its viscosity.

TIP

More viscous liquids move more slowly.

➡ Example

Of the room temperature liquids water (H_2O), ethylene glycol ($C_2H_4(OH)_2$), and glycerin ($C_3H_5(OH)_3$), which would exhibit the greatest viscosity?

Glycerin would show the greatest viscosity; it contains molecules between which the strongest intermolecular bonding is found. Although all three molecules exhibit hydrogen bonding in a noteworthy manner, glycerin exhibits the highest degree as it contains three hydroxyl (–OH) groups in its structure.

Capillary Action

Capillary action, the attraction of the surface of a liquid to the surface of a solid, is a property closely related to surface tension. A liquid will rise quite high in a very narrow tube if a strong attraction exists between the liquid molecules and the molecules that make up the surface of the tube. This attraction tends to pull the liquid molecules upward along the surface against the pull of gravity. This process continues until the weight of the liquid balances the gravitational force. Capillary action can occur between water molecules and paper fiber, causing the water molecules to rise up the paper. When a water soluble ink is placed on the paper, the ink moves up the paper and separates into its various colored components. This separation occurs because the water and the paper attract the molecules of the ink components differently. These phenomena are used in the separation process of **paper chromatography**, as shown in the paper chromatography experiment on page 307. Capillary action is at least partly responsible for the transportation of water from the roots of a plant to its leaves. The same process is responsible for the concave liquid surface, called a **meniscus**, that forms in a test tube or graduated cylinder. (See the drawing on page 49.)

Heat of Vaporization

The **heat of vaporization**, often symbolized ΔH_{vap}, is defined as the amount of heat required to vaporize a certain amount of a liquid at its boiling point. From a particle perspective, the process involves changing the clustery nature of a liquid into the random nature of a gas. As such, attractive interactions between the particles that cause the particles to cluster must be overcome to separate them into randomness. The separation of the particles does not make the particles move any faster, and so there is no increase in the kinetic energy throughout this process; the temperature remains constant. However, as energy is added to the system to overcome forces of attraction, the energy is stored by the substance, and the potential energy of the substance rises.

Once again, the magnitude of the heat of vaporization is dependent on the strength of the attractive forces, and so a higher heat of vaporization is generally associated with liquids that have a higher degree of intermolecular bond strength.

➡ Example _____

The heat of vaporization of water at its normal boiling point of 100.0°C is 41 kJ/mol. Since the unit quantity of substance referenced by this value is the mole, the value is commonly referred to as the **molar heat of vaporization**. Would the predicted molar heat of vaporization for carbon tetrachloride, CCl_4, be greater or less than this value at its normal boiling point of 77°C?

Compared to water, carbon tetrachloride would be predicted to have a lower molar heat of vaporization at its boiling point. This is so because carbon tetrachloride is a nonpolar molecular substance that exhibits only London dispersion forces between its molecules. Experimentally, the value is found to be 32 kJ/mol. Carbon tetrachloride's weaker intermolecular forces of attraction indicate that less energy would be needed to cause clusters of the liquid to be pulled apart into the random gas phase.

Equilibrium Vapor Pressure

Recall that the Maxwell-Boltzmann distribution of kinetic energy associated with a liquid (see Figure 22) indicates that most of the particles in a liquid possess a kinetic energy near the average value. Some, however, have values lower than the average, and others have values higher than the average. Now envision a situation where a high-energy particle finds itself on the surface of a liquid and possesses enough energy to break away from the cluster and turn into a gas. This is precisely what happens when **evaporation** occurs: as a certain fraction of the molecules in a liquid meet this threshold value, a portion of the liquid turns into **vapor** particles. It is important to note that this can happen below the boiling point of the liquid (where that process normally occurs) but only on its surface.

Since evaporation involves the loss of the high-energy particles from the liquid, it is commonly viewed as a **cooling process**. As the high-energy particles leave, the temperature of the liquid decreases, and the average kinetic energy necessarily goes down. The validity of this claim is easily recognized: consider how humans naturally sweat to cool themselves.

If a liquid is placed in a closed container and the temperature is held constant, evaporation will occur, and vapor particles will begin to fill the space above the liquid surface. Because the container is closed as time goes by, some of the vapor particles will have an opportunity to collide with the surface of the liquid; in the process they will perhaps lose a portion of their energy and be "captured" by the clusters of the liquid. This is precisely what happens during the process of **condensation** (*when a gas turns into a liquid*). Initially, the rate of evaporation is higher than that of condensation because of the fact that not many vapor particles are around. But as the number of vapor particles increases, so does the rate of condensation. It should now be easy to envision a point in time where the rates of evaporation and condensation are equal to each other and the number of vapor particles above the liquid becomes constant (see Figure 23). Situations like this, where two opposing processes are equal to each other in rate, are common in nature and are referred to as a **dynamic equilibrium**. Translating gaseous particles create pressure as the particles collide with their container walls (see Chapter 5). Consequently, the phenomenon described above is the reason why an **equilibrium vapor pressure** is associated with a given liquid.

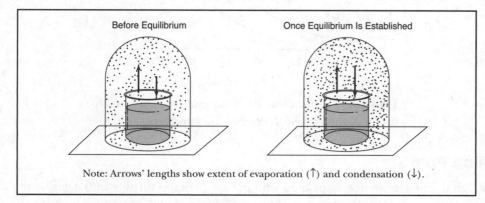

Note: Arrows' lengths show extent of evaporation (↑) and condensation (↓).

Figure 23. Closed System in Dynamic Equilibrium

As we have seen with regard to the other properties of liquids, the equilibrium vapor pressure of a liquid is also fundamentally related to the attractive force strength between the particles in the liquid—this time, however, in the opposite way. Higher attractive force strengths produce lower vapor pressures. This is true because a smaller fraction of high-energy particles in a given sample will possess the threshold energy needed to break away and turn into a gas on the surface of the liquid if the attractive force holding them to the cluster is stronger. The consequence is that equilibrium will be established with a smaller number of vapor particles above the liquid and hence a lower vapor pressure.

More about equilibrium is discussed in Chapter 10.

A review of the Maxwell-Boltzmann distribution (Figure 22) provides insight into how the vapor pressure of a liquid could change. Recall how at a higher temperature the distribution skews to the right and in so doing raises the average kinetic energy of the sample. As a result, a larger number of particles will have available the threshold energy needed to produce vapor particles through evaporation. If evaporation is enhanced, more vapor particles will be present when equilibrium is established, and the vapor pressure will be higher. Figure 24 shows the effect of temperature on vapor pressure for liquid water and carbon tetrachloride.

➥ Example

From Figure 24, determine the vapor pressure of CCl_4 and H_2O at 60°C and rationalize the relative magnitude of the values.

Reading from the graph, the vapor pressure of CCl_4 at 60°C is ~400 mm of Hg and that of H_2O reads ~200 mm of Hg. This is in agreement with the information presented in the last example, where CCl_4 was shown to have a lower heat of vaporization than water thanks to its lower intermolecular bond strength.

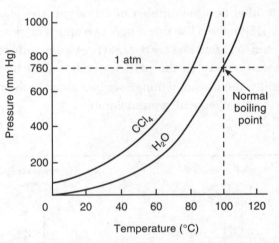

Figure 24. Vapor Pressure-Temperature Relationship for Carbon Tetrachloride and Water

Boiling Point

Boiling point is defined as the temperature at which the liquid's vapor pressure equals the atmospheric pressure.

Although the phase change associated with boiling (a liquid turning into a gas) is the same as seen with evaporation, boiling has several significantly different characteristics. As your own experience is likely to have verified, boiling occurs throughout the body of a liquid, not only on the surface, as happens with evaporation. This is shown by the presence of bubbles in the liquid during boiling. The requirement of bubble production creates another difference between boiling and evaporation. The creation of a bubble mandates that the vapor pressure of the gas making the bubble be high enough to withstand the pressure of the atmosphere above the liquid pushing down on the bubble to collapse it or inhibit its production altogether. Consequently, the temperature must be high enough to produce a vapor pressure for the liquid that is at least equal to the atmospheric pressure. This means that boiling occurs at a particular temperature, called the **boiling point**, where the vapor pressure of the liquid and the atmospheric pressure are the same. The ramification of that concept is that the boiling point of a liquid is variable: it is dependent on the atmospheric pressure under which the liquid exists. That is, if the atmospheric pressure on a liquid is lower (as it is above sea level, as may be seen in the mountains), the boiling point of the liquid is lower. This is so because the temperature to which the liquid must be raised to cause the vapor pressure of the liquid to be equal to the atmosphere will not be as high. The opposite is true for a liquid below sea level. The **normal boiling point** is the temperature at which the vapor pressure of a liquid is equal to 1.0 atmosphere (760 mm of Hg), the atmospheric pressure at sea level. The table below summarizes the differences and similarities between boiling and evaporation.

Process:	Evaporation	Boiling
Similarities:	Particles break away from liquid clusters to form a gas. The process is endothermic (requires energy).	
Differences:	Below the boiling point	At the boiling point
	Occurs on the surface of the liquid	Occurs throughout the body of the liquid (bubbles)

Since vapor pressure and boiling point are related to each other and vapor pressure is related to attractive force strength, boiling point must also be related to attractive force strength. Generally, higher boiling points indicate higher attractive force strength between the particles of a liquid. This makes sense: stronger bonds will require higher temperatures to make the vapor pressure high enough to equal that of the atmosphere.

➡ Example _____

From Figure 24, determine the normal boiling point of CCl_4 and compare it to that of H_2O. Also, rationalize the relative difference between the two values.

From the figure, CCl_4 has a boiling point slightly less than 80°C. This is so because the vapor pressure of CCl_4 is shown to be equal to 760 mm of Hg (normal atmospheric pressure) around that temperature. This agrees with the actual value of 77°C provided in a previous example using the substance carbon tetrachloride. As has been seen, carbon tetrachloride, a nonpolar molecule, has weaker intermolecular bond strength (only LDF) than water, a polar molecule that exhibits hydrogen bonding (H-bonding), whose normal boiling point is known to be 100°C, as the graph shows.

Critical Temperature

There is a limit to the attractive force that binds particles together in a liquid. It is only so strong. As a result, one can envision a temperature above which the kinetic energy of the particles will always be high enough to inhibit the clustering of gas particles into a liquid. This temperature, referred to as the **critical temperature**, represents a value above which the substance cannot be condensed. Since the attractive force limit dictates the temperature at which this occurs for various substances, one sees again that the magnitude of the attractive forces must dictate the magnitude of the critical temperature. Higher intermolecular bond strengths are associated with higher critical temperatures.

➡ Example _____

Carbon dioxide has a critical temperature of 31°C. Water has a critical temperature of 374°C. Why is carbon dioxide's critical temperature so much lower?

Carbon dioxide is a small nonpolar molecule and exhibits only weak London dispersion forces between its molecules. As noted before, water is a polar molecule and exhibits hydrogen bonding between its molecules. The attractions are stronger in water; hence, water has a higher critical temperature.

IMPORTANT PROPERTIES OF SOLIDS

Just as the attractive forces between the particles in liquids dictate the magnitude of the properties of liquids, the attractive forces between the particles in a solid do the same.

Heat of Fusion

The **heat of fusion**, often symbolized ΔH_{fus}, is defined as the amount of heat *required* to liquefy a certain amount of a solid at its melting point. As with heat of vaporization, the energy is used to overcome the attractive forces between the particles, not to make them move more. Consequently, the crystalline nature of a solid is transformed into the clustery nature of a liquid, and the temperature of the material does not rise. As seen previously, since the temperature is not on the rise, neither is the kinetic energy. What does increase is the potential energy of the material as the energy provided to the system is stored in the liquid phase. It is important to recognize that as a liquid is frozen (i.e., the opposite process of melting), this same quantity of energy is *released*. As is seen with heat of vaporization, here, too, higher attractive force strengths result in a higher heat of fusion for a given substance.

➡ **Example** _____

The **molar heat of fusion** for water at its normal melting point of 0.0°C is 6.0 kJ/mol. Would a larger amount of energy be released upon solidifying one mole of carbon tetrachloride at its freezing point than would occur with water?

No, as carbon tetrachloride is predicted to have a lower molar heat of fusion at its normal melting point of –23°C. This is so because carbon tetrachloride is a nonpolar molecular substance that exhibits only weaker London dispersion forces between its molecules. Experimentally, the value is found to be 2.5 kJ/mol. Although this is the energy *required* to melt one mole of CCl_4, that same amount of energy would be *released* upon freezing and is smaller than the 6.0 kJ/mol that is released when water normally freezes at 0.0°C.

Melting (and Freezing) Point

Earlier, the boiling point of a liquid was shown to depend upon the vapor pressure of the liquid. The melting and freezing points of a substance show a similar dependency although for a different reason. Just like liquids, solids exhibit an equilibrium vapor pressure if placed in a closed container. High-energy particles on the surface of the solid can break away from the crystal and enter the gaseous state. The process of a solid turning into a gas is called **sublimation**. When the rate of sublimation equals the rate of **vapor deposition** (the opposite of sublimation), equilibrium is established and the number of gaseous particles above the solid is constant, creating a certain pressure. Also, as with the liquid, the vapor pressure of the solid is dependent on the temperature of the material. Consequently, there is a temperature at which the vapor pressure of the solid is equal to the vapor pressure of the liquid. This temperature at which melting (or freezing) takes place is called the **melting** (or **freezing**)

point, depending on whether energy is being added to or taken away from the system. Since the vapor pressure of a substance is contingent upon the strength of the attractive forces between the particles, and the melting/freezing point is dictated by the liquid's and solid's vapor pressure, then the melting/freezing point depends on the strength of the attractive forces. Higher attractive force strength is indicative of a higher melting/freezing point.

➡ Example

The **normal melting point** of elemental magnesium (Mg) is 650°C. The normal melting point of magnesium oxide (MgO) is 2,825°C. Why is the melting point of MgO higher than the melting point of Mg?

Magnesium is a metallic solid, while magnesium oxide is ionic. Although fairly strong, metallic bonding is generally weaker than ionic bonding. The atoms of magnesium in a magnesium crystal are held tightly enough for magnesium to be stable as a solid up to 650°C. The same is true for ions of magnesium and oxygen in a magnesium oxide crystal, but its stronger ionic bonds make it stable to an even higher temperature.

Electrical Conductivity

In order for a substance to be electrically conductive, it must contain charged particles that can move translationally. For most types of solids, this is not practically possible because the particles, even if they are charged, are held in fixed positions. The only exception is for metallic solids in which the core of the metal atoms is held reasonably firm by the *metallic bond* but the valence electrons of those atoms are not. Recall from Chapter 3 how the metallic bond can be described as a *mobile* "sea of electrons" surrounding the remaining positive cores of the atoms.

➡ Example

Which substance, Mg or MgO, would be the most electrically conductive?

As stated in the previous example, magnesium oxide is an ionic substance. Despite containing charged particles (i.e. ions of Mg^{2+} and O^{2-}), the ions are not able to move translationally because of the strong ionic bonding. Magnesium, on the other hand, contains atoms held together by the metallic bond. The valence electrons act as mobile charge carriers and make magnesium electrically conductive.

Solubility

The solubility of solids varies with the type of bonding between the particles making up the solid and the kind of substance in which the solid will be dissolved. Although a more in-depth discussion on the solution process will be seen later in this chapter, a good rule of thumb concerning the degree to which a solute (what is being dissolved) will dissolve in a solvent (what is being dissolved into) is the phrase "*like dissolves like.*" Keep in mind that this is a *general rule* and that, for good reason, many exceptions exist. The "like dissolves like" rule refers to the nature of the materials involved with reference to their distribution of charged parts. For example, solids that have permanent charges (i.e., ionic or polar molecular substances) tend to dissolve in similar substances (i.e., polar molecular liquids). Likewise, nonpolar molecular solids tend to dissolve in nonpolar molecular liquids.

➥ Example _____

Paradichlorobenzene ($C_6H_4Cl_2$), a material used to produce mothballs, would best dissolve in which room temperature liquid: water or carbon tetrachloride?

   Paradichlorobenzene is an aromatic nonpolar molecular solid. (See the discussion in Chapter 14 on aromatic compounds to further familiarize yourself with this substance.) Using the rule "like dissolves like," paradichlorobenzene would best dissolve in carbon tetrachloride, as it was shown in previous examples to also be a nonpolar molecular liquid. Water is a polar molecular liquid and will not dissolve appreciable amounts of the substance.

TYPES OF SOLIDS

As seen in the many examples describing the properties of solids, the particles that make up a solid and the types of bonds with which they are held together are critical pieces of information needed in predicting the relative behavior of a material. Accordingly, solids can be categorized based on those characteristics. The table below summarizes the different types of solids and the properties generally associated with them.

Type of Solid	Examples	Pertinent Particles	Type of Bonds	General Properties
Macromolecular (Also known as a network solid)	C (as in diamond or graphite), SiC, SiO_2	Nonmetal atoms	Covalent	High heat of fusion and melting point; low electrical conductivity and solubility in water
Ionic	NaCl, $MgSO_4$, $K_2Cr_2O_7$	Metal and nonmetal ions	Ionic	High heat of fusion, melting point, and solubility in water*; low electrical conductivity
Metallic	Cu, Mg, Hg	Metal atoms	Metallic	Moderate to high heat of fusion and melting point; high electrical conductivity; low solubility in most solvents
Molecular	SO_2, NH_3, CH_4	Molecules	Intermolecular	Low heat of fusion, melting point, and electrical conductivity; variable solubility in water

*There are many ionic substances insoluble in water.

   There is a special type of solid that contains ions *and* molecules combined in a single substance; it is called a **hydrate**. Hydrates can be viewed as containing fully charged ions and polar water molecules surrounded by and attracted to each other in a crystal structure that appears dry despite the presence of the water. A hydrate's formula displays the formula for the regular ionic substance followed by the number of water molecules associated with it per formula unit. Examples include $CuSO_4 \cdot 5H_2O$ and $BaCl_2 \cdot 2H_2O$ (the • in the formula is read as "with"). Often, the heating of hydrated crystals releases their water of hydration and alters the crystal structure of the remaining "anhydrous" ionic material; this in turn causes a color change in the solid substance. For example, if a sample of $CuSO_4 \cdot 5H_2O$ (called copper (II) sulfate pentahydrate) is heated in the bottom of a test tube, the blue solid turns into a white powder with liquid water seen accumulating near the cooler top of the test tube. Certain hydrated substances—for example, $MgSO_4 \cdot 7H_2O$ (Epsom salt)—lose their water of hydration at room temperature without the need for additional heating in a process called **efflorescence**. Conversely, other ionic solids are so attracted to water molecules that they readily *become* hydrates by absorbing moisture from the air; these are referred to as **hygroscopic** materials. Calcium chloride is such a substance and is commonly used as a drying agent in the laboratory because of its desire to be associated with water. Some ionic substances are so hygroscopic that they can become wet and form "puddles" of ionic solutions with the water they accumulate. These ionic substances are additionally referred to as being **deliquescent**. Sodium hydroxide is an example of such a substance.

PHASE DIAGRAMS

The previously described properties of solids, liquids, and gases, as well as the processes relating to their change from one to another, all point toward the importance of temperature and pressure in determining the stable state of matter for a given substance. A **phase diagram** provides a graphical way to summarize the conditions of those parameters that dictate the phase the substance will primarily find itself in once equilibrium is established. Figure 25 is an example of a phase diagram of the substance water. It is important to note that many phase diagrams (like the one shown in Figure 25) are not shown to scale but are meant to convey important information about a substance. Analysis of Figure 25 shows that line BD is essentially the vapor pressure curve for water's liquid phase. Notice that when the pressure on a sample of water is 760 mm of Hg, the vapor pressure of the water matches that at 100°C, and the water will boil. However, if the pressure is raised, the boiling point temperature increases, and if the pressure is less than 760 mm of Hg, the boiling point decreases along the *BD* curve down to point *B*.

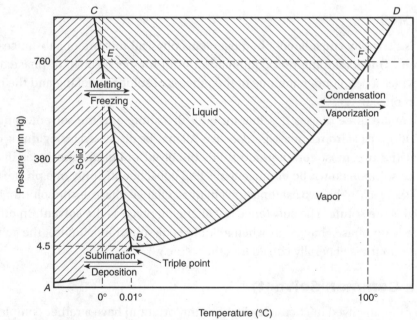

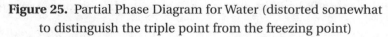

Figure 25. Partial Phase Diagram for Water (distorted somewhat to distinguish the triple point from the freezing point)

TIP

Know the significance of each gray area and boundary line.

TIP

The *triple point* is the only temperature and pressure at which all three phases of a substance can exist.

As line *BD* is essentially the vapor pressure curve for liquid water, line *AB* is the vapor pressure curve for solid water. Point *B* is one position where the vapor pressure (or VP) of the solid is equal to the vapor pressure of the liquid. As the pressure is increased, the temperature at which this remains true goes down, as seen by the negative slope of the line emanating upward from point *B*. This differs from most other substances, which display a positive slope in this condition, and has to do with the density of solid water (ice) being less than the density of liquid water (discussed earlier in this chapter). Recall too that when the VP of a solid is equal to the VP of a liquid, melting and freezing take place in equilibrium with each other. Consequently, the normal melting/freezing point for water (i.e., at 760 mm of Hg) is shown on the diagram to be 0.0°C. Again, this point is affected by pressure along line *BC* so that if the pressure is decreased, the melting/freezing point is slightly higher up to point *B*, or 0.01°C.

Point *B* on the diagram is unique in that not only are melting and freezing in equilibrium with each other but so are *all* other phase changes. Point *B*, commonly referred to as the **triple point**, describes the only combination of temperature and pressure where all three phases are simultaneously stable. Combinations of temperature and pressure that fall on the lines in these diagrams describe stable two-phase regions. All other combinations represent stable single-phase regions for the substance.

Another noteworthy point on a phase diagram is the end of the vapor pressure curve for the liquid. The phase diagram for water in Figure 25 cannot show this because of the figure's scale, but it is shown on many other diagrams. As earlier discussed in reference to the properties of liquids, this point, called the critical temperature, is the temperature above which a substance cannot exist as a liquid. That explains why the vapor pressure curve comes to an end. If the liquid cannot exist, there can be no vapor pressure measured for it. Note that if a substance is at its critical temperature, it can exist as a liquid, but in order to be so, it must also be at a high enough pressure. The pressure needed to liquefy a gas at its critical temperature is referred to as its **critical pressure**. The combination of critical temperature and critical pressure, often seen on phase diagrams, is called its **critical point**.

SOLUTIONS

A solution is generally considered a homogeneous physical mixture of substances, a solute and a solvent. Understanding that combination from the perspective of phase and phase change provides a deep understanding of the process of solution making and the properties of a solution once made.

Since there are three phases possible for both the solute and the solvent, nine possible kinds of solutions exist from a phase perspective. All may be discussed in beginning chemistry courses, but the two most common are solids dissolved in liquids and gases dissolved in liquids. Since solutions must be homogeneous and consequently only one phase, the material being dissolved (in the two examples given) must change phase as the solution is formed; this material is the **solute**. The substance not changing phase is the **solvent**. In other examples, if there is no phase change (as when a liquid is dissolved in a liquid), the substance in the smaller amount is generally considered the solute.

Water (a Common Solvent)

Water is so often involved in chemistry that it is important to have a rather complete understanding of this compound and its properties. Pure water has become a matter of national concern. Although commercial methods of purification will not be discussed here, the usual laboratory method of obtaining pure water, distillation, will be covered.

PURIFICATION OF WATER

The process of distillation involves the evaporation and condensation of the water molecules. The usual apparatus for the distillation of any liquid is shown in Figure 26.

This method of purification will remove any substance that has a boiling point higher than that of water. It cannot remove dissolved gases or liquids that boil off before water. These substances will be carried over into the condenser and subsequently into the distillate.

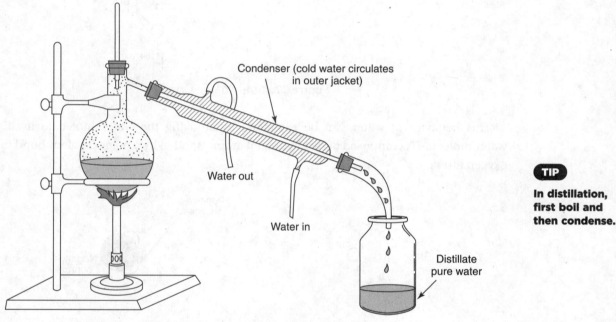

Figure 26. Distillation of Water

TIP

In distillation, first boil and then condense.

POLARITY AND HYDROGEN BONDING IN WATER

Water is different from most liquids in that it reaches its greatest density at 4°C and then its volume begins to expand. By the time water freezes at 0°C, its volume has expanded by about 9 percent. Most other liquids contract as they cool and change state to a solid because their molecules have less energy, move more slowly, and are closer together. This abnormal behavior of water can be explained as follows. X-ray studies of ice crystals show that H_2O molecules are bound into large molecules in which each oxygen atom is connected through **hydrogen bonds** to four other oxygen atoms as shown in Figure 27.

This rather wide open structure accounts for the low density of ice. As heat is applied and melting begins, this structure begins to collapse but not all the hydrogen bonds are broken. The collapsing increases the density of the water, but the remaining bonds keep the structure from completely collapsing. As heat is absorbed, the kinetic energy of the molecules breaks more of these bonds as the temperature rises from 0° to 4°C. At the same time this added kinetic energy tends to distribute the molecules farther apart. At 4°C these opposing forces are in balance—thus the greatest density. Above 4°C the increasing molecular motion again causes a decrease in density since it is the dominate force and offsets the breaking of any more hydrogen bonds.

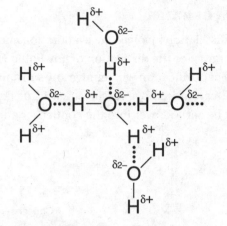

Figure 27. Study of Ice Crystal

This behavior of water can be explained by studying the water molecule itself. The water molecule is composed of two hydrogen atoms bonded by a polar covalent bond to one oxygen atom.

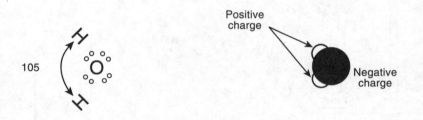

Because of the polar nature of the bond, the molecule exhibits the charges shown in the above drawing. It is this polar charge that causes the polar bonding discussed in Chapter 3 as the hydrogen bond. This bonding is stronger than the usual molecular attraction called van der Waals forces or dipole-dipole attractions.

WATER SOLUTIONS AND MIXTURES

To make molecules or ions of another substance go into solution, water molecules must overcome the forces that hold these molecules or ions together. The mechanism of the actual process is complex. To make sugar molecules go into solution, the water molecules cluster around the sugar molecules, pull them off, and disperse, forming the solution.

For an ionic crystal such as salt, the water molecules orient themselves around the ions (which are charged particles) and again must overcome the forces holding the ions together. Since the water molecule is polar, this orientation around the ion is an attraction of the polar ends of the water molecule. For example:

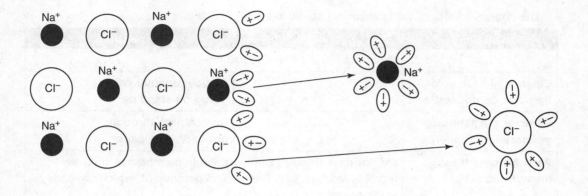

Once surrounded, the ion is insulated to an extent from other ions in solution because of the dipole property of water. The water molecules that surround the ion differ in number for various ions, and the whole group is called a **hydrated ion**.

In general, as stated in the preceding section, polar substances and ions dissolve in polar solvents and nonpolar substances such as fats dissolve in nonpolar solvents such as gasoline. The process of going into solution is **exothermic** if energy is released in the process, and **endothermic** if energy from the water is used up to a greater extent than energy is released in freeing the particle.

When two liquids are mixed and they dissolve each other, they are said to be completely **miscible**. If they separate and do not mix, they are said to be **immiscible**.

Two molten metals may be mixed and allowed to cool. This gives a "solid solution" called an **alloy**.

CONTINUUM OF WATER MIXTURES

Figure 28 shows the general sizes of the particles found in a water mixture.

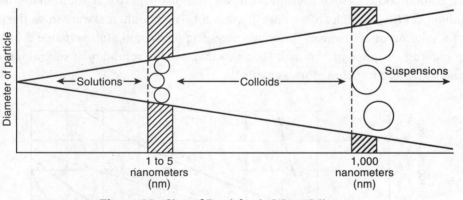

Figure 28. Size of Particles in Water Mixture

The basic difference between a colloid and a suspension is the diameter of the particles dispersed. All the boundaries marked in Figure 28 indicate only the general ranges in which the distinctions between solutions, colloids, and suspensions are usually made.

The characteristics of water mixtures are as follows:

Solutions	Colloids	Suspensions
. . . . 1 nm . . . . .1,000 nm		
Clear; may have color Particles do not settle.		Cloudy; opaque color Settle on standing
Particles pass through ordinary filter paper.		Do not pass through ordinary filter paper
Particles pass through membranes.	Do not pass through semipermeable membranes such as animal bladders, cellophane, and parchment, which have very small pores*	
Particles are not visible.	Visible in ultramicroscope	Visible with microscope or naked eye
	Show Brownian movement	No Brownian movement

Separation of a solution from a colloidal dispersion through a semipermeable membrane is called dialysis.

When a bright light is directed at right angles to the stage of an ultramicroscope, the individual reflections of colloidal particles can be observed to be following a random zigzag path. This is explained as follows: The molecules in the dispersing medium are in motion and continuously bumping into the colloidal particles, causing them to change direction in a random fashion. This motion is called **Brownian movement** after the Scottish botanist Robert Brown, who first observed it.

Solubility

The degree to which a solute can dissolve in a solvent, called **solubility**, varies with the solute/solvent combination, as seen previously (recall "like dissolves like"). Additionally, temperature influences the solubility of a solute in a solvent. This should make sense, as the production of a solution often involves the solute changing phase and phase change is influenced by temperature, as previously noted. Figure 29 displays the solubility of several ionic solids in liquid water as a function of temperature.

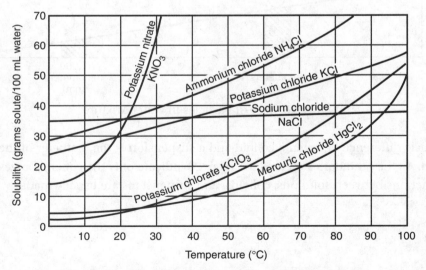

Figure 29. Solubility Curves of Some Common Salts

➥ Example

Describe how the solubility of ammonium chloride varies with temperature, and compare its solubility at 50°C to that of potassium chloride.

Figure 29 shows that at 0°C about 28 grams of ammonium chloride can be dissolved in 100.0 mL of water. At 50°C, about 48 grams of the substance is able to go into solution. Therefore, as is mostly the case for solid ionic substances dissolving in water, solubility increases with temperature. Potassium chloride shows lower solubility at 50°C as only about 39 grams of KCl can dissolve in 100.0 mL of water at that temperature.

Although it is most common for the solubility of solids dissolved in liquids to increase with temperature, the solubility of gases varies oppositely. Gases dissolve to a greater degree in a liquid as the temperature decreases. Recognition that a major step in the process of dissolving a gas solute in a liquid solvent requires the gas to condense (remember that solutes are the substances that change phase) suggests that lowering the temperature would aid in this process. In the same light, raising the pressure of a gaseous solute above a liquid solvent would also increase the solubility of the gas in the liquid as condensation is encouraged when the pressure of the gas is increased. Since solids and liquids are already condensed phases of matter, changes in pressure have little impact on the solubility of a solid in a liquid. The chart below summarizes these solubility trends.

Phases Involved:	Temperature Effect	Pressure Effect
Solid dissolved in a liquid	Solubility generally increases with temperature.	No significant influence on solubility.
Gas dissolved in a liquid	Solubility generally decreases with temperature.	Solubility generally increases as the pressure of the gas increases.

Factors That Affect Rate of Solution Making (How Fast They Go into Solution)

The following procedures increase the rate of solution making.

Pulverizing   increases surface exposed to solvent.

Stirring   brings more solvent that is unsaturated into contact with solute.

Heating   increases molecular action and gives rise to mixing by convection currents. (This heating affects the solubility as well as the rate of solution making.)

SEMI-QUANTITATIVE EXPRESSIONS OF CONCENTRATION

The solubility of a substance displayed by a solubility curve describes the maximum amount of solute that can be dissolved in a given amount of solvent at a particular temperature. Since solutions are fundamentally physical mixtures, there is no set ratio concerning the amount of solute and solvent in a solution. The solubility of a substance is recognition, however, that (in most cases) there are limits. When less than the amount of solute that can be dissolved in the solvent is dissolved, the solution is said to be **unsaturated**. Similarly, when the maximum amount of solute that can be dissolved in the solvent is dissolved, the solution is called **saturated**. Interestingly, there are, for certain solute/solvent combinations, situations where more than the typical amount of solute that can be dissolved in the solvent is dissolved; such a solution is described as **supersaturated**. With solids dissolved in liquids, this situation can occur in a quasi-stable manner when solvents are combined with more solute than can normally be dissolved and heated to get the undissolved solute to go into solution. Upon cooling, the "extra" solute can remain in solution, although somewhat tenuously.

Independent of the solubility of solute and solvent, the term **dilute** expresses that a small amount of solute is dissolved in the solvent for a given solution. When a large amount of solute is dissolved, the solution is described as **concentrated**. The exact amount of solute dissolved in each case is not specified, but chemists can easily apply appropriate terms to solutions as they gain experience in solution making and usage. It is common for inexperienced chemists to confuse dilute with unsaturated as well as concentrated with saturated, but they are not the same. A solution can be both saturated and dilute if the solubility of the solute in the solvent is low. If the solubility is low, large amounts of solute cannot be dissolved, and the solution is necessarily dilute even though the maximum amount of dissolvable solute has been reached. Conversely, solutes with high solubility in a solvent may not reach the maximum amount of dissolvable solute but still contain large amounts of solute and hence be described as both unsaturated and concentrated.

The term **soluble** is used to describe a solute when a reasonable amount of solute can dissolve in a solvent. **Insoluble** means this is not the case. Unlike the terms "dilute" and "concentrated," "soluble" and "insoluble" have a benchmark that differentiates them. Chemists generally recognize a solute as soluble if 0.10 moles of the solute can be dissolved in the amount of solvent needed to make 1.0 liter of solution.

RULES FOR SOLUBILITY (CONCERNING IONIC SOLIDS IN WATER AT ROOM TEMPERATURE)

TIP

You should be familiar with these general rules of the solubility of solids.

Compounds containing:

- nitrate, acetate, or chlorate are generally **soluble**.
- sodium, potassium, and ammonium are generally **soluble**.
- chlorides are generally **soluble**, with noteworthy exceptions being those also containing silver, mercury (I), and lead (II).
- sulfates are generally **soluble**, with noteworthy exceptions being those also containing lead (II), barium, strontium, and calcium.
- carbonate, phosphate, silicate, and sulfide are generally **insoluble**, with noteworthy exceptions being those also containing sodium, potassium, and ammonium.
- hydroxides are generally **insoluble**, with noteworthy exceptions being those also containing sodium, potassium, ammonium, calcium, barium, and strontium.

QUANTITATIVE EXPRESSIONS OF CONCENTRATION

The semi-quantitative terms referring to the amounts of solution parts present in a solution lack the precise information chemists often need about them. There are ways, however, to describe particularly the amounts of those solution parts useful in common chemical calculations.

Percent by mass considers the masses of the solution parts. With reference to the solute, it is defined as

$$\frac{\text{Grams of solute}}{\text{Grams of solution}} \times 100\% = \text{percent by mass (solute)}$$

➡️ **Example 1** _____

At 25°C, a saturated solution of potassium nitrate contains 45 grams of KNO_3 dissolved in 100.0 grams of water. What is the percent by mass (with regards to KNO_3) in this solution?

To solve this problem one needs to recognize that KNO_3 is the solute and has a mass of 45 grams and the mass of the solution (both solute and solvent) is 145 grams.

$$\frac{45\ g}{145\ g} \times 100\% = 31\%$$

➡️ **Example 2** _____

How many grams of KNO_3 and water are needed to prepare 300. grams of a 5% solution?

To solve this problem one must first find the mass of solute needed and then calculate the mass of solvent by difference.

$$\frac{5}{100} \times 300\ \text{grams} = 15\ \text{grams solute}$$

Since 15 grams of KNO_3 is needed, then 285 grams of water will be required to make 300. grams of solution.

Molarity (abbreviated M) considers the moles of solute dissolved and the volume of the overall solution; it is defined as

$$\frac{\text{Moles of solute}}{\text{Liter of solution}} = \text{Molarity}$$

➡️ **Example 1** _____

What is the molarity of a solution in which 10.5 grams of KNO_3 is dissolved in enough water to make 100.0 mL of solution?

In order to solve this problem, the mass of the solute needs to be converted into moles via dimensional analysis and then divided by the volume of the solution.

$$\frac{10.5\ \text{grams}}{1} \times \frac{1\ \text{mole}}{101\ \text{grams}} = 0.104\ \text{moles}$$

$$\frac{0.104\ \text{moles}}{0.1000\ \text{L}} = 1.04\ \text{M}$$

This molarity of the KNO_3 solution is also the molar concentration of the K^+ ion as well as the NO_3^- ion in solution because the formula of potassium nitrate indicates that there is one ion of potassium and one ion of nitrate per formula unit of KNO_3 dissolved. If the formula of an ionic substance shows multiple ions are present for each formula unit, then the subscript for each ion indicates the number the molarity should be multiplied by in order to describe the concentration of each ion in solution.

➡ Example 2

If a 47.60 g sample of $MgCl_2$, with a molar mass of 95.20 g/mol, is dissolved in enough water to make 250.0 mL of solution, determine the molarity (i.e., molar concentration) of each ion in the solution.

Using mental math, the molarity of the solution with regard to the $MgCl_2$ would be 2.000 M. This is because 47.60 g of $MgCl_2$ is equal to half a mole of magnesium chloride with that compound dissolved in a quarter liter of solution. Since *one-half* is twice as big as *one-quarter*, the molarity of the solution described is 2.000 M (considering the number of significant figures given). The molar concentration of the magnesium ion, Mg^{2+}, would be the same, as there is one magnesium ion per formula unit. The molar concentration of the chloride ion, Cl^-, would be 4.000 M, as there are two chloride ions per formula unit.

➡ Example 3

What mass of KNO_3 would be needed to produce 250 mL of solution with a concentration of 0.500 M, and how would this solution be produced in the laboratory?

This is a common calculation chemists do when solving problems associated with solutions. A rearrangement of the molarity equation above yields

$$\text{Molarity} \times \text{Volume (in liters)} = \text{Moles of solute}$$

Once moles of solute are determined, the amount must be converted into grams via dimensional analysis.

$$0.500 \text{ M} \times 0.250 \text{ L} = 0.125 \text{ moles of solute}$$

$$\frac{0.125 \text{ moles}}{1} \times \frac{101 \text{ grams}}{1 \text{ mole}} = 12.6 \text{ grams}$$

To produce this solution, 12.6 grams of KNO_3 would be placed in a **volumetric flask** and enough water added to bring the solution to a volume of 250 mL. A volumetric flask is a special device commonly used in the laboratory to make solutions. Precisely etched in the neck of the flask is *only one* graduation, which defines the volume of the solution in the container.

Dilution

Concentrated solutions are often bought or made as stock solutions in chemical laboratories for commonly dissolved substances. Solutions that are less concentrated (or more dilute) than those are also often needed, and so problems associated with dilution are often encountered. Solving a dilution problem, once again, takes advantage of the molarity formula as rearranged above, where molarity times volume equals the number of moles of solute in the solution. To solve a dilution problem, one simply needs to keep in mind that the moles of solute received from the more concentrated solution stay the same in the diluted solution; they simply exist in a greater volume. From this it should be easy to rationalize the dilution equation:

$$M_c V_c = M_d V_d$$

where M_c and V_c represent the molarity and volume in the concentrated case and M_d and V_d represent the molarity and volume in the diluted case. In either case, the number of moles remains the same (i.e., is equal) because all that is involved in dilution is taking a certain volume of the concentrated solution and adding some water to it to make it less concentrated. In other words, if the molarity goes down, the volume must go up to keep the number of moles the same. The equation is particularly useful because typically the molarity and volume for the desired, less concentrated solution is known, as is the molarity of the more concentrated solution. Therefore, the volume of original solution to be diluted is easily found and placed in a volumetric flask to which water is added to complete the dilution.

➡ Example

A solution of hydrochloric acid (HCl dissolved in H_2O) is commonly purchased in large quantities at a concentration where it is saturated, 12.1 M. Solutions of less than 12.1 M are typically used, however. Describe how to make 0.500 L of solution of 3.00 M HCl from the 12.1 M stock solution.

Values for three of the four variables found in the dilution equation are provided in the problem:

$$M_c = 12.1 \text{ M} \qquad V_c = ? \qquad M_d = 3.00 \text{ M} \qquad V_d = 0.500 \text{ L}$$

Plugging these values into the equation gives

$$12.1 \text{ M} \times V_c = 3.00 \text{ M} \times 0.500 \text{ L}$$

Solving for the volume of the original (more concentrated) solution needed gives

$$V_c = 0.124 \text{ L}$$

Therefore, in order to produce the diluted solution, 0.124 L (i.e., 124 mL) of the concentrated HCl solution should be transferred via a pipette to a 0.500 L (i.e., 500 mL) volumetric flask and enough water added to bring the solution up to the line on the flask.

COLLIGATIVE PROPERTIES OF SOLUTIONS

Properties of solutions that show predictable variation from those of the solvent itself and that are dependent primarily on the number of solute particles present in the solution (not particularly on the identity of those particles) are called **colligative properties**.

Vapor pressure is one of those properties. Whenever a solute is dissolved in a solvent, new interactions between the particles present ensue. Generally, in the case of a solid or liquid solute dissolved in a liquid solvent, the interactions are such that the solvent is inhibited from evaporating to the same degree it had in its pure state, and so the vapor pressure of the solution is lower than that of the pure solvent. This phenomenon, called **vapor pressure lowering**, is displayed graphically below as a portion of a phase diagram that has been modified to show the vapor pressure versus temperature for *both* pure water and a solution in which water is the solvent. Notice that the vapor pressure increases as the temperature does for both solution and pure solvent, but at any given temperature the vapor pressure of the solution is lower than that of the pure water. If you recall that boiling occurs when the vapor pressure of a liquid is equal to the atmospheric pressure, it becomes evident that the boiling point of a solution will be different from the boiling point of the pure liquid solvent because the vapor pressure has been lowered. This is also shown graphically below. The higher temperature required to make the vapor pressure of the solution equal to the atmospheric pressure (normally 760 mm of Hg) causes the boiling point to be raised. This phenomenon, called **boiling**

point elevation, is commonly taken advantage of to get liquids to boil at higher temperatures. Since foods cook faster when the water they are cooked in is hotter, chefs often put salt into the water to raise its temperature.

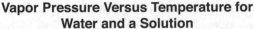

Vapor Pressure Versus Temperature for Water and a Solution

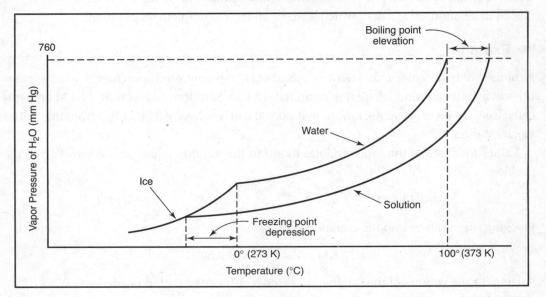

In addition to elevating the boiling point of a liquid solvent, added solutes change the freezing point of the solvent as well. This is also seen graphically above. Freezing occurs when the vapor pressure of a solid is equal to the vapor pressure of the liquid with which it is associated. Since the vapor pressure of the liquid solution is lowered relative to that of the pure solvent, the liquid will not freeze at the normal freezing point because the liquid and the solid no longer share a vapor pressure value at that temperature. Instead, the temperature must be lowered for this to be true. Upon lowering the temperature, the vapor pressure lowers in both the solid and liquid solution, but the solid's vapor pressure lowers to a greater extent with each degree drop in temperature. Consequently, there will be a temperature at which the vapor pressures are again equal and a new, lower freezing point is realized. One common practice that takes advantage of this phenomenon, called **freezing point depression**, is the wintertime use of salt to melt ice on streets and roads.

As previously mentioned, the extent to which solutes lower the vapor pressure, elevate the boiling point, and depress the freezing point of solvents is dependent simply on the number of particles dissolved in the solvent, not on the particular identity of the solute particles found in the solution. But the nature of the solute does make a difference in terms of the relative degree to which the effect is noticed. This is due to the fact that some solutes dissolve and remain principally intact as the particle they were before the dissolution process. Others dissociate upon dissolution and produce a larger number of particles having a greater impact on the colligative properties.

Most molecular solutes that dissolve in a solvent will not dissociate into multiple particles, and so the molecules remain whole and homogeneously mixed with the solvent particles upon dissolving. In the case of water as a solvent, most polar molecules tend to dissolve well but don't break apart to an appreciable degree. This means that the vapor pressure of the solution will be less modified from the pure water compared to when an equal number of formula units of an ionic substance might dissolve in the same amount of water. This is so because the formula units of ionic substances break apart into the individual ions that make up the formula unit when the ionic substance is dissolved.

➥ Example

Solutions made from 0.50 moles of both table sugar (sucrose) and table salt (sodium chloride) dissolved in 1.0 kg of water will exhibit higher boiling points than the water itself. Will both solutions have the same boiling point, and if not, which of the solutions will exhibit a higher boiling point?

Despite the fact that an equal number of moles of each solid are being dissolved in the same mass of water, the boiling points of these solutions will not be the same. While sucrose ($C_{12}H_{22}O_{11}$) is a molecular substance that dissolves in water because it is a polar molecule, it does not dissociate into ions. Therefore, there will be 0.50 moles of sucrose molecules dissolved in the 1.0 kg of water once the solution is made. Sodium chloride (NaCl) is an ionic substance that not only dissolves in water but, as a part of its solution process, dissociates into two ions, Na^+ and Cl^-. Therefore, there will be 1.0 mole of ions (0.50×2) present in the 1.0 kg of water once the solution is made. Since the salt solution has a larger number of particles dissolved in the same mass of water, the salt solution will have a higher boiling point. Colligative properties are independent of the identity of the dissolved particles and are dependent only on their total number.

CHAPTER SUMMARY

The following terms summarize all the concepts and ideas that were introduced in this chapter. You should be able to explain their meaning and how you would use them in chemistry. They appear in boldface type in this chapter to draw your attention to them. The boldface type also makes it easier for you to look them up if you need to. You could also use the Internet search engine *google.com* on your computer to get a quick and expanded explanation of these terms, laws, and formulas.

alloy	endothermic	molarity
boiling point	efflorescence	phase equilibrium
Brownian movement	exothermic	polarity
colligative property	heat of fusion	saturated
concentrated	heat of vaporization	solute
critical pressure	"heavy" water	solvent
critical temperature	hydrate	sublimation
crystal	hydrogen bonds	supersaturated
deliquescent	hygroscopic	surface tension
dilute	melting point	unsaturated
dynamic equilibrium	miscible/immiscible	viscosity

Online content that reinforces major concepts discussed in this chapter can be found at the following Internet addresses if they are still available. *Some may have been changed or deleted.*

The Chemistry of Water

http://www.nsf.gov/news/special_reports/water/index_low.jsp

This is a National Science Foundation site that offers a special report on the chemistry of water. It has beautiful artwork to supplement the topics.

Phase Diagrams

http://www.chemguide.co.uk/physical/phaseeqia/phasediags.html

This website offers clear explanations and interpretations of the types of phase diagrams that a first-year chemistry student should be familiar with.

1. The shape and volume of what phase are fixed and not dependent on the container housing the material?

   (A) Solid
   (B) Liquid
   (C) Gas
   (D) Plasma

2. When two different substances existing at the same temperature and pressure, one a solid and one a liquid, are compared, the average kinetic energy of

   (A) the liquid will be higher
   (B) the solid will be higher
   (C) both will be the same
   (D) both will be less than a gas under similar conditions

3. Which of the substances listed is likely to have the highest boiling point?

   (A) He
   (B) H_2
   (C) N_2
   (D) HF

4. When a list of substances is considered, the one with the smallest ΔH_{vap} is likely to be

   (A) the one with the highest viscosity
   (B) the one with the smallest vapor pressure
   (C) the one with the lowest boiling point
   (D) the one with the highest critical point

5. Copper, Cu, has a higher melting point than methane, CH_4, because

   (A) metallic bonds are generally stronger than ionic bonds
   (B) metallic bonds are generally weaker than ionic bonds
   (C) metallic bonds are generally weaker than covalent bonds
   (D) metallic bonds are generally stronger than intermolecular bonds for small molecules

6. Which substance is likely to be most soluble in liquid hexane?

   (A) Benzene
   (B) Sodium chloride
   (C) Hydrogen chloride
   (D) Ammonia

7. Which of the following has a temperature associated with it that cannot change for a substance?

   (A) Boiling point
   (B) Melting point
   (C) Triple point
   (D) Freezing point

8. Insoluble substances can

   (A) form only concentrated solutions
   (B) form only unsaturated solutions
   (C) form both saturated and dilute solutions
   (D) form only saturated solutions

9. The molar concentration of the nitrate ion in a 1.5 molar solution of aluminum nitrate is

(A) 6.0 M
(B) 4.5 M
(C) 3.0 M
(D) 1.5 M

10. A chemist who wanted to make a 150.0 mL sample of 0.30 M sucrose solution from a 0.45 M sucrose solution would require what volume of the higher concentration to do so?

(A) 30.0 mL
(B) 45.0 mL
(C) 75.0 mL
(D) 100.0 mL

11. The boiling point of a sample of water would rise to the highest degree with the addition of

(A) 0.50 mol NaCl
(B) 0.25 mol $CaCl_2$
(C) 0.75 mol $C_{12}H_{22}O_{11}$
(D) 0.35 mol $AlCl_3$

Answers and Explanations

1. **(A)** Solid substances have definite volumes and shapes and are independent of the device holding them. A liquid has a definite volume but a variable shape. Gases have both variable shape and volume.

2. **(C)** The average kinetic energy of both samples would be the same because the average kinetic energy of a material is simply dependent upon the temperature of the sample, which was described as the same for both the solid and the liquid.

3. **(D)** The boiling point of HF would be the highest because it exerts the highest attractive force strength between the pertinent particles. HF exhibits hydrogen bonding, the others all exhibit weaker intermolecular bonds.

4. **(C)** The smallest value for the heat of vaporization for a substance is associated with the substance having the weakest bonds. The substance having the weakest bonds is the one associated with the lowest boiling point. A low vapor pressure suggests high attractive force strength, as does a high viscosity and a high critical point.

5. **(D)** Metallic bonds are generally stronger than the intermolecular bonds between small molecules. Copper is a solid at room temperature, and methane is a gas. Metallic bonds *are* generally weaker than covalent bonds or ionic bonds (as stated in choices B and C), but neither covalent nor ionic bonds dictate the properties of methane gas. The properties of methane are dictated by London dispersion forces (a type of intermolecular bond).

6. **(A)** The general rule for solubility is "like dissolves like." Hexane is a nonpolar molecule and will dissolve benzene and other nonpolar molecules. All the other choices are polar molecules.

7. **(C)** The triple point is the *one* combination of temperature and pressure for a substance in which all three phases of the substance are in equilibrium. Boiling, melting, and freezing points are all temperatures that can vary with pressure for a given substance.

8. **(C)** When a substance is insoluble in a solvent, a solution with a reasonably appreciable concentration *cannot* be produced. "Dilute" implies only a small amount of solute is dissolved. "Saturated" means that the maximum amount of solute that can be dissolved is dissolved. Therefore, a dilute solution can also be saturated because of the solute's low solubility.

9. **(B)** Since the formula for aluminum nitrate is $Al(NO_3)_3$, there are 3 nitrate ions per formula unit. Since the molarity in reference to the formula unit is 1.5 M, the molar concentration of the nitrate would be three times that, or 4.5 M.

10. **(D)** This is a dilution problem. Using the equation $M_c V_c = M_d V_d$, the volume of the more concentrated solution needed, V_c, can be found.

$$(0.45 \text{ M})V_c = (0.30 \text{ M})(150.0 \text{ mL})$$

$$V_c = 100.0 \text{ mL}$$

11. **(D)** Boiling point elevation is a colligative property. Colligative properties are dependent on the number of solute particles present in the solution. Since $AlCl_3$ is a soluble ionic substance that dissociates into 4 particles, the molar concentration in reference to the ions turns out to be 1.40 M. The molar concentrations of the other substances, which concern dissolved particles, are lower.

Chemical Reactions and Thermochemistry

8

These skills are usually tested on the SAT Subject Test in Chemistry. You should be able to...

→ Identify the driving force for these four major types of chemical reactions, and write balanced equations for each: combination (or synthesis), decomposition, single replacement, and double replacement.

→ Identify and explain graphically enthalpy changes in exothermic and endothermic reactions.

→ Solve calorimetry/heating curve problems.

→ Use Hess's Law to show the additivity of heats of reactions.

This chapter will review and strengthen these skills. Be sure to do the Practice Exercises at the end of the chapter.

Many of the kinds of reactions you may encounter in a first-year chemistry course can be placed into four basic categories: synthesis, decomposition, single replacement, and double replacement.

The first type, **synthesis**, is also known as **direct combination**. A synthesis reaction involves the making of a more complex product from the union of two or more simpler substances. A *particular* type of synthesis is a **formation reaction**, in which one mole of a compound is made; that is, a more complex product is formed from the direct combination of its elements, the elements being simpler substances in the manner in which they standardly exist. All formation reactions are synthesis reactions, but not all synthesis reactions are formation reactions. Some examples of synthesis reactions (which also happen to be formation reactions) are

$$Zn(s) + S(s) \rightarrow ZnS(s)$$
$$Ca(s) + Cl_2(g) \rightarrow CaCl_2(g)$$
$$C(s) + O_2(g) \rightarrow CO_2(g)$$

The second type of reaction, **decomposition**, can also be referred to as **analysis**. This means the breakdown of one reactant, often a compound, to yield two or more products. These products can be simpler compounds or individual elements. In this regard, a decomposition reaction can be viewed as the reverse of a synthesis reaction. Some examples of this type are:

$$2H_2O(\ell) \rightarrow 2H_2(g) + O_2(g) \text{ (electrolysis of water)}$$
$$C_{12}H_{22}O_{11}(s) \rightarrow 12C(s) + 11H_2O(\ell)$$
$$2HgO(s) \rightarrow 2Hg(s) + O_2(g)$$

The third type of reaction is called **single replacement** or **single displacement**. In this type, one reactant (an element) is exchanged with an ion in a second reactant (an ionic compound). There are two types of single replacement reactions, the type depending on whether the elemental reactant is a metal or a nonmetal. The first two examples below represent *metal* single replacement reactions, and the third represents a *nonmetal* single replacement reaction. Some examples are:

$$Fe(s) + CuSO_4(aq) \rightarrow FeSO_4(aq) + Cu(s)$$
$$Zn(s) + H_2SO_4(aq) \rightarrow ZnSO_4(aq) + H_2(g)$$
$$Cl_2(g) + 2NaBr(aq) \rightarrow 2NaCl(aq) + Br_2(\ell)$$

The last type of reaction is called **double replacement** or **double displacement** because there is an actual exchange of "partners" to form new compounds. Some examples are:

$$AgNO_3(aq) + NaCL(aq) \rightarrow AgCl(s) + NaNO_3(aq)$$
$$H_2SO_4(aq) + NaOH(aq) \rightarrow Na_2SO_4(aq) + 2H_2O(\ell) \text{ (neutralization)}$$
$$BaCl_2(aq) + Na_2SO_4(aq) \rightarrow BaSO_4(s) + 2NaCl(aq)$$

THERMOCHEMISTRY

In general, chemical reactions either liberate or absorb heat. The energy changes in a reaction are due, to a large extent, to the changes in potential energy that accompany the alterations in the interactions between particles (particularly the breaking and making of chemical bonds). A more fundamental origin of chemical energy lies in the potential and kinetic energy of the atoms, molecules, and subatomic particles that make up a sample, in addition to the situation in which the sample exists—that is, with respect to pressure and volume. Each of these factors contributes to the total capacity these particles have to exchange heat; this property is referred to as **enthalpy**. Enthalpy is symbolized by the letter H. Although it is not possible to measure the enthalpy of a chemical system in an absolute manner, it is possible to measure the *change* in this property as a system changes; the change is consequently symbolized as ΔH. ΔH is often referred to as the heat of reaction. Changes in enthalpy for exothermic and endothermic reactions can be shown graphically, as in the examples below.

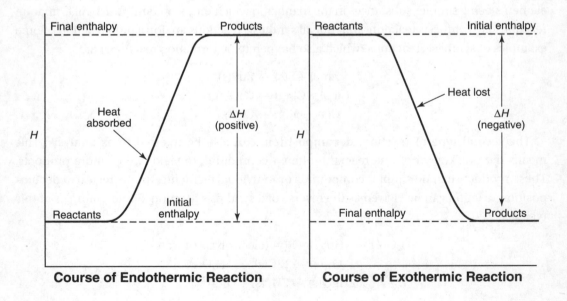

Course of Endothermic Reaction   **Course of Exothermic Reaction**

Notice that the ΔH for the endothermic reaction is positive, while that for the exothermic reaction is negative. This is so because chemists always take the perspective of the chemical system when describing change. In other words, in an exothermic process, the system is releasing energy (i.e., the energy of the system is going down). During an endothermic process, the system is absorbing energy (i.e., the energy of the system is going up). Changes in enthalpy are independent of the path, or series of steps, by which a reaction occurs; this form of change is known as a **state function**. ΔH simply describes the difference in the capacity substances have to exchange heat relative to the initial and final state of the system.

PREDICTING REACTIONS

One of the most important topics of chemistry deals with the reasons why reactions take place. Taking each of the previously mentioned reaction types and considering the above-mentioned information about thermochemistry, let's see how a prediction can be made concerning the driving force that makes these reactions occur.

1. Combination (Also Known as Synthesis)

A good source of predictive information (although technically not the only thing to consider) for whether a chemical combination will generate a reaction is a **heat of formation** table. Table A in the Appendix is such a table. The standard heat of formation, symbolized ΔH_f°, is a particular type of heat of reaction and is associated only with formation reactions. Table A shows the amount of energy released or absorbed by the chemical system when one mole of a compound is produced from its elements in their standard states. As can be seen in the table, the phase of the compound forming (solid, liquid, or gas) is important to note, as differences seen in the values depend on the compound's state of matter.

TIP

$-\Delta H$ = exothermic reaction

$+\Delta H$ = endothermic reaction

If the heat of formation is a relatively large negative value, the combination is likely to occur spontaneously, since one driving force for a reaction is for a chemical system to lower its energy. On the other hand, if the heat of formation is a small negative value or even a positive value, the reaction is likely to be non-spontaneous and also likely to require energy to proceed at any noticeable rate. The examples below are for synthesis reactions that are likely to occur.

➡ **Example 1** _____

$$Zn(s) + S(s) \rightarrow ZnS(s) + 202.7 \text{ kJ} \qquad \Delta H_f = -202.7 \text{ kJ}$$

This means that 1 mole of zinc (65 grams) reacts with 1 mole of sulfur (32 grams) to form 1 mole of zinc sulfide (97 grams) and releases 202.7 kilojoules of heat.

➡ **Example 2** _____

$$Mg(s) + \tfrac{1}{2}O_2(g) \rightarrow MgO(s) + 601.6 \text{ kJ} \qquad H_f = -601.6 \text{ kJ/mol}$$

indicates that the formation of 1 mole of magnesium oxide requires 1 mole of magnesium and $\frac{1}{2}$ mole of oxygen with the release of 601.6 kJ of heat. Notice the use of the fractional coefficient for oxygen. If the equation had been written with the usual whole-number coefficients, 2 moles of magnesium oxide would have been produced and twice the amount of energy would have been released.

$$2Mg(s) + O_2(g) \rightarrow 2MgO(s) + 1203.2 \text{ kJ}$$

2. Decomposition (Also Known as Analysis)

The prediction of decomposition reactions uses the same source of information, the heat of formation table. If the heat of formation of the reactant is a large exothermic (ΔH is negative) value, the compound will be difficult to decompose since this same quantity of energy must be returned to the compound. A relatively low and negative heat of formation indicates decomposition would not be difficult, such as the decomposition of mercuric oxide with a $\Delta H_f = -90.8$ kJ/mole:

$$2HgO(s) \rightarrow 2Hg(s) + O_2(g) \text{ (Priestley's method of preparation)}$$

A high positive heat of formation indicates extreme instability of a compound, which can explosively decompose.

3. Single Replacement

A prediction of the feasibility of this type of reaction can be based on a comparison of the heat of formation of the original compound and that of the compound to be formed. It is noteworthy that the heats of formation for the elements, in this type of reaction or others, are not included in the analysis. This is so because their ΔH_f° values, by definition, are zero and thus have no impact in the overall energy analysis. Some examples are

➡ Example 1 _____

$$Zn(s) + 2HCl(aq) \rightarrow ZnCl_2(aq) + H_2(g)$$

Since the value for the heat of formation of HCl is –92.3 kJ per mole and there are two moles of HCl shown as reactants, the value of $2 \times (-92.3$ kJ$)$ would have to be compared to the heat of formation of $ZnCl_2$, which is –415.5 kJ/mol. That comparison shows that the products possess an overall larger negative ΔH value (–415.5 kJ compared to –184.6 kJ). As the reaction is likely to occur with the difference of 230.9 kJ of heat being released, the energy of the chemical system is lowered.

➡ Example 2 _____

In this example, the heat of formation difference is –157.0 kJ, which is the excess to be given off as the reaction occurs:

$$
\begin{array}{cccc}
Fe(s) + & CuSO_4(aq) & \rightarrow & FeSO_4(aq) & + Cu(s) \\
& -771.4 \text{ kJ} & & -928.4 \text{ kJ}
\end{array}
$$

Another way of predicting the spontaneity of single replacement reactions is to check the relative positions of the two elements in the activity series on page 209. If the element that is to replace the other in the compound is higher on the chart, the reaction will occur. If it is below, there will be no reaction.

In predicting the replacement of hydrogen by zinc in hydrochloric acid, as seen in the example above, reference to the activity series shows that zinc will replace hydrogen. Again, predicting this reaction would occur:

$$Zn(s) + 2HCl(aq) \rightarrow ZnCl_2(aq) + H_2(g)$$

In fact, most metals in the activity series would replace hydrogen in an acid solution as hydrogen is fairly low in the activity series. If a metal such as copper were chosen, no reaction would occur, however, as copper is even lower then hydrogen in the series.

$$Cu(s) + HCl(aq) \rightarrow \text{no reaction}$$

A more fundamental reason driving these replacements, where these reactions are viewed as involving the transfer of electrons, is covered in Chapter 12.

Activity Series of Common Elements

		Activity of metals	Activity of halogen nonmetals
Most active	Li		F_2
	Rb	React with cold water and acids, replacing hydrogen. React with oxygen, forming oxides.	Cl_2
	K		Br_2
	Ba		I_2
	Sr		
	Ca		
	Na		
	Mg		
	Al	React with steam (but not cold water) and acids, replacing hydrogen. React with oxygen, forming oxides.	
	Mn		
	Zn		
	Cr		
	Fe		
	Cd		
	Co	Do not react with water. React with acids, replacing hydrogen. React with oxygen, forming oxides.	
	Ni		
	Sn		
	Pb		
	H_2		
	Sb	React with oxygen, forming oxides.	
	Bi		
	Cu		
	Hg		
	Ag	Fairly unreactive, forming oxides only indirectly	
	Pt		
Least active	Au		

4. Double Replacement

TIP

Know these three conditions for going to completion.

For double replacement reactions to go to completion, that is, proceed until the supply of one of the reactants is exhausted, one of the following conditions must be present: (1) an insoluble precipitate is formed, (2) a nonionizing substance is formed, or (3) a gaseous product is given off.

1. To predict the formation of an **insoluble precipitate**, you should have some knowledge of the solubilities of compounds as discussed in Chapter 7. Table 9 gives some general solubility rules.

TIP

Know the general solubility rules.

Table 9. Solubilities of Compounds

Soluble	Except
Na^+ NH_4^+ $\Big\}$ compounds K^+	
Acetates	
Bicarbonates	
Chlorates	
Chlorides..	Ag, Hg, Pb ($PbCl_2$, sol. in hot water)
Nitrates	
Sulfates...	Ba, Ca (slight), Pb
Insoluble	
Carbonates, phosphates..	Na, NH_4, K compounds
Sulfides, hydroxides..	Na, NH_4, K, Ba, Ca

An example of this type of reaction is found when a solution of potassium chloride reacts with a solution of silver nitrate.

$$KCl(aq) + AgNO_3(aq) \rightarrow AgCl(s) + KNO_3(aq)$$

When the reaction is given in its complete ionic form,

$$K^+(aq) + Cl^-(aq) + Ag^+(aq) + NO_3^-(aq) \rightarrow AgCl(s) + K^+(aq) + NO_3^-(aq)$$

it's easy to see that the silver ion combines with the chloride ion to make the insoluble precipitate silver chloride; the potassium and nitrate ions are not really involved in the reaction as they are not changing. They are referred to as *spectator ions*.

2. Another reason for a reaction of this type to go to completion is the formation of **a nonionizing product** such as water. This weak electrolyte keeps its component ions in molecular form and thus eliminates the possibility of reversing the reaction. All neutralization reactions are of this type.

$$H^+(aq) + Cl^-(aq) + Na^+(aq) + OH^-(aq) \rightarrow H_2O(\ell) + Na^+(aq) + Cl^-(aq)$$

This example shows the ions of the reactants, hydrochloric acid and sodium hydroxide, and the nonelectrolyte product water with sodium and chloride ions in solution. Since the water does not ionize to any extent, the reverse reaction cannot occur.

3. The third reason for double displacement to occur is the **evolution of a gaseous product.** An example of this is calcium carbonate reacting with hydrochloric acid:

$$CaCO_3(s) + 2HCl(aq) \rightarrow CaCl_2(aq) + H_2O(\ell) + CO_2(g)$$

Viewed from an exchange-of-partners perspective, the water and the carbon dioxide in the products above are the decomposition products of H_2CO_3, the compound that would have formed along with the $CaCl_2$ when the exchange of partners took place.

Another example of a double replacement reaction that evolves a gaseous product is the reaction of sodium sulfite with an acid:

$$Na_2SO_3(aq) + 2HCl(aq) \rightarrow 2NaCl(aq) + H_2O(\ell) + SO_2(g)$$

In general, acids combining with carbonates or sulfites are good examples of this type of reaction.

Entropy

Predicting the spontaneity of reactions with enthalpy change has been useful in many of the preceding reactions, as lowering the energy of a chemical system is *one* of the fundamental driving forces in nature that causes change. There is, however, another property of a system whose change *also* contributes to the spontaneity of a process. This property, called **entropy**, measures the disorder of a system. In this case, however, an increase in the property promotes change. Unlike enthalpy, the absolute entropy of a system *can* be measured. This is so because a benchmark for complete lack of disorder exists naturally in a perfect crystal at a temperature of 0 kelvins (absolute zero). The combination of the change in enthalpy and the change in entropy constitutes a complete view in which the spontaneity of a process can be determined. It is because entropy change was not considered in reference to most previous reactions that the word "likely" was used to preface the predictions. Chapter 10, page 240, gives a fuller, more quantitative treatment of entropy change, but the present focus will be on enthalpy changes in predicting the occurrence of reactions.

TIP

Entropy is a measure of the degree of disorder.

DETERMINING CHANGES IN ENTHALPY

There are two ways in which chemists routinely determine ΔH values. One is experimentally using calorimetry; the other is theoretically using Hess's Law.

Calorimetry

A **calorimeter** is a laboratory device used to measure the enthalpy change in a chemical or physical process. There are two types of calorimeters, one open, the other closed. In open calorimetry, the chemical system is open to the atmosphere and, as such, is subject to a constant pressure. In this case, the heat that flows into or out of a system is equal to the change in enthalpy. This is not the case in closed calorimetry. A simple open calorimeter consists of an insulating container (often just a Styrofoam cup) housing a known mass of water, along with a thermometer and stirring device. Since a **calorie** is a unit of energy (described in Chapter 1) and is defined as the amount of energy needed to raise the temperature of 1 gram of water by 1°C, a calorimeter, as its name implies, *measures energy* change. Based on this definition, the **specific heat** for water—that is, the particular amount of energy needed to raise a particular amount of water by a certain degree—is 1 cal/g°C. Since a joule, the SI unit for energy, is 4.186 times smaller than the calorie, the specific heat of water is also 4.186 J/g°C. Because of the known specific heat of water, the amount of energy change involved in a process can easily be calculated by monitoring the temperature of the water in the calorimeter present to the reaction to see how much it changes during the process. The calculation uses the formula

$$q = mc\Delta T$$

where q denotes the heat flow into or out of the water, m stands for the mass of the water, c represents the specific heat of the water, and ΔT its temperature change.

➡ Example 1

Find the change in enthalpy, ΔH, in kJ/mol for the dissolution of NaOH in water; assume that when 0.10 moles (4.0 g) of the sodium hydroxide is dissolved in 96.0 g of water, the temperature of the water rises from 25.0°C to 34.9°C. Also, assume the specific heat of the overall solution is the same as water and that the overall solution (not just the water) will be what absorbs or releases energy from or to the system. This is typically done in fairly dilute solutions.

Using the equation to find the heat flow *into* the solution (this is the case since the temperature of the solution went up) we find that

$$q = (100.0 \text{ g})(4.184 \text{ J/g°C})(9.9°C)$$
$$q = 4,100 \text{ J}$$

This value represents the energy *gained* by the solution (mostly the water) and therefore the amount of energy *lost* by the chemical system. This value also represents the change in enthalpy for the reaction involving the 4.0 grams of NaOH used (which is only 0.10 moles of NaOH). Recall, however, that heat-of-reaction values are typically reported in units of kJ per mole. Therefore, it would be typical to change the sign on the heat flow, q, from positive to negative and the value from joules to kilojoules. Finally, dividing that amount of energy by 0.10 moles allows for representation of the reaction in a typical manner:

$$NaOH(s) \rightarrow Na^+(aq) + OH^-(aq) + 41 \text{ kJ}$$

or

$$\Delta H = -41 \text{ kJ/mol}$$

The making of this solution would be an exothermic process because the heat of reaction is negative. As 41 kJ of energy would be released if it was to occur, it is therefore shown as a product. The production of this solution would feel warm to the touch.

➡ Example 2

0.050 L of a 1.0 M solution of NaOH is poured into 0.050 L of a 1.0 M solution of HCl already present in a calorimeter. The temperature of both solutions is 20.0°C initially, but once they are combined in the calorimeter, the temperature rises to 27.0°C. Find ΔH for this double replacement/acid-base/neutralization reaction. Assume the combination of both solutions produces a fairly dilute solution (mostly water) with a specific heat that would be the same as water. Also, assume that the density of all solutions is the same as water: 1 g/mL.

The mass of the total solution would be 100.0 grams, with the assumption above that each of the combined solutions is 50.0 g (as their volumes are 50.0 mL); therefore,

$$q = (100.0 \text{ g})(4.184 \text{ J/g°C})(7.0°C)$$
$$q = 2,900 \text{ J}$$

To associate this change in enthalpy with the chemical reaction, the sign would have to be changed to negative (as what the solution gained was lost by the chemical system), the units changed to kilojoules, and the value divided by the number of moles of NaOH or HCl used in the reaction (both of which are the limiting reactant). Recall from the previous chapter that according to the definition of molarity, the number of moles of a solute present in a solution is found by multiplying the molarity of the solution by its volume. Since each solution started out at 1.0 M and 0.050 L of it was present, the number of moles of either reactant used is 0.050 moles. Therefore,

$$\Delta H = -58 \text{ kJ/mol (i.e., } -2.9 \text{ kJ/0.050 mol)}$$

or

$$\text{NaOH(aq)} + \text{HCl(aq)} \rightarrow \text{NaCl(aq)} + \text{H}_2\text{O}(\ell) + 58 \text{ kJ}$$

Heating and Cooling Curves

Besides calorimetry, the equation $q = mc\Delta T$ is used in determining the amount of energy needed to raise the temperature of any substance, not just water in a calorimeter. This is typically the case when evaluating the amount of energy needed to take a substance in the solid state, below its freezing point, to the gaseous state, above its boiling point. Figure 30 provides a visual depiction of the process *for water* and is called a **heating curve**, as heat energy is constantly being supplied to the water as time goes by. If viewed in reverse, with energy being removed as time goes by and the gas caused to be turned into a solid, it would be called a **cooling curve**.

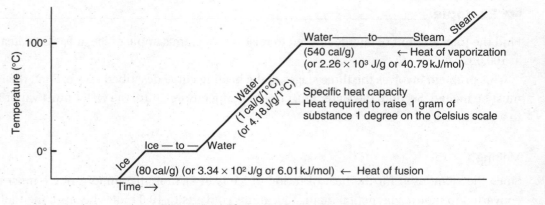

TIP

Water's heat of vaporization = 40.79 kJ/mol

TIP

Water's heat of fusion = 6.01 kJ/mol

Figure 30. Changing Ice to Steam

Ice changing to water and then to steam is not a process of continuous and constant change of temperature as time progresses; rather, it is accomplished in stages. Although there are really five stages to analyze when changing a solid below its freezing point to a gas above its boiling point, this graph supplies only numerical information to calculate the amount of energy needed to take ice *at* its freezing point to steam *at* its boiling point. Therefore, we will evaluate only these three energy changes for the system.

Starting at 0°C, water's normal melting point, the ice begins to melt. Notice that the temperature is not increasing despite the constant supply of heat as time passes. The lack of temperature increase indicates that the kinetic energy of the system is not changing; this is so because temperature and kinetic energy are directly related to each other (see Chapter 7). If the energy is not being used to make the particles move more, what is it doing? The answer is that it is overcoming the attractive forces between the solid particles that hold them in fixed crystalline positions and thus allowing them to exist as non-rigid clusters of particles characteristic of the liquid phase. Consequently, the potential energy of the system must be on the rise, as the energy is being used to overcome the attractive forces between the particles and accomplish the phase change. The energy required to do this, the **heat of fusion**, was discussed in Chapter 7 as an important property of a solid. The heating curve shows various values, utilizing various units, for the heat of fusion of water. Different substances have values different from those shown for water. An upcoming calculation will use the value 6.01 kJ/mol as the heat of fusion of ice. In light of what has been discussed in this chapter, it is simply another particular ΔH value.

Once the entire sample of the solid has melted, the liquid water's temperature begins to rise as more heat is pumped into the system. Since the temperature of the water is rising, the kinetic energy of the system is rising as well. The $q = mc\Delta T$ equation can be used to calculate the energy required to do this. The temperature will continue to rise until it reaches 100.0°C and the liquid water begins to normally boil.

At the normal boiling point, liquid water turns into gaseous water (steam); once again we see a plateau associated with the temperature change. As before, *KE* (kinetic energy) is not increasing, and so *PE* (potential energy) must be going up. This time, the clusters associated with the liquid are being ripped apart into randomness. The energy required to do this, the **heat of vaporization**, was discussed as well in Chapter 7 as an important property of a liquid. From the heating curve, one value for the heat of vaporization of water is 40.79 kJ/mol.

➡ Example _____

Find the total amount of energy needed to raise a 180. gram sample of ice at 0.0°C to steam at 100.0°C.

This problem involves the three stages of the heating curve described above. First, the ice must be melted. Second, the water must rise in temperature. Third, the water must vaporize to a gas.

Melting

Since the value used for the heat of fusion of ice is in kJ/mol, the amount of ice must be converted to moles. Via mental math, 180. grams converts to 10.0 moles because the molar mass of water is 18.0 grams. Using the heat of fusion, we find that

10.0 moles (6.01 kJ/mol) = 60.1 kJ (needed to melt the ice)

Since the specific heat term has grams in it, the 180-gram value (not moles) should be used for the amount of water in this calculation. ΔT will be 100.0°C as the temperature changes from the normal melting point of water to the normal boiling point of water. Using the q formula, we find that

$$q = (180. \text{ g})(4.18 \text{ J/g°C})(100.0°C) = 75{,}200 \text{ J} = 75.2 \text{ kJ}$$

Boiling

Again, moles will be used for the amount of water because the heat of vaporization noted above incorporates that unit. Using heat of vaporization, we find that

$$10.0 \text{ moles } (40.1 \text{ kJ/mol}) = 401 \text{ kJ (needed to boil the liquid)}$$

To find the total energy required, simply add the energy associated with the various stages investigated. The total energy required to change ice at its melting point to steam at its boiling point is 60.1 kJ + 75.2 kJ + 401 kJ or 535 kJ.

Hess's Law of Constant Heat Summation

As suggested previously, there is a theoretical way to determine the heat of reaction for a process. **Hess's Law** takes advantage of the **First Law of Thermodynamics**, which states that like matter, energy is conserved in chemical and physical changes. Consequently, knowledge of ΔH values for certain reactions allows you to find the ΔH value for a reaction whose heat of reaction you don't know. This is so because the change in enthalpy is constant—irrespective of whether a process takes place in one step or many—because energy is conserved. This idea complements previous ideas in this chapter, which stated that heat of reaction is pathway independent (and is called a state function) in that the total enthalpy change is dependent only on the initial and final enthalpies of the reactants and products.

Stated in a useful manner, Hess's Law allows for reactions about which you know ΔH values to be added to find ΔH values for reactions about which you don't know. This is seen in the examples below.

➡ Example _____

Find the heat of reaction, ΔH, for the process

$$C(s) + O_2(g) \rightarrow CO_2(g)$$

when the heats of reaction for the following processes are provided:

$$C(s) + \tfrac{1}{2} O_2(g) \rightarrow CO(g) \qquad \Delta H = -110.5 \text{ kJ}$$
$$CO(g) + \tfrac{1}{2} O_2(g) \rightarrow CO_2(g) \qquad \Delta H = -283.0 \text{ kJ}$$

Addition of the chemical processes above results in the reaction for which the heat of reaction is desired. The CO seen in both the products of the first reaction and the reactants of the second cancel out upon the addition of the reactions. Furthermore, the $\tfrac{1}{2} O_2(g)$ reactant, seen in both reactions, adds to make present 1 $O_2(g)$, as is found in the target reaction (i.e., the reaction for which the heat of reaction is not given). Carbon, C, is found in the reactants upon addition, and CO_2 is likewise found in the products. Hence, the addition of the two "steps" produces the target reaction. As stated before, energy change is constant; it is not contingent upon the number of steps it takes to accomplish a process. Since the two reactions add up to the target reaction, their ΔH values should add up to that of the overall reaction. The ΔH for the target reaction is then −393.5 kJ.

Sometimes reactions with known ΔH values are given as steps for an overall process but cannot be added together directly to achieve the target reaction. These may require some preliminary mathematical manipulation prior to the addition of the steps. The types of mathematical manipulation include

1. **FLIPPING A REACTION** (i.e., looking at the reaction in reverse). This will cause the substances previously in the reactants to now be in the products and vice versa. As a result, the enthalpy change will become the opposite of what it previously was from a thermodynamic perspective. This is easily shown by changing the sign on ΔH.

2. **MULTIPLYING OR DIVIDING A REACTION.** Since enthalpy is an extensive property, in which the amount of substance involved in a reaction is important to consider, proper numbers of moles of reactants and products need to be seen in the steps to achieve the target reaction upon addition. Whatever value the reaction is multiplied or divided by should also be applied to the ΔH value.

➡ **Example** _____

Find the heat of reaction, ΔH, for the process

$$W(s) + C(s) \rightarrow WC(s) \text{ (the target reaction)}$$

with the heats of reaction for the following processes provided:

$2W(s) + 3O_2(g) \rightarrow 2WO_3(s)$	$\Delta H = -1{,}680.6 \text{ kJ}$
$C(s) + O_2(g) \rightarrow CO_2(g)$	$\Delta H = -393.5 \text{ kJ}$
$2WC(s) + 5O_2(g) \rightarrow 2WO_3(s) + 2CO_2(g)$	$\Delta H = -2{,}391.6 \text{ kJ}$

Addition of the reactions above, as written, would not yield the target reaction, and so appropriate mathematical manipulation must be done. Analysis of the reactions shows that the pertinent substance in both the target reaction and the first step is tungsten, W. Since the target reaction indicates only 1 mole of tungsten reacting but the first step indicates 2 moles of tungsten reacting, the first step must be divided by 2 and rewritten as

$$W(s) + \frac{3}{2} O_2(g) \rightarrow WO_3(s) \qquad \Delta H = -840.3 \text{ kJ}$$

Further analysis of the reactions shows that the pertinent substance in both the target reaction and the second step is carbon, C. Since the target reaction indicates 1 mole of carbon is reacting and the second step indicates the same, no manipulation of step 2 is needed.

$$C(s) + O_2(g) \rightarrow CO_2(g) \qquad \Delta H = -393.5 \text{ kJ}$$

Final analysis of the reactions shows that the pertinent substance in both the target reaction and the third step is tungsten carbide, WC. Since the target reaction indicates 1 mole of tungsten carbide is produced and the third step shows 2 moles of tungsten carbide reacting, the third step must be divided by 2 and flipped.

$$WO_3(s) + CO_2(g) \rightarrow WC(s) + \frac{5}{2} O_2(g) \qquad \Delta H = 1{,}195.8 \text{ kJ}$$

Addition of the three manipulated equations produces the target reaction (after the CO_2, O_2, and WO_3 all cancel) and so the addition of the ΔH values associated with those reactions will give the ΔH for the target reaction.

$$W(s) + C(s) \rightarrow WC(s) \qquad\qquad \Delta H = -38 \text{ kJ}$$

Standard heats of formation (introduced at the beginning of this chapter) are values typically found in the appendix of most chemistry textbooks. They serve as a resource to predict the spontaneity of reactions like synthesis, decomposition, and single replacement. They can also be used as a resource to help determine $\Delta H°$ values for reactions as a less cumbersome alternative to the method shown above. Rooted in Hess's Law, the technique is based on the concept that a standard heat of reaction, $\Delta H°$, is equal to the difference between the total enthalpy of the reactants and the total enthalpy of the products. This can be expressed as follows:

$$\Delta H°_{reaction} = \sum n\Delta H°_{f\,(products)} - \sum n\Delta H°_{f\,(reactants)}$$

where $\sum$ = the summation of

n = the number of moles of the substances involved in the reaction

$\Delta H°_f$ = the standard heat of formation for the substance found in Table A in the Appendix.

➡ Example

Find the standard heat of reaction, $\Delta H°_{reaction}$, for the decomposition of sodium chlorate:

$$2NaClO_3(s) \rightarrow 2NaCl(s) + 3O_2(g)$$

Using Table A in the Appendix yields

$$\Delta H°_f \text{ for } NaClO_3(s) = -358.2 \text{ kJ/mol}$$
$$\Delta H°_f \text{ for } NaCl(s) = -410.5 \text{ kJ/mol}$$
$$\Delta H°_f \text{ for } O_2(g) = 0 \text{ kJ/mol}$$

Using the values above in the equation yields

$$\Delta H°_{reaction} = \sum n\Delta H°_{f\,(products)} - \sum n\Delta H°_{f\,(reactants)}$$
$$\Delta H°_{reaction} = (2 \text{ mol})(-410.5 \text{ kJ/mol}) - (2 \text{ mol})(-358.2 \text{ kJ/mol})$$
$$\Delta H°_{reaction} = -104.6 \text{ kJ}$$

Once known, the heat of reaction can be used to determine the energy released or absorbed in a process involving different amounts of substances relative to what is shown in the reaction equation to which the ΔH value is associated.

➡ Example

Having found the heat of reaction for the decomposition of sodium chlorate,

$$2NaClO_3(s) \rightarrow 2NaCl(s) + 3O_2(g) \qquad \Delta H° = -104.6 \text{ kJ}$$

find the total amount of energy released when a 106.5 gram sample of sodium chlorate is decomposed.

Since the reaction equation describes the amount of energy change associated with 2 moles of $NaClO_3$ decomposing, you must first find the number of moles of $NaClO_3$ represented by 106.5 grams of the substance. A quick check of the periodic table reveals that the molar mass of sodium chlorate is 106.5 grams. Therefore, the amount of $NaClO_3$ decomposing is simply 1.000 mol. Employing dimensional analysis, complete the calculation using the ΔH value associated with the reaction equation:

$$(1.000 \text{ mol } NaClO_3)(-104.6 \text{ kJ}/2 \text{ mol } NaClO_3) = -52.3 \text{ kJ}$$

Since the energy change is shown as a negative value, the information should be interpreted to mean that 52.3 kJ of energy will be released upon the decomposition of 106.5 g of sodium chlorate.

CHAPTER SUMMARY

The following terms summarize all the concepts and ideas that were introduced in this chapter. You should be able to explain their meaning and how you would use them in chemistry. They appear in boldface type in this chapter to draw your attention to them. The boldface type also makes it easier for you to look them up if you need to. You could also use the Internet search engine *google.com* on your computer to get a quick and expanded explanation of these terms, laws, and formulas.

combination reaction   heat of combustion
decomposition reaction   heat of formation
double replacement reaction   molar heat of formation
enthalpy   single replacement reaction
entropy   standard enthalpy of formation

First Law of Thermodynamics
Hess's Law of Heat Summation

INTERNET RESOURCES

Online content that reinforces major concepts discussed in this chapter can be found at the following Internet addresses if they are still available. *Some may have been changed or deleted.*

Classification of Chemical Reactions
http://boyles.sdsmt.edu/subhead/types_of_reactions.htm
This site gives a very good overview of various types of chemical reactions as well as video access to the actual reaction occurring.

Hess's Law
http://chemistry2.csudh.edu/lecture_help/Hesslaw.html
This web page gives you a chance to solve problems related to Hess's Law and receive instant feedback about the accuracy of your calculations.

1. A synthesis reaction will occur spontaneously after the activation energy is provided if the heat of formation of the product is

   (A) large and negative
   (B) small and negative
   (C) large and positive
   (D) small and positive

2. The reaction of aluminum with dilute H_2SO_4 can be classified as

   (A) synthesis
   (B) decomposition
   (C) single replacement
   (D) double replacement

3. For a metal atom to replace another kind of metallic ion in a solution, the metal atom must be

   (A) a good oxidizing agent
   (B) higher in the activity series than the metal in solution
   (C) lower in the electromotive chart than the metal in solution
   (D) equal in activity to the metal in solution

4. One reason for a double displacement reaction to go to completion is that

   (A) a product is soluble
   (B) a product is given off as a gas
   (C) the products can react with each other
   (D) the products are miscible

5. An endothermic reaction

   (A) has a ΔH value that's negative and would feel cold if it occurred in your hand
   (B) has a ΔH value that's positive and would feel cold if it occurred in your hand
   (C) has a ΔH value that's negative and would feel warm if it occurred in your hand
   (D) has a ΔH value that's positive and would feel warm if it occurred in your hand

6. The heat flowing into 50.0 g of water to produce a 10.0°C rise would be

   (A) 5.00 calories
   (B) 10.0 calories
   (C) 50.0 calories
   (D) 500. calories

7. During a phase change, the temperature of a substance

   (A) increases
   (B) decreases
   (C) stays the same
   (D) goes down and then up

8. Given the following reaction,

   $$H_2(g) + ½ O_2(g) \rightarrow H_2O(g)$$
   $$\Delta H = -242 \text{ kJ}$$

   if 64.0 grams of oxygen were to react with an excess of hydrogen,

   (A) 968 kJ of energy would be released
   (B) 968 kJ of energy would be absorbed
   (C) 484 kJ of energy would be released
   (D) 484 kJ of energy would be absorbed

9. Given the following reaction,

$$C(s) + O_2(g) \rightarrow CO_2(g)$$

$$\Delta H = -393.5 \text{ kJ}$$

the decomposition of 1.0 mole of gaseous CO_2 into the elements that compose it would

(A) release 393.5 kJ but would likely not occur
(B) absorb 393.5 kJ but would likely not occur
(C) release 393.5 kJ and would likely occur
(D) absorb 393.5 kJ and would likely occur

10. A $\Delta H_{reaction}$ of –100 kJ/mole indicates the reaction is

(A) endothermic
(B) unstable
(C) in need of a catalyst
(D) exothermic

Directions: Before attempting to answer the following questions (11–15), you may want to review the directions for this type of question, found on pages xv–xvi.

Every question below contains two statements, I in the left-hand column and II in the right-hand column. For each question, decide if statement I is true or false <u>and</u> whether statement II is true or false, and fill in the corresponding T or F ovals in the answer spaces. *<u>Fill in oval CE only if statement II is a correct explanation of statement I.</u>

	I		II
11.	If the heat of formation of a compound is a large number preceded by a minus sign, the reaction is exothermic	BECAUSE	the First Law of Thermodynamics states that a negative heat of formation is associated with an exothermic reaction.
12.	The burning of carbon with excess O_2 to form CO_2 will go to completion	BECAUSE	when a reaction results in the release of a gas that is allowed to escape, the reaction will go to completion.
13.	The heat of formation of a compound can be calculated by adding two or more thermal reaction equations	BECAUSE	Hess's Law states that a heat of reaction can be arrived at by the algebraic summation of two or more other thermal reactions.
14.	Entropy can be described as a measure of disorder of a system	BECAUSE	when high amounts of energy are released from a reaction, the reaction is said to be exothermic.
15.	The reaction in which HgO is heated to release O_2 is called a decomposition	BECAUSE	in a decomposition reaction, the original compound is broken apart into equal numbers of atoms.

Record your answers here:

	I	II	CE*
11.	T F	T F	◯
12.	T F	T F	◯
13.	T F	T F	◯
14.	T F	T F	◯
15.	T F	T F	◯

Answers and Explanations

1. **(A)** A large negative heat of formation indicates that the reaction will give off a large amount of energy and will self-perpetuate after getting the activation energy necessary for it to start, like burning paper.

2. **(C)** Because the aluminum replaces the hydrogen to form aluminum sulfate, this is classified as a single-replacement reaction.

3. **(B)** For a metal to replace another metallic ion in a solution, it must be higher in the activity series of elements than the metal in solution.

4. **(B)** For a double displacement to go to completion, a product or products must either deposit as a precipitate or leave the reaction as a gas.

5. **(D)** In endothermic reactions, the enthalpy of the system is rising, and so the sign on ΔH will be positive. In order for the system to go up in energy, the surroundings (you) would have to lose energy, and so you would feel cold.

6. **(D)** Using the equation $q = mc\Delta T$, the energy needed to flow into the water would be

$$q = (50.0 \text{ g})(1 \text{ cal/g°C})(10.0°C) = 500. \text{ cal}$$

7. **(C)** During a phase change, the potential energy of the substance is changing but the kinetic energy isn't. Therefore, the temperature, which is directly related to the average kinetic energy of the substance, is constant.

8. **(A)** Since 64.0 g of O_2 represents 2 moles of O_2, and the reaction equation describes 242 kJ of energy released for every ½ mole of O_2 reacting (because of the negative ΔH and the coefficient in front of oxygen in the equation), then 968 kJ of energy will be released.

9. **(B)** The reaction as shown is for the formation of CO_2. The decomposition of CO_2 into the elements that compose it would be the opposite reaction, and so the sign of the ΔH value would have to change. A positive ΔH means the reaction is endothermic, and the system would absorb the released energy. Since the value is fairly large, the reaction is unlikely to occur spontaneously.

10. **(D)** Because the heat of the reaction is negative, it indicates that energy (heat) is given off when this reaction occurs. It is therefore an exothermic reaction.

11. **(T, F)** While it is true that a negative heat of formation is associated with an exothermic reaction, the First Law of Thermodynamics states that the total energy of the universe is constant and cannot be easily created nor destroyed. It is a convention of chemistry that energy given off from a combination reaction is given a negative value.

12. **(T, T, CE)** When a combination reaction results in a gas that is allowed to be released, a precipitate that drops to the bottom of the container, or a nonionizing product, the reaction will go to completion.

13. **(T, T, CE)** Both statements are true and Hess's Law is the basis for arriving at a heat of reaction by algebraically adding two or more thermal equations.

14. **(T, T)** Both statements are true, but the second does not explain the first, and there is no cause and effect relationship. Entropy is a measure of a system's randomness or disorder. Enthalpy is related to the change of heat content in a reaction. Negative enthalpies indicate exothermic reactions.

15. **(T, F)** In the decomposition of HgO, Hg and O_2 are the products. The equation is $2HgO \rightarrow 2Hg + O_2(g)$

    It is not always true that the number of atoms of each product is equal. It happens in this reaction because there is an equal number of atoms of Hg and O in the formula. In decomposing water you have: $2H_2O \rightarrow 2H_2(g) + O_2(g)$ because there are twice as many atoms of hydrogen than oxygen in the original formula.

Rates of Chemical Reactions

9

These skills are usually tested on the SAT Subject Test in Chemistry. You should be able to...

→ **Explain how each of the following factors affects the rate of a chemical reaction: nature of the reactants, surface area exposed, concentrations, temperature, and the presence of a catalyst.**

→ **Draw reaction diagrams with and without a catalyst.**

→ **Explain the Law of Mass Action.**

→ **Describe the relationship between reaction mechanisms and rates of reaction.**

This chapter will review and strengthen these skills. Be sure to do the Practice Exercises at the end of the chapter.

The measurement of reaction rate is based on the rate of appearance of a product or disappearance of a reactant. It is usually expressed in terms of a change in concentration of one of the participants per unit time.

Experiments have shown that for most reactions the concentrations of all participants change most rapidly at the beginning of the reaction; that is, the concentration of the products shows the greatest rate of increase, and the concentrations of the reactants the highest rate of decrease, at this point. This means that the rate of a reaction changes with time. Therefore a rate must be identified with a specific time.

FACTORS AFFECTING REACTION RATES

Five important factors control the rate of a chemical reaction. These are summarized below.

TIP

Know the factors that affect reaction rates.

1. **The nature of the reactants.** In chemical reactions, some bonds break and others form. Therefore, the rates of chemical reactions should be affected by the nature of the bonds in the reacting substances. For example, reactions between ions in an aqueous solution may take place in a fraction of a second. Thus, the reaction between silver nitrate and sodium chloride is very fast. The white silver chloride precipitate appears immediately. In reactions where many covalent bonds must be broken, reaction usually takes place slowly at room temperatures. The decomposition of hydrogen peroxide into water and oxygen happens slowly at room temperatures. In fact, about 17 minutes is required for half the peroxide in a 0.50 M solution to decompose.

2. **The surface area exposed.** Since most reactions depend on the reactants coming into contact, the surface exposed proportionally affects the rate of the reaction.

3. **The concentrations.** The reaction rate is usually proportional to the concentrations of the reactants. The usual dependence of the reaction rate on the concentration of the reactants can simply be explained by theorizing that, if more molecules or ions of the

reactant are in the reaction area, then there is a greater chance that more reactions will occur. This idea is further developed in the collision theory discussed below.

4. **The temperature.** A temperature increase of 10°C above room temperature usually causes the reaction rate to double or triple. The basis for this generality is that, as the temperature increases, the average kinetic energy of the particles involved increases. As a result the particles move faster and have a greater probability of hitting other reactant particles. Because the particles have more energy, they can cause an effective collision, resulting in the chemical reaction that forms the product substance.

5. **The presence of a catalyst.** It is a substance that increases or decreases the rate of a chemical reaction without itself undergoing any permanent chemical change. The catalyst provides an alternative pathway by which the reaction can proceed and in which the activation energy is lower. It thus increases the rate at which the reaction comes to completion or equilibrium. Generally, the term is used for a substance that increases reaction rate (a positive catalyst). Some reactions can be slowed down by negative catalysts.

ACTIVATION ENERGY

Often a reaction rate may be increased or decreased by affecting the activation energy, that is, the energy necessary to cause a reaction to occur. This is shown graphically below for the forward reaction.

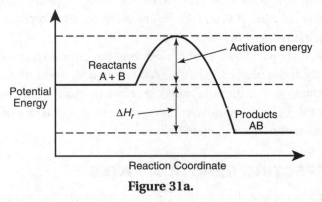

Figure 31a.

A **catalyst**, as explained in the preceding section, is a substance that is introduced into a reaction to speed up the reaction by changing the amount of activation energy needed. The effect of a catalyst used to speed up a reaction can be shown as follows:

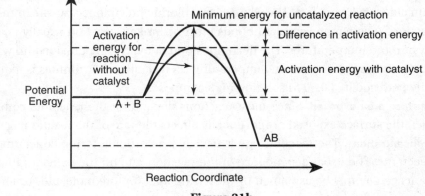

Figure 31b.

REACTION RATE LAW

The relationship between the rate of a reaction and the masses (expressed as concentrations) of the reacting substances is summarized in the **Law of Mass Action**. It states that the rate of a chemical reaction is proportional to the product of the concentrations of the reactants. For a general reaction between A and B, represented by

$$aA + bB \rightarrow \dots$$

the rate law expression is

$$r \propto [A]^a[B]^b$$

or, inserting a constant of proportionality that mathematically changes the expression to an equality, we have

$$r = k[A]^a[B]^b$$

Here k is called the **specific rate** constant for the reaction at the temperature of the reaction.

The exponents a and b may be added to give the total reaction order. For example:

$$H_2(g) + I_2(g) \rightarrow 2HI(g)$$
$$r = k[H_2]^1[I_2]^1$$

The sum of the exponents is $1 + 1 = 2$, and therefore we have a second-order reaction.

Reaction Mechanism and Rates of Reaction

The beginning of this chapter stated that the reaction rate is *usually* proportional to the concentrations of the reactants. This occurs because some reactions do not directly occur between the reactants but may go through intermediate steps to get to the final product. The series of steps by which the reacting particles rearrange themselves to form the products of a chemical reaction is called the **reaction mechanism**. For example:

Step 1	$A + B$	$\rightarrow$	I_1 (fast)
Step 2	$A + I_1$	$\rightarrow$	I_2 (slow)
Step 3	$C + I_2$	$\rightarrow$	D (fast)
Net equation	$2A + B + C$	$\rightarrow$	D

Notice that the reactions of steps 1 and 3 occur relatively fast compared with the reaction of step 2. Now suppose that we increase the concentration of C. This will make the reaction of step 3 go faster, but it will have little effect on the speed of the overall reaction since step 2 is the rate-determining step. If, however, the concentration of A is increased, then the overall reaction rate will increase because step 2 will be accelerated. Knowing the reaction mechanism provides the basis for predicting the effect of a concentration change of a reactant on the overall rate of reaction. Another way of determining the effect of concentration changes is actual experimentation.

CHAPTER SUMMARY

The following terms summarize all the concepts and ideas that were introduced in this chapter. You should be able to explain their meaning and how you would use them in chemistry. They appear in boldface type in this chapter to draw your attention to them. The boldface type also makes it easier for you to look them up if you need to. You could also use the Internet search engine *google.com* on your computer to get a quick and expanded explanation of these terms, laws, and formulas.

activation energy	factors affecting reaction rates
catalyst	Law of Mass Action
collision theory	reaction mechanism

INTERNET RESOURCES

Online content that reinforces major concepts discussed in this chapter can be found at the following Internet addresses if they are still available. *Some may have been changed or deleted.*

Factors Affecting Reaction Rates
http://www.chm.davidson.edu/ChemistryApplets/kinetics/index.html
This site offers you virtual experiments to demonstrate how factors affect reaction rates.

1. List the five factors that affect the rate of a reaction:

   1.
   2.
   3.
   4.
   5.

2. The addition of a catalyst to a reaction

   (A) changes the enthalpy
   (B) changes the entropy
   (C) changes the nature of the products
   (D) changes the activation energy

3. An increase in concentration

   (A) is related to the number of collisions directly
   (B) is related to the number of collisions inversely
   (C) has no effect on the number of collisions

4. At the beginning of a reaction, the reaction rate for the reactants is

   (A) largest, then decreasing
   (B) largest and remains constant
   (C) smallest, then increasing
   (D) smallest and remains constant

5. The reaction rate law applied to the reaction $a\mathrm{A} + b\mathrm{B} \rightarrow \mathrm{AB}$ gives the expression

   (A) $rate = [\mathrm{A}]^b[\mathrm{B}]^a$
   (B) $rate = [\mathrm{AB}]^a[\mathrm{A}]^b$
   (C) $rate = [\mathrm{B}]^a[\mathrm{AB}]^b$
   (D) $rate = [\mathrm{A}]^a[\mathrm{B}]^b$

Answers and Explanations

1. (1) Nature of the reactants
   (2) Surface area exposed
   (3) Concentrations
   (4) Temperature
   (5) Presence of a catalyst

2. **(D)** When a catalyst is added to a reaction, it is used to either increase or decrease the activation energy needed for the reaction to occur. When the activation energy is decreased, it allows the reaction to occur at a faster rate; conversely, increasing the activation energy slows down the forward reaction.

3. **(A)** When the concentration of reactants is increased, the number of collisions between reactants is directly affected because there are more reactant units (ions, atoms, or molecules) available.

4. **(A)** At the beginning of a reaction, the reaction rate of the reactants is the highest because their concentration is the highest. As the reaction progresses, the concentration of the reactants decreases while the concentration of the products increases.

5. **(D)** For the reaction $a\text{A} + b\text{B} \rightarrow \text{AB}$, the reaction rate law would give the expression of r as proportional to $[\text{A}]^a[\text{B}]^b$.

Chemical Equilibrium

10

These skills are usually tested on the SAT Subject Test in Chemistry. You should be able to...

→ **Explain the development of an equilibrium condition and how it is expressed as an equilibrium constant, and use it mathematically.**

→ **Describe Le Châtelier's Principle and how changes in temperature, pressure, and concentrations affect an equilibrium.**

→ **Solve problems dealing with ionization of water, finding the pH, solubility products, and the common ion effect.**

→ **Explain the relationship of enthalpy and entropy as driving forces in a reaction and how they are combined in the Gibbs equation.**

This chapter will review and strengthen these skills. Be sure to do the Practice Exercises at the end of the chapter.

In some reactions no product is formed to allow the reaction to go to completion; that is, the reactants and products can still interact in both directions. This can be shown as follows:

$$A + B \rightleftharpoons C + D$$

The double arrow indicates that C and D can react to form A and B, while A and B react to form C and D.

REVERSIBLE REACTIONS AND EQUILIBRIUM

The reaction is said to have reached **equilibrium** when the forward reaction rate is equal to the reverse reaction rate. Notice that this is a dynamic condition, NOT a static one, although in appearance the reaction *seems* to have stopped. An example of an equilibrium is a crystal of copper sulfate in a saturated solution of copper sulfate. Although to the observer the crystal seems to remain unchanged, there is actually an equal exchange of crystal material with the copper sulfate in solution. As some solute comes out of solution, an equal amount is going into solution.

To express the rate of reaction in numerical terms, we can use the **Law of Mass Action**, discussed in Chapter 9, which states: The rate of a chemical reaction is proportional to the product of the concentrations of the reacting substances. The concentrations are expressed in moles of gas per liter of volume or moles of solute per liter of solution. Suppose, for example, that 1 mole/liter of gas A_2 (diatomic molecule) is allowed to react with 1 mole/liter of another diatomic gas, B_2, and they form gas AB; let R be the rate for the forward reaction

TIP

Equilibrium is reached when the forward and reverse reaction rate are equal.

forming AB. The bracketed symbols $[A_2]$ and $[B_2]$ represent the concentrations in moles per liter for these diatomic molecules. Then $A_2 + B_2 \rightarrow 2AB$ has the rate expression

$$R \propto [A_2] \times [B_2]$$

where $\propto$ is the symbol for "proportional to." When $[A_2]$ and $[B_2]$ are both 1 mole/liter, the reaction rate is a certain constant value (k_1) at a fixed temperature.

$$R = k_1 \ (k_1 \text{ is called the rate constant})$$

For any concentrations of A and B, the reaction rate is

$$R = k_1 \times [A_2] \times [B_2]$$

If $[A_2]$ is 3 moles/liter and $[B_2]$ is 2 moles/liter, the equation becomes

$$R = k_1 \times 3 \times 2 = 6k_1$$

The reaction rate is six times the value for a 1 mole/liter concentration of each reactant.

At the fixed temperature of the forward reaction, AB molecules are also decomposing. If we designate this reverse reaction as R', then, since

$$2AB \text{ (or } AB + AB) \rightarrow A_2 + B_2$$

two molecules of AB must decompose to form one molecule of A_2 and one of B_2. Thus the reverse reaction in this equation is proportional to the square of the molecular concentration of AB:

$$R' \propto [AB] \times [AB]$$
$$\text{or } R' \propto [AB]^2$$
$$\text{and } R' \propto k_2 \times [AB]^2$$

where k_2 represents the rate of decomposition of AB at the fixed temperature. Both reactions can be shown in this manner:

$$A_2 + B_2 \rightleftharpoons 2AB \text{ (note double arrow)}$$

When the first reaction begins to produce AB, some AB is available for the reverse reaction. If the initial condition is only the presence of A_2 and B_2 gases, then the forward reaction will occur rapidly to produce AB. As the concentration of AB increases, the reverse reaction will increase. At the same time, the concentrations of A_2 and B_2 will be decreasing and consequently the forward reaction rate will decrease. Eventually the two rates will be equal, that is, $R = R'$. At this point, equilibrium has been established, and

$$k_1[A_2] \times [B_2] = k_2[AB]^2$$

or

$$\frac{k_1}{k_2} = \frac{[AB]^2}{[A_2] \times [B_2]} = K$$

The convention is that k_1 (forward reaction) is placed over k_2 (reverse reaction) to get this expression. Then k_1/k_2 can be replaced by K_{eq}, which is called the **equilibrium constant** for this reaction under the particular conditions.

In another general example:

$$a\text{A} + b\text{B} \rightleftharpoons c\text{C} + d\text{D}$$

the reaction rates can be expressed as

$$R = k_1[\text{A}]^a \times [\text{B}]^b$$
$$R' = k_2[\text{C}]^c \times [\text{D}]^d$$

Note that the values of k_1 and k_2 are different, but that each is a constant for the conditions of the reaction. At the start of the reaction, [A] and [B] will be at their greatest values, and R will be large; [C], [D], and R' will be zero. Gradually R will decrease and R' will become equal. At this point the reverse reaction is forming the original reactants just as rapidly as they are being used up by the forward reaction. Therefore, no further change in R, R', or any of the concentrations will occur.

If we set R' equal to R, we have:

$$k_2 \times [\text{C}]^c \times [\text{D}]^d = k_1 \times [\text{A}]^a \times [\text{B}]^b$$

or

$$\frac{[\text{C}]^c \times [\text{D}]^d}{[\text{A}]^a \times [\text{B}]^b} = \frac{k_1}{k_2} = K$$

This process of two substances, A and B, reacting to form products C and D and the reverse can be shown graphically to represent what happens as equilibrium is established. The hypothetical equilibrium reaction is described by the following general equation:

$$a\text{A} + b\text{B} \rightleftharpoons c\text{C} + d\text{D}$$

At the beginning (time t_0), the concentrations of C and D are zero and those of A and B are maximum. The graph below shows that over time the rate of the forward reaction decreases as A and B are used up. Meanwhile, the rate of the reverse reaction increases as C and D are formed. When these two reaction rates become equal (at time t_1), equilibrium is established. The individual concentrations of A, B, C, and D no longer change if conditions remain the same. To an observer, it appears that all reaction has stopped; in fact, however, both reactions are occurring at the same rate.

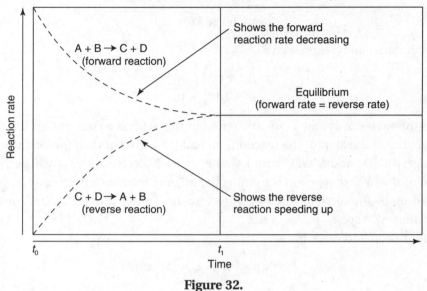

Figure 32.

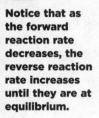
We see that, for the given reaction and the given conditions, K_{eq} is a constant, called the **equilibrium constant**. If K_{eq} is large, this means that equilibrium does not occur until the concentrations of the original reactants are small and those of the products large. A small value of K_{eq} means that equilibrium occurs almost at once and relatively little product is produced.

The equilibrium constant, K_{eq}, has been determined experimentally for many reactions, and the values are given in chemical handbooks.

Suppose we find K_{eq} for reacting H_2 and I_2 at 490°C to be equal to 45.9. Then the equilibrium constant for this reaction

$$H_2(g) + I_2(g) \rightleftharpoons 2HI(g) \text{ at } 490°C$$

is

$$K = \frac{[HI]^2}{[H_2][I_2]} = 45.9$$

➡ **Example 1**

Three moles of $H_2(g)$ and 3.00 moles of $I_2(g)$ are introduced into a 1-liter box at a temperature of 490°C. Find the concentration of each substance in the box when equilibrium is established.

Initial conditions:

$$[H_2] = 3.00 \text{ mol/L}$$
$$[I_2] = 3.00 \text{ mol/L}$$
$$[HI] = 0 \text{ mol/L}$$

The reaction proceeds to equilibrium and

$$K = \frac{[HI]^2}{[H_2][I_2]} = 45.9$$

At equilibrium, then,

$[H_2] = (3.00 - x)$ mol/L   (where x is the number of moles of H_2 that are in the form of HI at equilibrium)

$[I_2] = (3.00 - x)$ mol/L   (the same x is used since 1 mol of H_2 requires 1 mol of I_2 to react to form 2 mol of HI)

$[HI] = 2x$ mol/L

Then

$$K = \frac{(2x)^2}{(3.00 - x)(3.00 - x)} = 45.9$$

If

$$\frac{(2x)^2}{(3 - x)^2} = 45.9$$

then taking the square root of each side gives

$$\frac{2x}{3.00 - x} = 6.77$$

Solving for x:

$$x = 2.32$$

Note: On the SAT Subject Test in Chemistry, calculators are not permitted. So answers to questions like this would be easily calculated "perfect squares."

Substituting this x value into the concentration expressions at equilibrium we have

$$[H_2] = (3.00 - x) = 0.68 \text{ mol/L}$$
$$[I_2] = (3.00 - x) = 0.68 \text{ mol/L}$$
$$[HI] = 2x = 4.64 \text{ mol/L}$$

The crucial step in this type of problem is setting up your concentration expressions from your knowledge of the equation. Suppose that this problem had been:

➡ Example 2 _____

Find the concentrations at equilibrium for the same conditions as in the preceding example except that only 2.00 moles of HI are injected into the box.

$$[H_2] = 0 \text{ mol/L}$$
$$[I_2] = 0 \text{ mol/L}$$
$$[HI] = 2.00 \text{ mol/L}$$

At equilibrium

$$[HI] = (2.00 - x) \text{ mol/L}$$

(For every mole of HI that decomposes, only 0.5 mole of H_2 and 0.5 mole of I_2 are formed.)

$$[H_2] = 0.5x$$

$$[I_2] = 0.5x$$

$$K = \frac{(2-x)^2}{(x/2)^2} = 45.9$$

Solving for x gives:

$$x = 0.456$$

Then, substituting into the equilibrium conditions,

$$[HI] = 2 - x = 1.54 \text{ mol/L}$$

$$[I_2] = \frac{1}{2}x = 0.228 \text{ mol/L}$$

$$[H_2] = \frac{1}{2}x = 0.228 \text{ mol/L}$$

LE CHÂTELIER'S PRINCIPLE

TIP

Know Le Châtelier's Principle.

A general law, **Le Châtelier's Principle**, can be used to explain the results of applying any change of condition (stress) on a system in equilibrium. *It states that if a stress is placed upon a system in equilibrium, the equilibrium is displaced in the direction that counteracts the effect of the stress.* An increase in concentration of a substance favors the reaction that uses up that substance and lowers its concentration. A rise in temperature favors the reaction that absorbs heat and so tends to lower the temperature. These ideas are further developed below.

EFFECTS OF CHANGING CONDITIONS

Effect of Changing the Concentrations

TIP

At equilibrium, K_{eq} stays the same at a given temperature.

When a system at equilibrium is disturbed by adding or removing one of the substances (thus changing its concentration), all the concentrations will change until a new equilibrium point is reached with the same value of K_{eq}.

If the concentration of a reactant in the forward action is increased, the equilibrium is displaced to the right, favoring the forward reaction. If the concentration of a reactant in the reverse reaction is increased, the equilibrium is displaced to the left. Decreases in concentration will produce effects opposite to those produced by increases.

Effect of Temperature on Equilibrium

TIP

Know how each factor affects equilibrium.

If the temperature of a given equilibrium reaction is changed, the reaction will shift to a new equilibrium point. If the temperature of a system in equilibrium is raised, the equilibrium is shifted in the direction that absorbs heat. Note that the shift in equilibrium as a result of temperature change is actually a change in the value of the equilibrium constant. This is different from the effect of changing the concentration of a reactant; when concentrations are changed, the equilibrium shifts to a condition that maintains the same equilibrium constant.

Effect of Pressure on Equilibrium

A change in pressure affects only equilibria in which a gas or gases are reactants or products. Le Châtelier's Law can be used to predict the direction of displacement. If it is assumed that the total space in which the reaction occurs is constant, the pressure will depend on the total number of molecules in that space. An increase in the number of molecules will increase pressure; a decrease in the number of molecules will decrease pressure. If the pressure is increased, the reaction that will be favored is the one that will lower the pressure, that is, decrease the number of molecules.

An example of the application of these principles is the Haber process of making ammonia. The reaction is

$$N_2(g) + 3H_2(g) \rightleftharpoons 2NH_3(g) + heat \text{ (at equilibrium)}$$

If the concentrations of the nitrogen and hydrogen are increased, the forward reaction is increased. At the same time, if the ammonia produced is removed by dissolving it into water, the forward reaction is again favored.

Because the reaction is exothermic, the addition of heat must be considered with care. Increasing the temperature causes an increase in molecular motion and collisions, thus allowing the product to form more readily. At the same time, the equilibrium equation shows that the reverse reaction is favored by the increased temperature, so a compromise temperature of about 500°C is used to get the best yield.

An increase in pressure will cause the forward reaction to be favored since the equation shows that four molecules of reactants are forming two molecules of products. This effect tends to reduce the increase in pressure by the formation of more ammonia.

EQUILIBRIA IN HETEROGENEOUS SYSTEMS

The examples so far have involved systems made up of only gaseous substances. Expression of the K values of systems changes when other phases are present.

Equilibrium Constant for Systems Involving Solids

If the experimental data for this reaction are studied:

$$CaCO_3(s) \rightleftharpoons CaO(s) + CO_2(g)$$

it is found that at a given temperature an equilibrium is established in which the concentration of CO_2 is constant. It is also true that the concentrations of the solids have no effect on the CO_2 concentration as long as both solids are present. Therefore, the K_{eq}, which would conventionally be written like this:

$$K = \frac{[CaO][CO_2]}{[CaCO_3]}$$

can be modified by incorporating the concentrations of the two solids. This can be done since the concentration of solids is fixed. It becomes a new constant K, known as:

$$K = [CO_2]$$

Any heterogeneous reaction involving gases does not include the concentrations of pure solids. As another example, K for the reaction

$$NH_4Cl(s) \rightleftharpoons NH_3(g) + HCl(g)$$

is

$$K = [NH_3][HCl]$$

Acid Ionization Constants

When a weak acid does not ionize completely in a solution, an equilibrium is reached between the acid molecule and its ions. The mass action expression can be used to derive an equilibrium constant, called the **acid dissociation constant**, for this condition. For example, an acetic acid solution ionizing is shown as

$$HC_2H_3O_2(aq) + H_2O(l) \rightleftharpoons H_3O^+(aq) + C_2H_3O_2^-(aq)$$

$$K = \frac{[H_3O^+][C_2H_3O_2^-]}{[HC_2H_3O_2][H_2O]}$$

TIP

K_a incorporates the concentration of water.

The concentration of water in moles/liter is found by dividing the mass of 1 liter of water (which is 1,000 g at 4°C) by its gram-molecular mass, 18 grams, giving H_2O a value of 55.6 moles/liter. Because this number is so large compared with the other numbers involved in the equilibrium constant, it is practically constant and is incorporated into a new equilibrium constant, designated as K_a. The new expression is

$$K_a = \frac{[H_3O^+][C_2H_3O_2^-]}{[HC_2H_3O_2]}$$

Ionization constants have been found experimentally for many substances and are listed in chemical tables. The ionization constants of ammonia and acetic acid are about 1.8×10^{-5}. For boric acid $K_a = 5.8 \times 10^{-10}$, and for carbonic acid $K_a = 4.3 \times 10^{-7}$.

If the concentrations of the ions present in the solution of a weak electrolyte are known, the value of the ionization constant can be calculated. Also, if the value of K_a is known, the concentrations of the ions can be calculated.

A small value for K_a means that the concentration of the un-ionized molecule must be relatively large compared with the ion concentrations. Conversely, a large value for K_a means that the concentrations of ions are relatively high. Therefore, the smaller the ionization constant of an acid, the weaker the acid. Thus, of the three acids referred to above, the ionization constants show that the weakest is boric acid, and the strongest, acetic acid. It should be remembered that, in all cases where ionization constants are used, the electrolytes must be weak in order to be involved in ionic equilibria.

Ionization Constant of Water

Because water is a very weak electrolyte, its ionization constant can be expressed as follows:

$$2H_2O(l) \rightleftharpoons H_3O^+(aq) + OH^-(aq)$$

TIP

K_w incorporates the $[H_2O]^2$.

(Equilibrium constant) $K = \dfrac{[H_3O^+][OH^-]}{[H_2O]^2}$ (since $[H_2O]^2$ remains relatively constant, it is incorporated into K_w)

(Dissociation constant) $K_w = [H_3O^+][OH^-] = 1 \times 10^{-14}$ at 25°C

From this expression, we see that for distilled water $[H_3O^+] = [OH^-] = 1 \times 10^{-7}$. Therefore, the pH, which is $-\log[H_3O^+]$, is

$$pH = -\log[1 \times 10^{-7}]$$
$$pH = -[-7] = 7 \text{ for a neutral solution}$$

The pH range of 1 to 6 is acid, and the pH range of 8 to 14 is basic. See the chart below.

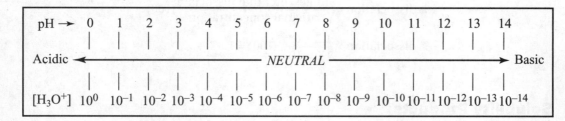

Know the meaning of the pH scale.

➥ Example

(This example incorporates the entire discussion of dissociation constants, including finding the pH.)

Calculate (a) the $[H_3O^+]$, (b) the pH, and (c) the percentage dissociation for 0.10 M acetic acid at 25°C. The symbol K_a is used for the acid dissociation constant. K_a for $HC_2H_3O_2$ is 1.8×10^{-5}.

(a) For this reaction

$$H_2O(l) + HC_2H_3O_2(l) \rightleftharpoons H_3O^+(aq) + C_2H_3O_2^-(aq)$$

and

$$K_a = \frac{[H_3O^+][C_2H_3O_2^-]}{[HC_2H_3O_2]} = 1.8 \times 10^{-5}$$

Let x = number of moles/liter of $HC_2H_3O_2$ that dissociate and reach equilibrium. Then

$$[H_3O^+] = x, \ [C_2H_3O_2^-] = x, \ [HC_2H_3O_2] = 0.10 - x$$

Substituting in the expression for K_a gives

$$K_a = 1.8 \times 10^{-5} = \frac{(x)(x)}{0.10 - x}$$

Because a weak acid, such as acetic, at concentrations of 0.01 M or greater dissociates very little, the equilibrium concentration of the acid is very nearly equal to the original concentration, that is,

$$0.10 - x \cong 0.10$$

Therefore, the expression can be changed to

$$1.8 \times 10^{-5} = \frac{(x)(x)}{0.10}$$
$$x^2 = 1.8 \times 10^{-6}$$
$$x = 1.8 \times 10^{-3} = [H_3O^+]$$

(b) Substituting this result in the pH expression gives

$$pH = -\log[H_3O^+] = -\log[1.3 \times 10^{-3}]$$
$$pH = 3 - \log 1.3$$
$$\underline{\underline{pH = 2.9}}$$

(c) The percentage of dissociation of the original acid may be expressed as

$$\% \text{ dissociation} = \frac{\text{moles/liter that dissociate}}{\text{original concentration}}(100)$$

$$\% \text{ dissociation} \times \frac{1.3 \times 10^{-3}}{1.0 \times 10^{-1}}(100) = \underline{\underline{1.3\%}}$$

Solubility Products

A saturated solution of a substance has been defined as an equilibrium condition between the solute and its ions. For example:

$$AgCl(s) \rightleftharpoons Ag^+(aq) + Cl^-(aq)$$

TIP

K_{sp} incorporates the concentration of the solute.

The equilibrium constant would be:

$$K = \frac{[Ag^+][Cl^-]}{[AgCl]}$$

Since the concentration of the solute remains constant for that temperature, the [AgCl] is incorporated into the K to give the K_{sp}, called the **solubility constant**:

$$K_{sp} = [Ag^+][Cl^-] = 1.2 \times 10^{-10} \text{ at } 25°C$$

This setup can be used to solve problems in which the ionic concentrations are given and the K_{sp} is to be found or the K_{sp} is given and the ionic concentrations are to be determined.

➡ Example 1 _____

Finding the K_{sp}.

By experimentation it is found that a saturated solution of $BaSO_4$ at 25°C contains 3.9×10^{-5} mole/liter of Ba^{2+} ions. Find the K_{sp} of this salt.

Since $BaSO_4$ ionizes into equal numbers of Ba^{2+} and SO_4^{2-}, the barium ion concentration will equal the sulfate ion concentration. Then the solution is

$$BaSO_4(s) \rightleftharpoons Ba^{2+}(aq) + SO_4^{2-}(aq)$$

and

$$K_{sp} = [Ba^{2+}][SO_4^{2-}]$$

Therefore

$$K_{sp} = (3.9 \times 10^{-5})(3.9 \times 10^{-5}) = \underline{\underline{1.5 \times 10^{-9}}}$$

➥ Example 2

Finding the solubility.

If the K_{sp} of radium sulfate, $RaSO_4$, is 4×10^{-11}, calculate the solubility of the compound in pure water. Let x = moles of $RaSO_4$ that dissolve per liter of water. Then, in the saturated solution,

$$[Ra^{2+}] = x \, mol/L$$
$$[SO_4^{2-}] = x \, mol/L$$
$$RaSO_4(s) \rightleftharpoons Ra^{2+}(aq) + SO_4^{2-}(aq)$$
$$[Ra^{2+}][SO_4^{2-}] = K_{sp} = 4 \times 10^{-11}$$

Let $x = [Ra^{2+}]$ and $[SO_4^{2-}]$. Then

$$(x)(x) = 4 \times 10^{-11} = 40 \times 10^{-12}$$
$$\underline{x = 6 \times 10^{-6} \, mol/L}$$

Thus the solubility of $RaSO_4$ is 6×10^{-6} mole/liter of water, for a solution 6×10^{-6} M in Ra^{2+} and 6×10^{-6} M in SO_4^{2-}.

➥ Example 3

Predicting the formation of a precipitate.

In some cases, the solubility products of solutions can be used to predict the formation of a precipitate.

Suppose we have two solutions. One solution contains 1.00×10^{-3} mole of silver nitrate, $AgNO_3$, per liter. The other solution contains 1.00×10^{-2} mole of sodium chloride, NaCl, per liter. If 1 liter of the $AgNO_3$ solution and 1 liter of the NaCl solution are mixed to make a 2-liter mixture, will a precipitate of AgCl form?

In the $AgNO_3$ solution, the concentrations are:

$$[Ag^+] = 1.00 \times 10^{-3} \, mol/L \text{ and } [NO_3^-] = 1.00 \times 10^{-3} \, mol/L$$

In the NaCl solution, the concentrations are:

$$[Na^+] = 1.00 \times 10^{-2} \, mol/L \text{ and } [Cl^-] = 1.00 \times 10^{-2} \, mol/L$$

When 1 liter of one of these solutions is mixed with 1 liter of the other solution to form a total volume of 2 liters, the concentrations will be halved.

In the mixture then, the initial concentrations will be:

$$[Ag^+] = 0.50 \times 10^{-3} \quad or \quad 5.0 \times 10^{-4} \, mol/L$$
$$[Cl^-] = 0.50 \times 10^{-2} \quad or \quad 5.0 \times 10^{-3} \, mol/L$$

For the K_{sp} of AgCl,

$$[Ag^+][Cl^-] = [5.0 \times 10^{-4}][5.0 \times 10^{-3}]$$
$$[Ag^+][Cl^-] = 25 \times 10^{-7} \quad or \quad \underline{2.5 \times 10^{-6}}$$

This is far greater than 1.7×10^{-10}, which is the K_{sp} of AgCl. These concentrations cannot exist, and Ag^+ and Cl^- will combine to form solid AgCl precipitate. Only enough Ag^+ ions and Cl^- ions will remain to make the product of the respective ion concentrations equal 1.7×10^{-10}.

COMMON ION EFFECT

When a reaction has reached equilibrium, and an outside source adds more of one of the ions that is already in solution, the result is to cause the reverse reaction to occur at a faster rate and reestablish the equilibrium. This is called the **common ion effect**. For example, in this equilibrium reaction:

$$NaCl(s) \rightleftharpoons Na^+(aq) + Cl^-(aq)$$

the addition of concentrated HCl (12M) adds H^+ and Cl^- both at a concentration of 12 M. This increases the concentration of the Cl^- and disturbs the equilibrium. The reaction will shift to the left and cause some solid NaCl to come out of solution.

The "common" ion is the one already present in an equilibrium before a substance is added that increases the concentration of that ion. The effect is to reverse the solution reaction and to decrease the solubility of the original substance, as shown in the above example.

DRIVING FORCES OF REACTIONS

Relation of Minimum Energy (Enthalpy) to Maximum Disorder (Entropy)

Some reactions are said to go to completion because the equilibrium condition is achieved when practically all the reactants have been converted to products. At the other extreme, some reactions reach equilibrium immediately with very little product being formed. These two examples are representative of very large K values and very small K values, respectively. There are essentially two driving forces that control the extent of a reaction and determine when equilibrium will be established. These are the drive to the lowest heat content, or **enthalpy**, and the drive to the greatest randomness or disorder, which is called **entropy**. Reactions with negative ΔH's (enthalpy or heat content) are exothermic, and reactions with positive ΔS's (entropy or randomness) are proceeding to greater randomness.

The **Second Law of Thermodynamics** states that the entropy of the universe increases for any spontaneous process. This means that the entropy of a system may increase or decrease but that, if it decreases, then the entropy of the surroundings must increase to a greater extent so that the overall change in the universe is positive. In other words,

$$\Delta S_{universe} = \Delta S_{system} + \Delta S_{surroundings}$$

The following is a list of conditions in which ΔS is positive for the system:

TIP

When the ΔS is positive for the system, it means greater disorder.

1. When a gas is formed from a solid, for example,

$$CaCO_3(s) \rightarrow CaO(s) + CO_2(g).$$

2. When a gas is evolved from a solution, for example,

$$Zn(s) + 2H^+(aq) \rightarrow H_2(g) + Zn^{2+}(aq).$$

3. When the number of moles of gaseous product exceeds the moles of gaseous reactant, for example,

$$2C_2H_6(g) + 7O_2(g) \rightarrow 4CO_2(g) + 6H_2O(g).$$

4. When crystals dissolve in water, for example,

$$NaCl(s) \rightarrow Na^+(aq) + Cl^-(aq).$$

Looking at specific examples, we find that in some cases endothermic reactions occur when the products provide greater randomness or positive entropy. This reaction is an example:

$$CaCO_3(s) \rightleftharpoons CaO(s) + CO_2(g)$$

The production of the gas and thus greater entropy might be expected to take this reaction almost to completion. However, this does not occur because another force is hampering this reaction. It is the absorption of energy, and thus the increase in enthalpy, as the $CaCO_3$ is heated.

The equilibrium condition, then, at a particular temperature, is a compromise between the increase in entropy and the increase in enthalpy of the system.

The Haber process of making ammonia is another example of this compromise of driving forces that affect the establishment of an equilibrium. In this reaction

$$N_2(g) + 3H_2(g) \rightleftharpoons 2NH_3(g) + heat$$

the forward reaction to reach the lowest heat content and thus release energy cannot go to completion because the force to maximum randomness is driving the reverse reaction.

Change in Free Energy of a System—the Gibbs Equation

Free energy, ΔG, depends on ΔH (enthalpy) and ΔS (entropy).

These factors, enthalpy and entropy, can be combined in an equation that summarizes the change of **free energy** in a system. This is designated as ΔG. The relationship is

$$\Delta G = \Delta H - T\Delta S \quad (T \text{ is temperature in kelvins})$$

and is called the **Gibbs free-energy equation**.

The sign of ΔG can be used to predict the spontaneity of a reaction at constant temperature and pressure. If ΔG is negative, the reaction is (probably) spontaneous; if ΔG is positive, the reaction is improbable; and if ΔG is 0, the system is at equilibrium and there is no net reaction.

The ways in which the factors in the equation affect ΔG are shown in this table:

Know these relationships.

ΔH	ΔS	ΔG	Will It Happen?	Comment
Exothermic (–)	+	Always –	Yes	No exceptions
Exothermic (–)	–	– At lower temperatures	Probably	At low temperature
Endothermic (+)	+	– At higher temperatures	Probably	At high temperature
Endothermic (+)	–	Never –	No	No exceptions

This drive to achieve a minimum of free energy may be interpreted as the driving force of a chemical reaction.

CHAPTER SUMMARY

The following terms summarize all the concepts and ideas that were introduced in this chapter. You should be able to explain their meaning and how you would use them in chemistry. They appear in boldface type in this chapter to draw your attention to them. The boldface type also makes it easier for you to look them up if you need to. You could also use the Internet search engine *google.com* on your computer to get a quick and expanded explanation of these terms, laws, and formulas.

acid ionization constant	equilibrium constant
common ion effect	free energy
enthalpy (ΔH)	Gibbs free-energy equation
entropy (ΔS)	solubility product constant
equilibrium	

Le Châtelier's Principle
Second Law of Thermodynamics

INTERNET RESOURCES

Online content that reinforces major concepts discussed in this chapter can be found at the following Internet addresses if they are still available. *Some may have been changed or deleted.*

Le Châtelier's Principle

http://www.ausetute.com.au/lechatsp.html

This site gives an explanation of the principle. This site also provides specific examples of how the principle can be used in different circumstances to show the effects of changing temperature, pressure, and concentration on an equilibrium system.

Equilibrium Equations and Constants

http://www.800mainstreet.com/7/0007-007-Equi_exp_k.html

This site offers tutorials and exercises on how to work with equilibrium equations and constants along with exercises and answers to equilibrium problems.

1. For the reaction $A + B \rightleftharpoons C + D$, the equilibrium constant can be expressed as:

   (A) $K = \dfrac{[A][B]}{[C][D]}$

   (B) $K = \dfrac{[C][B]}{[A][D]}$

   (C) $K = \dfrac{[C][D]}{[A][B]}$

   (D) $K = \dfrac{[A][C]}{[B][D]}$

2. The concentrations in an expression of the equilibrium constant are given in

   (A) mol/mL
   (B) g/L
   (C) gram-equivalents/L
   (D) mol/L

3. In the equilibrium expression for the reaction

   $$BaSO_4(s) \rightleftharpoons Ba^{2+}(aq) + SO_4^{2-}(aq)$$

   K_{sp} is equal to

   (A) $[Ba^{2+}][SO_4^{2-}]$

   (B) $\dfrac{[Ba^{2+}][SO_4^{2-}]}{BaSO_4}$

   (C) $\dfrac{[Ba^{2+}][SO_4^{2-}]}{[BaSO_4]}$

   (D) $\dfrac{[BaSO_4]}{[Ba^{2+}][SO_4^{2-}]}$

4. The K_w of water at 298 K is equal to

   (A) 1×10^{-7}
   (B) 1×10^{-17}
   (C) 1×10^{-14}
   (D) 1×10^{-1}

5. The pH of a solution that has a hydrogen ion concentration of 1×10^{-4} mole/liter is

   (A) 4
   (B) −4
   (C) 10
   (D) −10

6. The pH of a solution that has a hydroxide ion concentration of 1×10^{-4} mole/liter is

   (A) 4
   (B) −4
   (C) 10
   (D) −10

7. A small value for K, the equilibrium constant, indicates that

   (A) the concentration of the un-ionized molecules must be relatively small compared with the ion concentrations
   (B) the concentration of the ionized molecules must be larger than the ion concentrations
   (C) the substance ionizes to a large degree
   (D) the concentration of the un-ionized molecules must be relatively large compared with the ion concentrations

8. In the Haber process for making ammonia, an increase in pressure favors

   (A) the forward reaction
   (B) the reverse reaction
   (C) neither reaction

9. A change in which of these conditions will change the K of an equilibrium given as a starting point?

   (A) Temperature
   (B) Pressure
   (C) Concentration of reactants
   (D) Concentration of products

10. If $Ca(OH)_2$ is dissolved in a solution of NaOH, its solubility, compared with that in pure water, is

    (A) increased
    (B) decreased
    (C) unaffected

The following questions are in the format that is used on the **SAT Subject Area Test in Chemistry**. If you are not familiar with these types of questions, study pages xiii–xiv before doing the remainder of the review questions.

Directions: The following set of lettered choices refers to the numbered questions immediately below it. For each numbered item, choose the one lettered choice that fits it best. Every choice in the set may be used once, more than once, or not at all.

Questions 11–15

(A) the free energy is always negative
(B) the free energy is negative at lower temperatures
(C) the free energy is negative at high temperatures
(D) the free energy is never negative
(E) the system is at equilibrium, and there is no net reaction

Complete the sentence with the appropriate choice.

11. When enthalpy is negative and entropy is positive,

12. When enthalpy is positive and entropy is positive,

13. When enthalpy is negative and entropy is negative,

14. When enthalpy is positive and entropy is negative,

15. When ΔG, free energy, is zero,

Answers and Explanations

1. **(C)** The correct setup of K_{eq} is the product of the concentration of the products over the products of the reactants.

2. **(D)** The concentrations in an expression of the equilibrium constant are given in moles/liter (mol/L).

3. **(A)** This reaction is the equilibrium between a precipitate and its ions in solution. Because $BaSO_4$ is a solid, it does not appear in the solubility product expression. Therefore, $K_{sp} = [Ba^{+2}]\ [SO_4^{2-}]$.

4. **(C)** This is the K expression of water.

5. **(A)** Because the pH is defined as the negative of the log of the H^+ concentration, it is $-(\log$ of $10^{-4})$, which is $-(-4)$ or 4.

6. **(C)** $pH + pOH = 14$. In this problem, the $pOH = -\log [OH] = -\log [10^{-4}] = -(-4) = 4$. Placing this value in the equation, you have $pH + 4 = 14$. Solving for pH, you get 10.

7. **(D)** For K_{eq} to be a small value, the numerator of the expression must be small compared with the denominator. Because the numerator is the product of the concentrations of the ion concentrations and the denominator is the product of the un-ionized molecules, the concentration of the un-ionized molecules must be relatively large compared with the ion concentrations.

8. **(A)** In the reaction for the formation of ammonia, $N_2 + 3H_2 \leftrightarrow 2NH_3 +$ heat (at equilibrium), an increase in pressure will cause the forward reaction to be favored, because the equation shows that four molecules of reactants are forming two molecules of products. This effect tends to reduce this increase in pressure by the formation of more ammonia.

9. **(A)** Only the changing of the temperature of the equilibrium reaction will change the K of the equilibrium given at the starting point.

10. **(B)** Because there already is a concentration of $(OH)^-$ ions from the NaOH in solution, this common ion effect will decrease the solubility of the $Ca(OH)_2$.

11. **(A)** When enthalpy is negative and entropy is positive, the free energy is always negative.

12. **(C)** When enthalpy is positive and entropy is positive, the free energy is negative at high temperatures.

13. **(B)** When enthalpy is negative and entropy is negative, the free energy is negative at lower temperatures.

14. **(D)** When enthalpy is positive and entropy is negative, the free energy is never negative.

15. **(E)** When ΔG, free energy, is zero, the system is at equilibrium and there is no net reaction.

11

Acids, Bases, and Salts

These skills are usually tested on the SAT Subject Test in Chemistry. You should be able to...

→ Describe the properties of an Arrhenius acid and base, and know the name, formula, and degree of ionization of common acids and bases.

→ Explain the Brønsted-Lowry Theory of acids and conjugate bases.

→ Determine pH and pOH of solutions.

→ Solve titration problems and the use of indicators in the process.

→ Describe how a buffer works.

→ Explain the formation and naming of salts.

→ Explain amphoteric substances in relation to acid–base theory.

This chapter will review and strengthen these skills. Be sure to do the Practice Exercises at the end of the chapter.

What defines an acid? A base? A salt? You must know the different ways that these interact. You must also know how to determine the pH of a substance and how to neutralize that same substance.

DEFINITIONS AND PROPERTIES

Acids

There are some characteristic properties by which an acid may be defined. The most important are:

1. *Water (aqueous) solutions of acids conduct electricity.* The degree of conduction depends on the acid's degree of ionization. A few acids ionize almost completely, while others ionize to only a slight degree. Table 10 indicates some common acids and their degrees of ionization.

2. *Acids will react with metals that are more active than hydrogen ions (see page 209) to liberate hydrogen.* (Some acids are also strong oxidizing agents and will not release hydrogen. Somewhat concentrated nitric acid is such an acid.)

3. *Acids have the ability to change the color of indicators.* Some common indicators are litmus and phenolphthalein. **Litmus** is a dyestuff obtained from plant life. When litmus is added to an acidic solution, or paper impregnated with litmus is dipped into an acid, the neutral purple color changes to pink-red. **Phenolphthalein** is pink in a basic solution and becomes colorless in a neutral or acid solution.

TIP

Learn these characteristics of common acids.

TIP

Learn the names
and formulas of
these common
acids.

Table 10. Degrees of Ionization of Common Acids

Completely or Nearly Completely Ionized	Moderately Ionized	Slightly Ionized
Nitric HNO_3	Oxalic $H_2C_2O_4$	Hydrofluoric HF
Hydrochloric HCl	Phosphoric H_3PO_4	Acetic $HC_2H_3O_2$
Sulfuric H_2SO_4	Sulfurous H_2SO_3	Carbonic H_2CO_3
Hydriodic HI		Hydrosulfuric H_2S
Hydrobromic HBr		(Most others)

4. *Acids react with bases so that the properties of both are lost to form water and a salt.*
   This is called **neutralization**. The general equation is:

$$\text{Acid} + \text{Base} \rightarrow \text{Salt} + \text{Water}$$

An example is:

$$Mg(OH)_2(aq) + H_2SO_4(aq) \rightarrow MgSO_4(aq) + 2H_2O(\ell)$$

5. *Acids react with carbonates to release carbon dioxide.* An example:

$$CaCO_3(s) + 2HCl(aq) \rightarrow CaCl_2(aq) + H_2CO_3 \text{(unstable and decomposes)}$$
$$ \longrightarrow H_2O(\ell) + CO_2(g)$$

The most common theory used in first-year chemistry is the **Arrhenius Theory,** which states that an acid is a substance that yields hydrogen ions in an aqueous solution. Although we speak of these hydrogen ions in the solution, they are really not separate ions but become attached to the oxygen of the polar water molecule to form the H_3O^+ ion (the hydronium ion). Thus, it is really this hydronium ion we are concerned with in an acid solution.

The general reaction for the dissociation of an acid, HX, is commonly written as

$$HX \rightleftharpoons H^+ + X^-$$

To show the formation of the hydronium ion, H_3O^+, the complete equation is:

$$HX + H_2O \rightleftharpoons H_3O^+ + X^-$$

A list of common acids and their formulas is given in Chapter 4, Table 7; an explanation of the naming procedures for acids precedes Table 7.

Bases

Bases may also be defined by some operational definitions that are based on experimental observations. Some of the important ones are as follows:

1. *Bases are conductors of electricity in an aqueous solution.* Their degrees of conduction depend on their degrees of ionization. The degrees of ionization of some common bases are shown in Table 11.

Table 11. Degrees of Ionization of Common Bases

Completely or Nearly Completely Ionized	Slightly Ionized
Potassium hydroxide KOH	Ammonium hydroxide $NH_4(OH)$
Sodium hydroxide NaOH	(All others)
Barium hydroxide $Ba(OH)_2$	
Strontium hydroxide $Sr(OH)_2$	
Calcium hydroxide $Ca(OH)_2$	

TIP

Learn the names and formulas of these common bases.

2. ***Bases cause a color change in indicators.*** Litmus changes from red to blue in a basic solution, and phenolphthalein turns pink from its colorless state.

3. ***Bases react with acids to neutralize each other and form a salt and water.***

4. ***Bases react with fats to form a class of compounds called soaps.*** Earlier generations used this method to make their own soap.

5. ***Aqueous solutions of bases feel slippery, and the stronger bases are very caustic to the skin.***

The Arrhenius Theory defines a base as a substance that yields hydroxide ions (OH^-) in an aqueous solution.

Some common bases have familiar names, for example:

Sodium hydroxide	= lye, caustic soda
Potassium hydroxide	= caustic potash
Calcium hydroxide	= slaked lime, hydrated lime, limewater
Ammonium hydroxide	= ammonia water, household ammonia

Much of the sodium hydroxide produced today comes from the Hooker cell electrolysis apparatus. The electrolysis process for the decomposition of water was discussed in Chapter 7. When an electric current is passed through a saltwater solution, hydrogen, chlorine, and sodium hydroxide are the products. The formula for this equation is:

$$2NaCl(aq) + 2HOH(\ell) \xrightarrow[\text{energy}]{\text{electrical}} H_2(g) + Cl_2(g) + 2NaOH(aq)$$

Broader Acid–Base Theories

Besides the common Arrhenius Theory of acids and bases discussed for aqueous solutions, two other theories, the Brønsted-Lowry Theory and the Lewis Theory, are widely used.

The Brønsted-Lowry Theory (1923) defines acids as proton donors and bases as proton acceptors. This definition agrees with the aqueous solution definition of an acid giving up hydrogen ions in solution, but goes beyond to other cases as well.

An example is the reaction of dry HCl gas with ammonia gas to form the white solid NH_4Cl.

$$HCl(g) + NH_3(g) \rightarrow NH_4Cl(s)$$

The HCl is the proton donor or acid, and the ammonia is a Brønsted-Lowry base that accepts the proton.

Conjugate Acids and Bases

In an acid–base reaction, the original acid gives up its proton to become a **conjugate base**. In other words, after losing its proton, the remaining ion is capable of gaining a proton, thus qualifying as a base. The original base accepts a proton, so it now is classified as a **conjugate acid** since it can release this newly acquired proton and thus behave like an acid.

Some examples are given below:

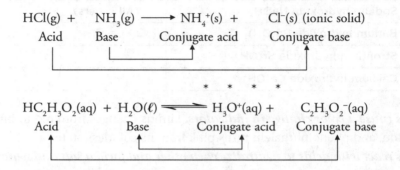

Strength of Conjugate Acids and Bases

The extent of the reaction between a Brønsted-Lowry acid and base depends on the relative strengths of the acids and bases involved. Consider the following example. Hydrochloric is a strong acid. It gives up protons readily. It follows that the Cl–ion has little tendency to attract and retain a proton. Consequently, the Cl–ion is an extremely weak base.

$$HCl(g) + H_2O(\ell) \rightarrow H_3O^+(aq) + Cl^-(aq)$$
$$\text{strong acid} \qquad \text{base} \qquad\qquad \text{acid} \qquad \text{weak base}$$

This observation leads to an important conclusion: the stronger an acid is, the weaker its conjugate base; the stronger a base is, the weaker its conjugate acid. This concept allows strengths of different acids and bases to be compared to predict the outcome of a reaction. As an example, consider the reaction of perchloric acid, $HClO_4$, and water.

$$HClO_4(aq) + H_2O(\ell) \rightarrow H_3O^+(aq) + ClO_4^-(aq)$$
$$\text{stronger acid} \qquad \text{stronger base} \qquad \text{weaker acid} \qquad \text{weaker base}$$

Another important general conclusion is that proton-transfer reactions favor the production of the weaker acid and the weaker base. For a reaction to approach completion, the reactants must be much stronger as an acid and as a base than the products.

The Lewis Theory (1916) defines acids and bases in terms of the electron-pair concept, which is probably the most generally useful concept of acids and bases. According to the Lewis definition, an acid is an electron-pair acceptor; and a base is an electron-pair donor. An example is the formation of ammonium ions from ammonia gas and hydrogen ions.

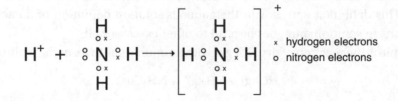

Notice that the hydrogen ion is in fact accepting the electron pair of the ammonia, so it is a Lewis acid. The ammonia is donating its electron pair, so it is a Lewis base.

Another example is boron trifluoride. It is an excellent Lewis acid. It forms a fourth covalent bond with many molecules and ions. Its reaction with a fluoride ion is shown below.

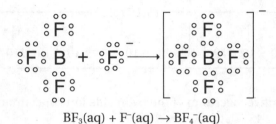

$$BF_3(aq) + F^-(aq) \rightarrow BF_4^-(aq)$$

The acid–base systems are summarized below.

Type	Acid	Base
Arrhenius	H^+ or H_3O^+ producer	OH^- producer
Brønsted-Lowry	proton (H^+) donor	proton (H^+) acceptor
Lewis	electron-pair acceptor	electron-pair donor

Acid Concentration Expressed as pH

Frequently, acid and base concentrations are expressed by means of the **pH** system. The pH can be defined as $-\log [H^+]$, where $[H^+]$ is the concentration of hydrogen ions expressed in moles per liter. The logarithm is the exponent of 10 when the number is written in the base 10. For example:

$$100 = 10^2 \text{ so logarithm of } 100, \text{ base } 10 = 2$$
$$10{,}000 = 10^4 \text{ so logarithm of } 10{,}000, \text{ base } 10 = 4$$
$$0.01 = 10^{-2} \text{ so logarithm of } 0.01, \text{ base } 10 = -2$$

REMEMBER

pH = –log[H⁺]

The logarithms of more complex numbers can be found in a logarithm table. An example of a pH problem is:

Find the pH of a 0.1 molar solution of HCl.

1st step.   Because HCl ionizes almost completely into H^+ and Cl^-, $[H^+] = 0.1$ mole/liter.
2nd step.   By definition
              $pH = -\log [H^+]$
       so
              $pH = -\log [10^{-1}]$
3rd step.   The logarithm of 10^{-1} is -1
       so
              $pH = -(-1)$
4th step.   The pH then is $= 1$.

Because water has a normal H^+ concentration of 10^{-7} mole/liter because of the slight ionization of water molecules, the water pH is 7 when the water is neither acid nor base. The normal pH range is from 0 to 14.

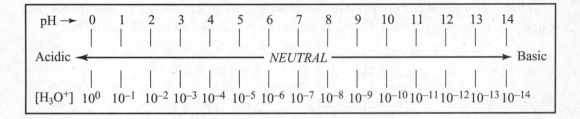

The pOH is the negative logarithm of the hydroxide ion concentration:

$$pOH = -\log\ [OH^-]$$

If the concentration of the hydroxide ion is 10^{-9} M, then the pOH of the solution is +9. From the equation

$$[H^+][OH^-] = 1.0 \times 10^{-14} \text{ at } 298 \text{ K}$$

the following relationship can be derived:

$$pH + pOH = 14.00$$

In other words, the sum of the pH and pOH of an aqueous solution at 298 K must always equal 14.00. For example, if the pOH of a solution is 9.00, then its pH must be 5.00.

REMEMBER

pH + pOH = 14

➥ Example

What is the pOH of a solution whose pH is 3.0?

Substituting 3.0 for pH in the expression

$$pH + pOH = 14.0$$

gives

$$3.0 + pOH = 14.0$$
$$pOH = 14.0 - 3.0$$
$$pOH = 11.0$$

INDICATORS

Some indicators can be used to determine pH because of their color changes somewhere along this pH scale. Some common indicators and their respective color changes are given below.

TIP

Notice that each indicator has its own range of color change.

Indicator	pH Range of Color Change	Color Below Lower pH	Color Above Higher pH
Methyl orange	3.1–4.4	Red	Yellow
Bromthymol blue	6.0–7.6	Yellow	Blue
Litmus	4.5–8.3	Red	Blue
Phenolphthalein	8.3–10.0	Colorless	Red/Pink

Here is an example of how to read this chart: At pH values below 4.5, litmus is red; above 8.3, it is blue. Between these values, it is a mixture of the two colors.

In choosing an indicator for a titration, we need to consider if the solution formed when the end point is reached has a pH of 7. Depending on the type of acid and base used, the resulting hydrolysis of the salt formed may cause it to be slightly acidic, slightly basic, or neutral. If the titration is of a strong acid and a strong base, the end point will be at pH 7 and practically any indicator can be used. This is because the addition of 1 drop of either reagent will change the pH at the end point by about 6 units. For titrations of strong acids and weak bases, we need an indicator, such as methyl orange, that changes color between 3.1 and 4.4 in the acid region. In the titration of a weak acid and a strong base, we should use an indicator that changes in the basic range. Phenolphthalein is a suitable choice for this type of titration because it changes color in the pH 8.3 to 10.0 range.

TITRATION—VOLUMETRIC ANALYSIS

Knowledge of the concentrations of solutions and the reactions they take part in can be used to determine the concentrations of "unknown" solutions or solids. The use of volume measurement in solving these problems is called **titration**.

A common example of a titration uses acid–base reactions. If you are given a base of known concentration, that is, a standard solution, let us say 0.10 M NaOH, and you want to determine the concentration of an HCl solution, you could **titrate** the solutions in the following manner.

First, introduce a measured quantity, 25.0 milliliters, of the NaOH into a flask by using a pipet or burette in a setup like the one in the accompanying diagram. Next, introduce 2 drops of a suitable indicator. Because NaOH and HCl are considered a strong base and a strong acid, respectively, an indicator that changes color in the middle pH range would be appropriate. Litmus solution would be one choice. It is blue in a basic solution but changes to red when the solution becomes acidic. Slowly introduce the HCl until the color change occurs. This point is called the **end point**. The point at which enough acid is added to neutralize all the standard solution in the flask is called the **equivalence point**.

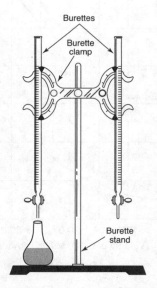

Figure 33. Burette Setup for Titration

Suppose 21.5 milliliters of HCl was needed to produce the color change. The reaction that occurred was

$$H^+(aq) + OH^-(aq) \rightarrow H_2O(\ell)$$

until all the OH^- was neutralized; then the excess H^+ caused the litmus paper to change color.

To solve the question of the concentration of NaOH, this equation is used:

$$M_{acid} \times V_{acid} = M_{base} \times V_{base}$$

Substituting the known amounts in this equation gives

$$x\,M_{acid} \times 21.5\ \text{mL} = 0.1\,\text{M} \times 25.0\ \text{mL}$$
$$x = 0.116\,\text{M}$$

TIP

For a titration:
$M_{acid} \times V_{acid} =$
$M_{Base} \times V_{Base}$

In choosing an indicator for a titration, we need to consider whether the solution formed when the end point is reached has a pH of 7. Depending on the types of acid and base used, the resulting hydrolysis of the salt formed may cause the solution to be slightly acidic, slightly basic, or neutral. If a strong acid and a strong base are titrated, the end point will be at pH 7, and practically any indicator can be used because adding 1 drop of either reagent will change the pH at the end point by about 6 units. For titrations of strong acids and weak bases, we need an indicator, such as methyl orange, that changes color between 3.1 and 4.4 in the acid region. When titrating a weak acid and a strong base, we should use an indicator that changes in the basic range. Phenolphthalein is the suitable choice for this type of titration because it changes color in the pH 8.3 to 10.0 range.

The process of the neutralization reaction can be represented by a titration curve like the one below, which shows the titration of a strong acid with a strong base.

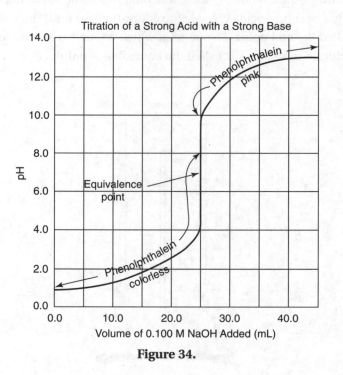

Figure 34.

➡ Example 1 _____

Find the concentration of acetic acid in vinegar if 21.6 milliliters of 0.20 M NaOH is needed to titrate a 25-milliliter sample of the vinegar.

Solution

Using the equation $M_{acid} \times V_{acid} = M_{base} \times V_{base}$, we have

$$x\, M_{acid} \times 25\ \text{mL} = 0.20 \times 21.6\ \text{mL}$$
$$x = \underline{\underline{0.17\, M_{acid}}}$$

Another type of titration problem involves a solid and a titrated solution.

➡ Example 2 _____

A solid mixture contains NaOH and NaCl. If 10.0 milliliters of 0.100 M HCl is required to titrate a 0.100-gram sample of this mixture to its end point, what is the percent of NaOH in the sample?

Solution

Since

$$\text{Molarity} = \frac{\text{No. of moles}}{\text{Liters of solution}}$$

then

$$M \times V = \frac{\text{No. of moles}}{\text{Liters of solution}} \times \text{Liters of solution} = \text{No. of moles}$$

Substituting the HCl information in the equation, we have

$$\frac{0.100\ \text{mol}}{\text{Liters of solution}} \times 0.0100\ \text{L of solution} = 0.001\ \text{mol}$$
(Note: this is 10 mL expressed in liters.)

Since 1 mol of HCl neutralizes 1 mol of NaOH, 0.001 mol of NaOH must be present in the mixture. Since 1 mol NaOH = 40.0 g
then

$$0.01\ \text{mol} \times 40.0\ \text{g/mol} = 0.04\ \text{g NaOH}$$

Therefore, 0.04 g of NaOH was in the 0.100-g sample of the solid mixture. The percent is 0.04 g/0.100 g × 100 = 40%.

In the explanations given to this point, the reactions that took place were between monoprotic acids (single hydrogen ions) and monobasic bases (one hydroxide ion per base). This means that each mole of acid had 1 mole of hydrogen ions available, and each mole of base had 1 mole of hydroxide ions available, to interact in the following reaction until the end point was reached:

$$H^+(aq) + OH^-(aq) \rightarrow 2H_2O(\ell)$$

This is not always the case, however, and it is important to know how to deal with acids and bases that have more than one hydrogen ion and more than one hydroxide ion per formula. The following is an example of such a problem.

➤ Example 3 _____

If 20.0 milliliters of an aqueous solution of calcium hydroxide, $Ca(OH)_2$, is used in a titration, and an appropriate indicator is added to show the neutralization point (end point), the few drops of indicator that are added can be ignored in the volume considerations. Therefore, if 25.0 milliliters of standard 0.050 M HCl is required to reach the end point, what was the original concentration of the $Ca(OH)_2$ solution?

The balanced equation for the reaction gives the relationship between the number of moles of acid reacting and the number of moles of base:

$$\underbrace{2HCl}_{2\,mol}(aq) + \underbrace{Ca(OH)_2}_{1\,mol}(aq) \rightarrow CaCl_2(aq) + 2H_2O(\ell)$$

The mole relationship here is that the number of moles of acid is twice the number of moles of base:

$$\text{No. of moles of acid} = 2 \times \text{No. of moles of base}$$
$$\uparrow \textbf{ mole factor}$$

Since the molar concentration of the acid times the volume of the acid gives the number of moles of acid:

$$M_a \times V_a = \text{moles of acid}$$

and the molar concentration of the base times the volume of the base gives the number of moles of base:

$$M_b \times V_b = \text{moles of base}$$

then, substituting these products into the mole relationship, we get

$$M_a V_a = 2M_b V_b$$

Solving for M_b gives

$$M_b = \frac{M_a V_a}{2V_b}$$

Substituting values, we get

$$M_b = \frac{0.050\,mol/L \times 0.0250\,\cancel{L}}{2 \times 0.0200\,\cancel{L}}$$
$$M_b = 0.0312\,mol/L \text{ or } 0.0312\,M$$

BUFFER SOLUTIONS

Buffer solutions are equilibrium systems that resist changes in acidity and maintain constant pH when acids or bases are added to them. A typical laboratory buffer can be prepared by mixing equal molar quantities of a weak acid such as $HC_2H_3O_2$ and its salt, $NaC_2H_3O_2$. When a small amount of a strong base such as NaOH is added to the buffer, the acetic acid reacts (and consumes) most of the excess OH^- ion. The OH^- ion reacts with the H^+ ion from the acetic acid, thus reducing the H^+ ion concentration in this equilibrium:

$$HC_2H_3O_2(aq) \rightleftharpoons H^+(aq) + C_2H_3O_2^-(aq)$$

This reduction of H^+ causes a shift to the right, forming additional $C_2H_3O_2^-$ ions and H^+ ions. For practical purposes, each mole of OH^- added consumes 1 mole of $HC_2H_3O_2$ and produces 1 mole of $C_2H_3O_2^-$ ions.

When a strong acid such as HCl is added to the buffer, the H^+ ions react with the $C_2H_3O_2^-$ ions of the salt and form more undissociated $HC_2H_3O_2$. This does not alter the H^+ ion concentration. Proportional increases and decreases in the concentrations of $C_2H_3O_2^-$ and $HC_2H_3O_2$ do not significantly affect the acidity of the solution.

SALTS

A salt is an ionic compound containing positive ions other than hydrogen ions and negative ions other than hydroxide ions. The usual method of preparing a particular salt is by neutralizing the appropriate acid and base to form the salt and water.

Five methods for preparing salts are as follows:

1. **Neutralization reaction.** An acid and a base neutralize each other to form the appropriate salt and water. For example:

There are five methods for preparing salts.

$$2HCl(aq) + Ca(OH)_2(aq) \rightarrow CaCl_2(aq) + 2H_2O(\ell)$$
$$\text{acid} \qquad + \quad \text{base} \qquad \rightarrow \quad \text{salt} \qquad + \text{water}$$

2. **Single replacement reaction.** An active metal replaces hydrogen in an acid. For example:

$$Mg(s) + H_2SO_4(aq) \rightarrow MgSO_4(aq) + H_2(g)$$

3. **Direct combination of elements.** An example of this method is the combination of iron and sulfur. In this reaction small pieces of iron are heated with powdered sulfur:

$$Fe(s) + S(s) \rightarrow FeS(s)$$
$$\textbf{iron (II) sulfide}$$

4. **Double replacement.** When solutions of two soluble salts are mixed, they form an insoluble salt compound. For example:

$$AgNO_3(aq) + NaCl(aq) \rightarrow NaNO_3(aq) + AgCl(s)$$

5. **Reaction of a metallic oxide with a nonmetallic oxide.** For example:

$$MgO(s) + SiO_2(s) \rightarrow MgSiO_3(s)$$

The naming of salts is discussed on page 117.

AMPHOTERIC SUBSTANCES

Some substances, such as the HCO_3^- ion, the HSO_4^- ion, the H_2O molecule, and the NH_3 molecule, can act as either proton donors (acids) or proton receivers (bases), depending upon which other substances they come into contact with. These substances are said to be **amphoteric**. Amphoteric substances donate protons in the presence of strong bases and accept protons in the presence of strong acids.

Examples are the reactions of the bisulfate ion, HSO_4^-:

With a strong acid, HSO_4^- accepts a proton:

$$HSO_4^-(aq) + H^+(aq) \rightarrow H_2SO_4(aq)$$

With a strong base, HSO_4^- donates a proton:

$$HSO_4^-(aq) + OH^-(aq) \rightarrow H_2O(\ell) + SO_4^{2-}(aq)$$

ACID RAIN—AN ENVIRONMENTAL CONCERN

TIP

Acid rain is the result of the formation of sulfuric acid from sulfur oxides reacting with water. Nitrogen oxides are also involved.

Acid rain is currently a subject of great concern in many countries around the world because of the widespread environmental damage it reportedly causes. It forms when the oxides of sulfur and nitrogen combine with atmospheric moisture to yield sulfuric and nitric acids—both known to be highly corrosive, especially to metals. Once formed in the atmosphere, these acids can be carried long distances from their source before being deposited by rain. The pollution may also take the form of snow or fog or be precipitated in dry form. This dry form is just as damaging to the environment as the liquid form.

The problem of acid rain can be traced back to the beginning of the industrial revolution, and it has been growing ever since. The term "acid rain" has been in use for more than a century and is derived from atmospheric studies made in the region of Manchester, England.

In 1988, as part of the Long-Range Transboundary Air Pollution Agreement sponsored by the United Nations and the United States, along with 24 other countries, a protocol freezing the rate of nitrogen oxide emissions at 1987 levels was ratified. The 1990 amendments to the Clean Air Act of 1967 put in place regulations to reduce the release of sulfur dioxide from power plants to 10 million tons per year by 2000. That achieved a 20 percent decrease in sulfur dioxide. The attempts continue through international organizations to further clean the air.

These equations show the most common reactions of sulfur- and nitrogen-containing gases with rainwater. The sulfur dioxide reacts with rainwater to form sulfuric acid solutions:

$$2SO_2(g) + O_2(g) \rightarrow 2SO_3(g)$$
$$SO_3(g) + H_2O(\ell) \rightarrow H_2SO_4(aq)$$

The oxides of nitrogen react to form nitrous and nitric acid:

$$2NO_2(g) + H_2O(\ell) \rightarrow HNO_2(aq) + HNO_3(aq)$$

CHAPTER SUMMARY

The following terms summarize all the concepts and ideas that were introduced in this chapter. You should be able to explain their meaning and how you would use them in chemistry. They appear in boldface type in this chapter to draw your attention to them. The boldface type also makes it easier for you to look them up if you need to. You could also use the Internet search engine *google.com* on your computer to get a quick and expanded explanation of these terms, laws, and formulas.

acid	conjugate base	neutralization
amphoteric	end point	pH
base	equivalence point	salt
buffer solution	indicator	titration
conjugate acid	litmus	volumetric analysis
Arrhenius Theory	Brønsted-Lowry Theory	Lewis Theory

Online content that reinforces major concepts discussed in this chapter can be found at the following Internet addresses if they are still available. *Some may have been changed or deleted.*

Acids and Bases

http://chemwiki.ucdavis.edu/Core/Physical_Chemistry/Acids_and_Bases/Acid
This site summarizes the important ways in which acids and bases can be defined. The role of water is also described in the acid/base process.

Titration Experiments and More

http://chemistry.about.com/od/workedchemistryproblems/a/titrationexampl.htm
This site has a step-by-step explanation about how to solve a titration problem. It also has links to other informative sites.

Acid Rain

http://www3.epa.gov/acidrain/index.html
This Environmental Protection Agency site offers a wealth of information concerning the major environmental problem called acid rain.

1. The difference between HCl and $HC_2H_3O_2$ as acids is

    (A) the first has less hydrogen in solution
    (B) the second has more ionized hydrogen
    (C) the first is highly ionized
    (D) the second is highly ionized

2. The hydronium ion is represented as

    (A) H_2O^+
    (B) H_3O^+
    (C) HOH^+
    (D) H^-

3. H_2SO_4 is a strong acid because it is

    (A) slightly ionized
    (B) unstable
    (C) an organic compound
    (D) highly ionized

4. The common ionic reaction of an acid with a base involves ions of

    (A) hydrogen and hydroxide
    (B) sodium and chloride
    (C) hydrogen and hydronium
    (D) hydroxide and nitrate

5. Which pH is an acid solution?

    (A) 3
    (B) 7
    (C) 9
    (D) 10

6. The pH of a solution with a hydrogen ion concentration of 1×10^{-3} is

    (A) +3
    (B) −3
    (C) ±3
    (D) 1 + 3

7. According to the Brønsted-Lowry Theory, an acid is

    (A) a proton donor
    (B) a proton acceptor
    (C) an electron donor
    (D) an electron acceptor

8. A buffer solution

    (A) changes pH rapidly with the addition of an acid
    (B) does not change pH at all
    (C) resists changes in pH
    (D) changes pH only with the addition of a strong base

9. The point at which a titration is complete is called the

    (A) end point
    (B) equilibrium point
    (C) calibrated point
    (D) chemical point

10. If 10. mL of 1 M HCl was required to titrate a 20. mL NaOH solution of unknown concentration to its end point, what was the concentration of the NaOH?

    (A) 0.5 M
    (B) 1.5 M
    (C) 2 M
    (D) 2.5 M

Answers and Explanations

1. **(C)** The strength of an Arrhenius acid is determined by the degree of ionization of the hydrogens in the formula. The HCl ionizes to a great degree and is considered a strong acid, whereas $HC_2H_3O_2$, acetic acid, which is found in vinegar, ionizes only to a small degree and is considered a weak acid.

2. **(B)** The hydronium ion is written as H_3O^+.

3. **(D)** H_2SO_4 is a strong acid because it is highly ionized.

4. **(A)** The basic reaction between an acid and a base is $H^+(aq) + OH^-(aq) \rightarrow H_2O(\ell)$ or $H_3O^+(aq) + OH^-(aq) \rightarrow 2H_2O(\ell)$.

5. **(A)** The pH scale is 0 to 14; the numbers below 7 are acid and those above 7 are increasingly basic.

6. **(A)** Because the pH is defined as the negative of the log of the H^+ concentration, it is $-(\log$ of $10^{-3})$, which is $-(-3)$ or $+3$.

7. **(A)** By definition, a Brønsted-Lowry acid is a proton donor.

8. **(C)** A buffer solution resists the changes in pH.

9. **(A)** The point in a titration when the "unknown" solution has been neutralized—in the case of an acid/base titration—by the "standard" solution of known concentration is called the end point or equivalence point.

10. **(A)** Use the equation $M_{acid} \times V_{acid} = M_{base} \times V_{base}$ and change mL to L by dividing by 1,000 mL/L to change 10. mL into .01 L and 20. mL into .02 L. Then substituting, you get $1\ M_{acid} \times .01\ L_{acid} = xM_{base} \times .02\ L_{base}$. Solving for x, you get $x = 0.5\ M_{base}$.

Oxidation-Reduction

12

These skills are usually tested on the SAT Subject Test in Chemistry. You should be able to...

→ **Assign oxidation states to elements in compounds.**

→ **Describe the process of oxidation and reduction.**

→ **Recognize when a substance is being oxidized or reduced.**

→ **Apply the appropriate terms to substances involved in redox reactions.**

→ **Use the concepts of oxidation-reduction to better describe combustion reactions.**

This chapter will review and strengthen these skills. Be sure to do the Practice Exercises at the end of the chapter.

In Chapter 8, the reaction category called **single replacement** was discussed. Recall that this type of reaction is characterized by a compound reacting with an element producing a new compound and a new element. It was shown that reactions such as these could be predicted to occur (i.e., be spontaneous) if the heat of formation of the compound in the products is negative and reasonably bigger than that of the compound in the reactants. These situations produce exothermic reactions, those in which the system under study becomes lower in energy, which nature tends to want to see occur.

Delving deeper into single replacement reactions allows one to understand why the energy of the system goes down in terms of the chemical process that is occurring. Take the reaction between a solution of silver nitrate and copper metal.

$$2AgNO_3(aq) + Cu(s) \rightarrow Cu(NO_3)_2(aq) + 2Ag(s)$$

The solution of silver nitrate is actually composed of silver ions (Ag^+) and nitrate ions (NO_3^-) surrounded by water molecules in solution. Likewise, the solution of copper(II) nitrate in the products contains ions of Cu^{2+} and NO_3^-. The nitrate ions, then, are found in both the reactants and the products. As discussed in Chapter 4, they can be referred to as **spectator ions**. Because spectator ions can be canceled out, the reaction can be simplified and written as the net ionic equation:

$$2Ag^+(aq) + Cu(s) \rightarrow Cu^{2+}(aq) + 2Ag(s)$$

Analysis of the reaction shows that for the process to occur, the positively charged silver ion in the reactants must turn into a silver atom in the products. Likewise, the copper atom must become a positively charged copper ion. How do these changes occur? The answer is simple and represents a major driving force for chemical change—the **transfer of electrons** from a substance that wants them less to a substance that wants them more.

The gain of 2 electrons

$$2Ag^+(aq) + Cu(s) \rightarrow Cu^{2+}(aq) + 2Ag(s)$$

The loss of 2 electrons

Whenever the transfer of electrons occurs during a chemical process, one substance must lose electrons in order for the other to gain them. The loss of electrons has a name in chemistry: **oxidation**. Accordingly, the gaining of electrons is named too: **reduction**. Since one process can't happen without the other, these processes are often intertwined into one distinctly chemical term called **redox**. Redox reactions are those in which oxidation and reduction occur in complementary ways. Redox is a major reaction category in chemistry. All single replacement reactions are redox reactions; however, all redox reactions are not single replacement reactions. Before we move on to other reaction types that can be seen as redox, let's take a further look at the single replacement category from a redox perspective.

The activity series of metals, whose use is described in Chapter 8 as another way to predict the spontaneity of single replacement reactions, can now be seen as a measure of the desire of certain metals to lose electrons compared to other metals. In the single replacement reaction example previously discussed, the process can be said to occur because the copper atom has a greater desire to lose electrons than does the silver atom. In other words, the reaction will proceed in the forward direction as written (as opposed to going backward) because there is a natural push for copper to lose electrons more than for the silver to lose electrons. Elements higher on the activity series then are simply those that have a stronger desire to lose electrons than the ones below them.

➥ **Example** _____

Using the activity series for metals found in Chapter 8, predict which of the following reactions is likely to occur as written. Also, for any reaction that does occur, describe the substances involved in the oxidation process as well as those involved in the reduction process.

$$Au(s) + CuCl_2(aq) \rightarrow Cu(s) + AuCl_2(aq)$$
$$3Zn(s) + 2FeCl_3(aq) \rightarrow 3ZnCl_2(aq) + 2Fe(s)$$

The first reaction will not occur (spontaneously). The activity of gold is lower than that of copper, so gold will not replace copper in a compound. In other words, gold has a lesser desire to lose electrons than does copper. The second reaction will occur. Zinc is more active than iron. In this reaction the zinc will be oxidized, or lose electrons, as the reaction occurs. The recipient of electrons lost by the zinc is the iron. Therefore, the iron will be reduced.

OXIDATION STATES

Because the transfer of electrons in chemical reactions is so prevalent, chemists have developed a manner to keep track of their movement. It's important to note that for processes involving ionic species, of the type we have looked at so far, it is somewhat straightforward to distinguish redox from non-redox: The sign and magnitude of the charge on the ions involved allows for an easy view of what is taking place. Using the spontaneous reaction between zinc and iron(III) chloride referenced in the previous section, the zinc started as an atom with no net charge and turned into an ion with a 2+ charge. During chemical reactions, in order for the charge on particles to increase, negative electrons have to be lost and the process of oxidation occurs. A similar analysis of the iron in the reaction shows that the iron began as a 3+ ion (ferric) and ended as a neutral iron atom in the products. In other words, the iron went *down* in charge. Particles that go down in charge do so by gaining negative electrons. This description gives insight to why the process is referred to as *reduction* despite the fact that electrons are being *gained*. Reduction refers to the *change in charge* involving ionic species, not to the change in the number of electrons possessed by a particle as a result of the process taking place. Since reduction involves a decrease in charge, oxidation must involve an increase in charge (when ionic species are involved). Using the term *oxidation* may seem like a strange way to describe this process, but its use will be better understood from a historical context when **combustion** reactions are looked at as redox processes in an upcoming section of this chapter.

As discussed, reactions involving ions are relatively simple to recognize as redox (or not) because it is straightforward to identify changes in charges if they occur. It's not quite so simple when redox reactions involve molecular species. For these types of reactions, as well as those involving ionic species, a related but noteworthily different term is used to describe the responsibility particles have for the electrons around them as a chemical reaction unfolds. To keep track of the transfer of electrons in all formulas (ionic or molecular), chemists have devised a system of electron bookkeeping called **oxidation states** (or **oxidation numbers**). In this system, an oxidation state is assigned to each member of a formula using rules that recognize the degree to which electrons *practically belong* to a particular element in a given substance from an ionic bonding perspective. Basically, elements with high electronegativities are given responsibility for the electrons in a bond, ionic or covalent, and the change in this responsibility will be noted by changes in their oxidation states. In this regard, the oxidation states system assumes an ionic perspective for *all* bonding, i.e., electrons are not shared but *belong* to one element or the other in a chemical bond. Although we know that electrons *are* shared in a large number of chemical bonds, particularly those described as covalent (or molecular), assigning responsibility for the electrons in this way allows for easy recognition of how the accountability for electrons changes during all chemical processes.

Oxidation states are designated by a small number superscript *preceded* by a plus or minus sign. This is not to be confused with the ionic charges we have been using thus far that are shown as plus or minus signs *to the right* of the magnitude of ionic charge as a superscript.

The Rules for Assigning an Oxidation State

The basic rules for assigning an oxidation state to an element in a substance's formula are given below. By applying these simple rules, you can assign oxidation states to elements in practically all substances you may encounter as a beginning chemistry student. To apply these rules, remember that the **sum of the oxidation states must be zero for an electrically neutral compound**. For a polyatomic ion, the **sum of the oxidation states must be equal to the charge on the ion**.

1. The oxidation state of an element of an atom in an element is zero. Examples: 0 for $Na(s)$, $O_2(g)$, and $Hg(\ell)$.
2. The oxidation state of a monatomic ion is the same as its charge. Examples: +1 for Na^+ and –1 for Cl^-.

3. The oxidation state for fluorine is –1 in its compounds. Examples: HF, hydrogen fluoride, has one H at +1 and one F at –1. PF_3 has one P at +3 and three F's at –1 each. (Note that in each compound the sum of the oxidation states is equal to zero.)

4. The oxidation state of oxygen is usually –2 in its compounds. Example: H_2O has two H's at +1 each and one O at –2. (Exceptions occur when the oxygen is bonded to fluorine and the oxidation state of fluorine takes precedence, and in peroxide compounds where the oxidation state is assigned the value of –1.)

5. The oxidation state of hydrogen in most compounds is +1. Examples: H_2O, HCl, and NH_3. (In hydrides, where hydrogen acts like an anion compounded with a metal, there is an exception, however. In this case, hydrogen is assigned the value of –1. Examples: LiH and KH.)

Some examples of determining the oxidation states of other elements in chemical formulas follow.

➥ Example 1

In Na_2SO_4, what is the oxidation state of sulfur?

The first thing to recognize is that this compound is an ionic substance. Ionic substances have two parts—the cation and the anion. In this case, the cation is monatomic and the anion is polyatomic. The cation is Na^+ and the anion is SO_4^{2-} (sulfate). By Rule #2, the oxidation state of the sodium is +1 because the oxidation states of monatomic ions are the same as their charges. By Rule #4, the oxidation state of the oxygen in the sulfate is –2. Now you can look at the complete formula and calculate the oxidation state of the sulfur.

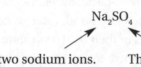

There are two sodium ions.
$2 \times (+1) = +2$

There are four oxygen atoms.
$4 \times (-2) = -8$

Since the positive sum and the negative sum must equal 0,

$$(+2) + x + (-8) = 0$$

The sulfur must have a +6 oxidation state.

➥ Example 2

What is the oxidation state of carbon in the *molecule* CO_2?

By Rule #4, the oxidation state of oxygen is –2, and since there are two oxygen atoms in this formula the total negative sum is –4. Since the positive and negative sums must add to zero, the oxidation state of carbon is +4 in this compound. Keep in mind that oxidation states are not "real" charges and carbon dioxide is not an ionic substance. In this case, the oxidation state of +4 for carbon indicates that *for electron bookkeeping purposes*, carbon will not be responsible for the electrons it is sharing in the bond between the carbon and oxygen atoms and the responsibility will lie with the oxygen. When CO_2 is involved in a chemical process and carbon's responsibility for electrons changes (i.e., its oxidation state changes), a redox reaction will be recognized.

➡ Example 3 _____

In the polyatomic ion dichromate ($Cr_2O_7^{2-}$), what is the oxidation state of chromium?

In a polyatomic ion, the algebraic sum of the positive and negative oxidation states of all the elements must equal the charge on the ion.

$$Cr = 2 \times (x) = 2x$$
$$O = 7 \times (-2) = -14$$

Since the sum of these values must equal –2 (the charge on the polyatomic ion)

$$2x + (-14) = -2$$
$$x = +6$$

The oxidation state of chromium in dichromate is +6.

Using Oxidation States to Recognize Redox Reactions

Once oxidation states can be assigned to elements in the substances involved in a chemical reaction, recognition of the process as being redox or not is straightforward. If the oxidation states change, then a transfer of electrons is taking place. If the oxidation states remain the same, then a redox reaction is not occurring.

Consider these two reactions:

$$Na_2S(aq) + CuSO_4(aq) \rightarrow Na_2SO_4(aq) + CuS(s)$$
$$\text{and}$$
$$2Na(\ell) + Cl_2(g) \rightarrow 2NaCl(s)$$

The oxidation states of each element in all the substances can be determined as shown above each of them here:

$$\overset{+1\ -2}{Na_2S}(aq) + \overset{+2\ +6\ -2}{CuSO_4}(aq) \rightarrow \overset{+1\ +6\ -2}{Na_2SO_4}(aq) + \overset{+2\ -2}{CuS}(s)$$
$$\text{and}$$
$$2\overset{0}{Na}(\ell) + \overset{0}{Cl_2}(g) \rightarrow 2\overset{+1\ -1}{NaCl}(s)$$

Because the oxidation states of the elements in the first reaction don't change, this reaction is not a redox process. You should recognize this as a precipitation reaction, as described in Chapter 8. The second reaction does exhibit a change in oxidation states and should be viewed as a redox reaction. In this reaction the sodium is being oxidized and the chlorine is being reduced. The chlorine could not be reduced (i.e., gain electrons) if the sodium wasn't being oxidized (i.e., losing electrons.) In one regard, the sodium is acting as a facilitator for the reduction of the chlorine. As such, it is referred to as a **reducing agent**. Likewise, the sodium would not lose its electron if it had nowhere to go, so the chlorine is referred to as the **oxidizing agent** in this reaction. Oxidizing agents, then, contain elements that are capable of being reduced by other substances (reducing agents) that contain elements that are capable of being oxidized. Sometimes the terms **oxidizer** and **reducer** are used as labels on the bottles of substances with the tendency to act as oxidizing and reducing agents, respectively.

➥ Example 1 _____

Identify the elements that are being oxidized and reduced in the following reaction. Also, name the oxidizing and reducing agents.

$$2H_2O(\ell) + 2MnO_4^-(aq) + I^-(aq) \rightarrow 2MnO_2(aq) + IO_3^-(aq) + 2OH^-(aq)$$

The oxidation states of the hydrogen and oxygen are not changing in the reaction. The oxidation states of the manganese and the iodine are changing, as shown below.

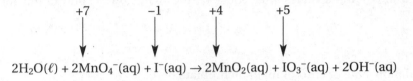

$$2H_2O(\ell) + 2MnO_4^-(aq) + I^-(aq) \rightarrow 2MnO_2(aq) + IO_3^-(aq) + 2OH^-(aq)$$

Since the manganese is changing from an oxidation state of +7 to that of +4, the manganese is being reduced. The iodine is being oxidized from a value of –1 to a value of +5. Manganese is gaining electron responsibility while iodine is losing it. In that light, then, the permanganate ion (MnO_4^-), which contains the manganese, is facilitating the process in which iodine is being oxidized and so is referred to as the oxidizing agent. A typical source of the permanganate ion in chemical reactions like this one is from the compound potassium permanganate, $KMnO_4$, whose bottle is generally labeled with the term "oxidizer." Along a line of similar thinking, the iodide ion, I^-, can be called the reducing agent in this reaction.

➥ Example 2 _____

Completely analyze the following reaction from a redox perspective.

$$3CO(g) + Fe_2O_3(s) \rightarrow 3CO_2(g) + 2Fe(s)$$

The change in the oxidation states of the elements is shown.

$$3\overset{+2\ -2}{CO}(g) + \overset{+3\ -2}{Fe_2O_3}(s) + 3\overset{+4\ -2}{CO_2}(g) + 2\overset{0}{Fe}(s)$$

Since the oxidation state of carbon is increasing from +2 to +4, it is being oxidized, its responsibility for electrons is decreasing, and the compound of which it is a part, carbon monoxide, is called the reducing agent. Carbon monoxide is a common reducer because of its chemical desire to turn into carbon dioxide when oxygen becomes available from other substances. On the other hand, the oxidation state of iron is decreasing from +3 to 0, it is being reduced, its responsibility for electrons is going up, and the compound of which it is a part, iron(III) oxide, is called the oxidizing agent.

COMBUSTION REACTIONS

Combustion reactions are those chemical processes in which substances (called *fuels*) are rapidly oxidized, accompanied by the release of heat and usually light. Combustion is also referred to as **burning**. Typical, common, and historically well-known combustion reactions involve using oxygen as the oxidizing agent—hence, the name for the process of losing electrons has became known as *oxidation*. When a combustion reaction is complete, the elements in the burning fuel form compounds with the oxidizing agent and the responsibility for electrons changes.

The reaction between magnesium and oxygen is a common example of combustion. When the reaction occurs

$$2Mg(s) + O_2(g) \rightarrow 2MgO(s)$$

the oxidation states of magnesium and oxygen change from 0 each to +2 and –2, respectively. A blinding light and large quantity of heat is released by the system as the reaction unfolds. The amount of heat released when 1 mole of a fuel burns is referred to as its *heat of combustion* and is symbolized ΔH_c. The ΔH_c for Mg is 602 kJ/mol. The *change in enthalpy* associated with the burning of 1 mole of carbon

$$C(s) + O_2(g) \rightarrow CO_2(g) + 393.5 \text{ kJ}$$

is –393.5 kJ, also written as $\Delta H_c = -393.5$ kJ, and it represents the heat released in another combustion/redox reaction. The heats of combustion for many common substances can easily be found in reference tables like the kind found in Appendix B of this book.

Hydrocarbon Fuel Combustion Reactions

Another common combustion reaction involves the burning of hydrocarbon fuels. Methane, CH_4, is a typical hydrocarbon fuel. It is the main component of **natural gas**. Analysis of the reaction

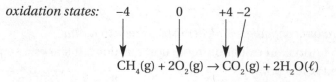

shows that carbon is being oxidized and the oxygen is being reduced. This fuel is used to heat homes and cook food and should be familiar to all chemistry students as the gas used in their laboratory burners.

➡ Example 1 _____

Describe the reaction that occurs when propane, C_3H_8 (the hydrocarbon fuel used in backyard barbecues) combusts, releasing 2,221 kJ when one mole is burned.

It should be noted that when hydrocarbon fuels combust completely in the presence of oxygen, the products of the reaction are carbon dioxide and water. Therefore, the reaction is:

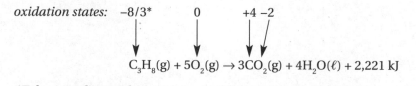

Take note that oxidation states may be expressed in fractions

Once again, the carbon is oxidized from the fraction –8/3 to +4 (increasing from negative to positive) and the oxygen is reduced from 0 to –2. Oxygen is the oxidizing agent and propane is the reducing agent. When this reaction occurs light and heat are released and the heat of combustion, ΔH_c, is –2,221 kJ.

The following terms summarize all the concepts and ideas that were introduced in this chapter. You should be able to explain their meaning and how you would use them in chemistry. They appear in boldface type in this chapter to draw your attention to them. The boldface type also makes it easier for you to look them up if you need to. You could also use the Internet search engine *google.com* on your computer to get a quick and expanded explanation of these terms, laws, and formulas.

anions	ionization	redox
burning	natural gas	reducing agent
cations	oxidation	reduction
combustion	oxidation states	
dissociation	oxidizing agent	

INTERNET RESOURCES

Online content that reinforces major concepts discussed in this chapter can be found at the following Internet addresses if they are still available. *Some may have been changed or deleted.*

Redox Reactions
http://chemistry.about.com/od/generalchemistry/ss/redoxbal.htm
This site offers a tutorial on oxidation-reduction reactions. It shows a step-by-step solution to balancing a redox reaction.

Questions 1–4 refer to the following reaction types:

(A) Single Replacement
(B) Decomposition
(C) Acid–Base
(D) Combustion
(E) Synthesis

1. The reaction type that can be viewed as redox often involving hydrocarbon fuels and oxygen releasing heat and light

2. The reaction type that can be viewed as redox involving an element and a compound becoming a new element and a new compound

3. The reaction type that does not generally involve the transfer of electrons and is not viewed as a redox reaction

4. The reaction type that could possibly be viewed as a redox reaction in which only one substance is present in the reactants

Questions 5–9

Use the following choices to indicate the oxidation state of the underlined symbol in the given formulas

(A) +1
(B) +2
(C) +4
(D) +5
(E) +6

5. $Na_2\underline{S}O_3$

6. $\underline{P}O_4^{3-}$

7. $\underline{Ca}CO_3$

8. $\underline{S}O_2$

9. $K_2\underline{Cr}O_4$

10. Which of the following is true?

(A) Fluorine is assigned the oxidation state of +1 in all compounds.
(B) In the formula of a compound, the algebraic sum of the oxidation states is never zero.
(C) Oxygen's oxidation state in most compounds is −1.
(D) Oxygen can have an oxidation state of −1.
(E) Fluorine can have an oxidation state of −2.

11. Reducing agents are substances that

(A) make reactions take place faster
(B) contain elements that are reduced
(C) make reactions take place slower
(D) contain elements that are oxidized
(E) reduce the transfer of electrons

12. When elemental bromine (Br_2) reacts with a solution of sodium iodide (NaI)

(A) bromine will be reduced
(B) bromine will be oxidized
(C) iodine will be reduced
(D) sodium will be reduced
(E) sodium will be oxidized

Questions 13–15

The following elements are listed in order of decreasing activity as they appear on the activity series

Ca, Na, Mg, Zn, Fe, H, Cu, Hg, Ag, Au

13. The element that is the best reducing agent and the easiest to oxidize is

(A) Ca
(B) Au
(C) H
(D) Fe
(E) Cu

14. Of the following, the element that does NOT react with hydrochloric acid to produce hydrogen gas is

(A) Zn
(B) Fe
(C) Hg
(D) Ca
(E) Mg

15. Which of the following statements is NOT true?

(A) Magnesium has a stronger desire to lose electrons than does mercury.
(B) Gold is often naturally found in its elemental state.
(C) Iron will not replace zinc in a single replacement reaction.
(D) Calcium is often naturally found in its elemental state.
(E) Iron will replace copper in a single replacement reaction.

16. In the combustion reaction between butane (C_4H_{10}) and oxygen, the oxidation state of carbon goes from

(A) +4 to +1, so the carbon is reduced
(B) +2.5 to +4, so the carbon is reduced
(C) +2.5 to +4, so the carbon is oxidized
(D) –2.5 to –4, so the carbon is oxidized
(E) –2.5 to +4, so the carbon is oxidized

17. In the decomposition reaction of a hydrogen peroxide solution:

$$2H_2O_2(aq) \rightarrow 2H_2O(\ell) + O_2(g)$$

(A) the hydrogen is being oxidized
(B) the hydrogen is being reduced
(C) the oxygen is being oxidized
(D) the oxygen is being reduced
(E) both (C) and (D)

Answers and Explanations

1. **(D)** Hydrocarbon combustion reactions can be seen as redox reactions due to the fact that an element, O_2 (with an oxidation state of 0), enters into two compounds, CO_2 and H_2O (with an oxidation state of –2). Elements on one side of a reaction equation that are found in compounds on the other side of the reaction equation mandate a transfer of electrons.

2. **(A)** Generally, reactions in which elements and compounds react to form new elements and compounds are called single replacement. All single replacement reactions are redox.

3. **(C)** Acid–base reactions do not involve the transfer of electrons. From the Brønsted-Lowry perspective, they involve the transfer of protons (H^+ ions).

4. **(B)** Decomposition reactions are characterized by having only one reactant that chemically changes into "simpler" substances. Often those simpler substances formed are elements, so a change in oxidation state takes place (changing from some value other than zero to that value). Therefore, many decomposition reactions are redox. The simpler substances in the products of decomposition do not have to be elements, however. Therefore, decompositions don't have to be redox.

5. **(C)** Since the oxidation state of sodium is +1 and the oxidation state of oxygen is –2, the oxidation state of the sulfur must be +4.

6. **(D)** Since the oxidation state of oxygen is –2 and the sum of the oxidation states of the elements in a polyatomic ion must equal the charge on the ion, the oxidation state of phosphorous must be +5.

7. **(B)** Calcium carbonate is an ionic substance. The cation in the compound is the monatomic ion calcium with a 2+ charge. The charges on monatomic ions are also their oxidation states.

8. **(C)** Since the oxidation state of oxygen is –2, the oxidation state of sulfur must be +4 so that the sum of the oxidation states of the elements in the molecule (a neutral particle) is zero.

9. **(E)** Potassium chromate is an ionic substance. The charge on the potassium is its oxidation state, +1. Since oxygen's oxidation state is –2, the oxidation state of the chromium must be +6.

10. **(D)** Although oxygen's oxidation state is generally –2, it can have an oxidation state of –1 in peroxide compounds.

11. **(D)** By definition, reducing agents contain substances that are oxidized because in order to reduce another substance electrons must be made available.

12. **(A)** The reaction described is a single replacement reaction in which bromine reacts with sodium iodide to produce sodium bromide and iodine.

$$Br_2(\ell) + 2NaI(aq) \rightarrow 2NaBr(aq) + I_2(aq)$$

The bromine is being reduced from 0 to –1. The iodine is being oxidized from –1 to 0 while the oxidation state of the sodium isn't changing.

13. **(A)** Calcium has the highest activity on the list of elements shown. It has a desire to lose electrons more so than any other element on the list.

14. **(C)** Mercury is lower than hydrogen in the activity series. It will not spontaneously replace hydrogen in a compound.

15. **(D)** Calcium is unlikely to be found naturally as an element. It has a high activity level and thus a strong desire to be in compound with other elements. Gold has a low activity level, causing it to be found elementally in nature.

16. **(E)** The reaction of butane and oxygen produces carbon dioxide and water. The oxidation state of carbon in butane is -2.5 since hydrogen is a $+1$. The oxidation state of carbon in carbon dioxide is $+4$ since the oxidation state of oxygen is -2. Because the oxidation state of the carbon is increasing, the carbon is being oxidized.

17. **(E)** The oxidation state of oxygen in peroxide compounds is -1. The oxidation state of the oxygen in water is -2 and in elemental oxygen is 0. Therefore, the oxygen is being both reduced and oxidized. This process is referred to as *disproportionation*. In the reaction, the hydrogen is not being reduced or oxidized.

Some Representative Groups and Families

13

These skills are usually tested on the SAT Subject Test in Chemistry. You should be able to...

→ Describe the properties, both physical and chemical, of the major members of each group and family, and the common compounds formed by the sulfur, halogen, and nitrogen families and by major metals and their alloys.

→ Write equations for major reactions involving these elements.

This chapter will review and strengthen these skills. Be sure to do the Practice Exercises at the end of the chapter.

In the following section, a brief description is given of some of the important and representative groups of elements usually discussed in most first-year chemistry courses.

SULFUR FAMILY

Since we discussed oxygen in Chapter 5, the most important element in this family left to discuss is sulfur.

Sulfur is found free in the volcanic regions of Japan, Mexico, and Sicily. It is removed from the rock mixtures by heating in retorts or furnaces.

Table 12. Allotropic Forms of Sulfur

Characteristic	Rhombic	Allotropic Form Monoclinic	Amorphous
Shape	Rhombic or octahedral crystals	Needle-shaped monoclinic crystals	Noncrystalline
Color	Pale-yellow, opaque, brittle	Yellow, waxy, translucent, brittle	Dark, tough, elastic

Sulfuric Acid

IMPORTANT PROPERTIES OF SULFURIC ACID. Sulfuric acid ionizes in two steps:

$$H_2SO_4(\ell) + H_2O(\ell) \rightleftharpoons H_3O^+(aq) + HSO_4^-(aq) \qquad K_{a_1} \text{ is very large}$$
$$HSO_4^-(aq) + H_2O(\ell) \rightleftharpoons H_3O^+(aq) + SO_4^{2-}(aq) \qquad K_{a_2} \text{ is somewhat small}$$

to form a strong acid solution. The ionization is more extensive in a dilute solution. Most hydronium ions are formed in the first step. Salts formed with the HSO_4^- (bisulfate ion) are called **acid salts**; the SO_4^{2-} (sulfate ion) forms **normal salts**.

Sulfuric acid reacts like other acids, as shown below:

With active metals: $Zn(s) + H_2SO_4(aq) \rightarrow ZnSO_4(aq) + H_2(g)$
(for dilute H_2SO_4)

With bases: $2NaOH(aq) + H_2SO_4(aq) \rightarrow Na_2SO_4(aq) + 2H_2O(\ell)$

With metal oxides: $MgO(s) + H_2SO_4(aq) \rightarrow MgSO_4(aq) + H_2O(\ell)$

With carbonates: $CaCO_3(s) + H_2SO_4(aq) \rightarrow CaSO_4(aq) + H_2O(\ell) + CO_2(g)$

Sulfuric acid has other particular characteristics.

As an oxidizing agent:

$$Cu(s) + 2H_2SO_4(aq) \text{ (concentrated)} \rightarrow CuSO_4(aq) + SO_2(g) + 2H_2O(\ell)$$

As a dehydrating agent with carbohydrates:

$$C_{12}H_{22}O_{11}(\text{sugar}) \xrightarrow[H_2SO_4]{\text{Conc.}} 12C(s) + 11H_2O(\ell)$$

Other Important Compounds of Sulfur

Hydrogen sulfide is a colorless gas having an odor of rotten eggs. It is fairly soluble in water and is poisonous in rather small concentrations. It can be prepared by reacting ferrous sulfide with an acid, such as dilute HCl:

$$FeS(s) + 2HCl(aq) \rightarrow FeCl_2(aq) + H_2S(g)$$

Hydrogen sulfide burns in excess oxygen to form compounds of water and sulfur dioxide. If insufficient oxygen is available, some free sulfur will form. It is only a weak acid in a water solution. Hydrogen sulfide is used widely in qualitative laboratory tests since many metallic sulfides precipitate with recognizable colors. These sulfides are sometimes used as paint pigments. Some common sulfides and their colors are:

ZnS—White
CdS—Bright yellow
As_2S_3—Lemon yellow
Sb_2S_3—Orange
CuS—Black
HgS—Black
PbS—Brown-black

Another important compound of sulfur is **sulfur dioxide**. It is a colorless gas with a suffocating odor.

The structure of sulfur dioxide is a good example of **resonance structures**. Its molecule is depicted in Figure 35.

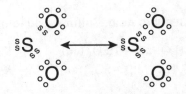

Figure 35. Sulfur Dioxide Molecule

You will notice in Figure 35 that the covalent bonds between sulfur and oxygen are shown in one drawing as single bonds and in the other as double bonds. This signifies that the bonds between the sulfur and oxygens have been shown by experimentation to be neither single nor double bonds, but "hybrids" of the two. Sulfur trioxide, shown below, also has resonance structures.

Resonance is a hybrid of the two structures shown.

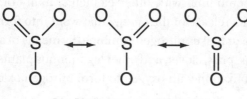

"——" indicates a covalent bond

HALOGEN FAMILY

The common members of the halogen family are shown in Table 13 with some important facts concerning them.

Table 13. Halogen Family

Item	Fluorine	Chlorine	Bromine	Iodine
Molecular formula	F_2	Cl_2	Br_2	I_2
Atomic number	9	17	35	53
Activity	Most active	←——— to ———→		Least active
Outer energy level structure	7 electrons	7 electrons	7 electrons	7 electrons
State and color at room temperature	Gas—pale yellow	Gas—green	Liquid—dark red	Solid—purplish black crystals

Because each halogen lacks one electron in its outer principal energy level, these elements usually are acceptors of electrons (oxidizing agents). Fluorine is the most active nonmetal in the periodic chart.

Some Important Halides and Their Uses

Hydrochloric acid —common acid prepared in the laboratory by reacting sodium chloride with concentrated sulfuric acid. It is used in many important industrial processes.

Silver bromide and silver iodide —halides used on photographic film. Light intensity is recorded by developing as black metallic silver those portions of the film upon which the light fell during exposure.

Hydrofluoric acid —acid used to etch glass by reacting with SiO_2 to release silicon fluoride gas. Also used to frost lightbulbs.

Fluorides —used in drinking water and toothpaste to reduce tooth decay.

NITROGEN FAMILY

The most common member of this family is nitrogen itself. It is a colorless, odorless, tasteless, and rather inactive gas that makes up about four-fifths of the air in our atmosphere. The inactivity of N_2 gas can be explained by the fact that the two atoms of nitrogen are bonded by three covalent bonds (:N ⋮ ⋮ N:) that require a great deal of energy to break. Since nitrogen must be "pushed" into combining with other elements, many of its compounds tend to decompose violently with a release of the energy that went into forming them.

Nature "fixes" nitrogen, or makes nitrogen combine, by means of a nitrogen-fixing bacteria found in the roots of beans, peas, clover, and other leguminous plants. Discharges of lightning also cause some nitrogen fixation with oxygen to form nitrogen oxides.

Nitric Acid

An important compound of nitrogen is nitric acid. This acid is useful in making dyes, celluloid film, and many of the lacquers on cars.

Its physical properties are: it is a colorless liquid (when pure), it is one and one-half times as dense as water, it has a boiling point of 86°C, the commercial form is about 68% pure, and it is miscible with water in all proportions.

Its outstanding chemical properties are: the dilute acid shows the usual properties of an acid except that it rarely produces hydrogen when it reacts with metals, and it is quite unstable and decomposes as follows:

$$4HNO_3(aq) \rightarrow 2H_2O(\ell) + 4NO_2(g) + O_2(g)$$

Because of this ease of decomposition, nitric acid is a good oxidizing agent. When it reacts with metals, the nitrogen product formed will depend on the conditions of the reaction, especially the concentration of the acid, the activity of the metal, and the temperature. If the nitric acid is concentrated and the metal is copper, the principal reduction product will be nitrogen dioxide (NO_2), a heavy, red-brown gas with a pungent odor.

$$Cu(s) + 4HNO_3(aq) \rightarrow Cu(NO_3)_2(aq) + 2NO_2(g) + 2H_2O(\ell)$$

With dilute nitric acid, this reaction is:

$$3Cu + 8HNO_3 \rightarrow 3Cu(NO_3)_2 + 4H_2O + 2NO(g)$$

The product NO, called nitric oxide, is colorless and is immediately oxidized in air to NO_2 gas.

With still more dilute nitric acid, considerable quantities of nitrous oxide (N_2O) are formed; with an active metal like zinc, the product may be the ammonium ion (NH_4^+).

When nitric acid is mixed with hydrochloric acid, the mixture is called **aqua regia** because of its ability to dissolve gold.

METALS
Properties of Metals

Some physical properties of metals are: they have metallic luster, they can conduct heat and electricity, they can be pounded into sheets (are malleable), they can be drawn into wires (are ductile), most have a silvery color, and none is soluble in any ordinary solvent without a chemical change.

The general chemical properties of metals are: they are electropositive, and the more active metallic oxides form bases, although some metals form amphoteric hydroxides that can react as both acids and bases.

Some Important Reduction Methods of Iron Ore

Iron ore is refined by reduction in **a blast furnace**, that is, a large, cylinder-shaped furnace charged with iron ore (usually hematite, Fe_2O_3), limestone, and coke. A hot air blast, often enriched with oxygen, is blown into the lower part of the furnace through a series of pipes called tuyeres. The chemical reactions that occur can be summarized as follows:

Burning coke:	$2C + O_2 \rightarrow 2CO(g)$
	$C + O_2 \rightarrow CO_2(g)$
Reduction of CO_2:	$CO_2 + C \rightarrow 2CO(g)$
Reduction of ore:	$Fe_2O_3 + 3CO \rightarrow 2Fe + 3CO_2(g)$
	$Fe_2O_3 + 3C \rightarrow 2Fe + 3CO(g)$
Formation of slag:	$CaCO_3 \rightarrow CaO + CO_2(g)$
	$CaO + SiO_2 \rightarrow CaSiO_3$

The molten iron from the blast furnace is called **pig iron**.

From pig iron, the molten metal may undergo one of three steel-making processes that burn out impurities and set the contents of carbon, manganese, sulfur, phosphorus, and silicon. Often nickel and chromium are alloyed in steel to give the particular properties of hardness needed for tool parts. The three most important means of making steel involve the basic oxygen, the open-hearth, and the electric furnaces. The first two methods are the most common.

The **basic oxygen furnace** uses a lined "pot" into which the molten pig iron is poured. Then a high-speed jet of oxygen is blown from a water-cooled lance into the top of the pot. This "burns out" impurities to make a batch of steel rapidly and cheaply.

The **open-hearth furnace** is a large oven containing a dish-shaped area to hold the molten iron, scrap steel, and other additives with which it is charged. Alternating blasts of flame are directed across the surface of the melted metal until the proper proportions of additives are established for that "heat" so that the steel will have the particular properties needed by the customer. The tapping of one of these furnaces holding 50 to 400 tons of steel is a truly beautiful sight.

The final method of making steel involves the **electric arc furnace**. This method uses enormous amounts of electricity through graphite cathodes that are lowered into the molten iron to purify it and produce a high grade of steel.

Alloys

An **alloy** is a mixture of two or more metals. In a mixture certain properties of the metals involved are affected. Three of these are:

1. Melting point   The melting point of an alloy is lower than that of its components.
2. Hardness   An alloy is usually harder than the metals that compose it.

3. Crystal structure   The size of the crystalline particles in the alloy determines many of the physical properties. The size of these particles can be controlled by heat treatment. If the alloy cools slowly, the crystalline particles tend to be larger. Thus, by heating and cooling an alloy, its properties can be altered considerably.

Common alloys are:

1. Brass, which is made up of copper and zinc.
2. Bronze, which is made up of copper and tin.
3. Steel, which has controlled amounts of carbon, manganese, sulfur, phosphorus, and silicon, is alloyed with nickel and chromium.
4. Sterling silver, which is alloyed with copper.

Metalloids

In the preceding sections, representative metals and nonmetals have been reviewed, along with the properties of each. Some elements, however, are difficult to classify as one or the other. One example is carbon. The diamond form of carbon is a poor conductor, yet the graphite form conducts fairly well. Neither form looks metallic, so carbon is classified as a nonmetal.

Silicon looks like a metal. However, its conductivity properties are closer to those of carbon.

Since some elements are neither distinctly metallic nor clearly nonmetallic, a third class, called the **metalloids**, is recognized.

The properties of metalloids are intermediate between those of metals and those of nonmetals. Although most metals form ionic compounds, metalloids as a group may form ionic or covalent bonds. Under certain conditions pure metalloids conduct electricity, but do so poorly, and are thus termed **semiconductors**. This property makes the metalloids important in microcircuitry.

The metalloids are located in the periodic table along the heavy dark line that starts alongside boron and drops down in steplike fashion between the elements found lower in the table (see Figure 36).

	13 IIIA	14 IVA	15 VA	16 VIA	17 VIIA
	5 B	6 C	7 N	8 O	9 F
12 IIB	13 Al	14 Si	15 P	16 S	17 Cl
30 Zn	31 Ga	32 Ge	33 As	34 Se	35 Br
48 Cd	49 In	50 Sn	51 Sb	52 Te	53 I
80 Hg	81 Ti	82 Pb	83 Bi	84 Po	85 At

Figure 36. Location of Metalloids

CHAPTER SUMMARY

The following terms summarize all the concepts and ideas that were introduced in this chapter. You should be able to explain their meaning and how you would use them in chemistry. They appear in boldface type in this chapter to draw your attention to them. The boldface type also makes it easier for you to look them up if you need to. You could also use the Internet search engine *google.com* on your computer to get a quick and expanded explanation of these terms, laws, and formulas.

acid salt	bronze	pig iron
allotropic form	electric arc furnace	resonance structure
alloy	metalloid	rhombic, monoclinic,
basic oxygen furnace	normal salt	amorphous
blast furnace	open-hearth furnace	semiconductor
brass		

INTERNET RESOURCES

Online content that reinforces major concepts discussed in this chapter can be found at the following Internet addresses if they are still available. *Some may have been changed or deleted.*

Essays On All of the Chemical Elements
http://pubs.acs.org/cen/80th/elements.html
This site, taken from *Chemical & Engineering News*, displays a periodic table with links to essays on all of the chemical elements by experts in their field.

1. The most active nonmetallic element is

   (A) chlorine
   (B) fluorine
   (C) oxygen
   (D) sulfur

2. The order of decreasing activity of the halogens is

   (A) Fl, Cl, I, Br
   (B) F, Cl, Br, I
   (C) Cl, F, Br, I
   (D) Cl, Br, I, F

3. A light-sensitive substance used on photographic films has the formula

   (A) AgBr
   (B) CaF_2
   (C) CuCl
   (D) $MgBr_2$

4. Sulfur dioxide is the anhydride of

   (A) hydrosulfuric acid
   (B) sulfurous acid
   (C) sulfuric acid
   (D) hyposulfurous acid

5. The charring action of sulfuric acid is due to its being

   (A) a strong acid
   (B) an oxidizing agent
   (C) a reducing agent
   (D) a dehydrating agent

6. Ammonia is prepared commercially by the

   (A) decomposition of salts
   (B) arc process
   (C) combining of hydrogen and nitrogen gases (Haber process)
   (D) contact process

7. A nitrogen compound that has a color is

   (A) nitric oxide
   (B) nitrous oxide
   (C) nitrogen dioxide
   (D) ammonia

8. If a student heats a mixture of ammonium chloride and calcium hydroxide in a test tube, he will detect

   (A) no reaction
   (B) the odor of ammonia
   (C) the odor of rotten eggs
   (D) nitric acid fumes

9. The difference between ammonia and the ammonium ion is

   (A) an electron
   (B) a neutron
   (C) a proton
   (D) hydroxide

10. An important ore of iron is

    (A) bauxite
    (B) galena
    (C) hematite
    (D) smithsonite

11. A reducing agent used in the blast furnace is

    (A) $CaCO_3$
    (B) CO
    (C) O_2
    (D) SiO_2

12. The metal with the electron shell configuration of [Ar] $3d^{10}s^1$ is

    (A) Cu
    (B) Ag
    (C) Au
    (D) Zn
    (E) Al

13. The placement of the halogen family in the Periodic Table explains which of the following statements?

   I. The most active nonmetallic element in the periodic table is fluorine.

   II. The normal physical state of the halogens goes from a solid to a gaseous state as you go down the family.

   III. The halogen elements become ions by filling the outermost *d* orbital.

(A) I only

(B) II only

(C) I and II

(D) II and III

(E) I and III

14. Which of the following properties are attributed to metals?

   I. They are conductors of heat and electricity.

   II. They are malleable and ductile.

   III. They are *all* solids at room temperature.

(A) I only

(B) II only

(C) I and II

(D) I and III

(E) I, II, and III

Answers and Explanations

1. **(B)** Fluorine is the most active nonmetallic element because of its atomic structure. It needs only one electron to complete its outer shell and has the highest electronegativity.

2. **(B)** The order of activity of the halogens is from the smallest atomic radii to the largest straight down the family on the Periodic Table.

3. **(A)** Silver bromide, like many of the halogen salts, is light sensitive and is used in photographic films.

4. **(B)** The reaction of SO_2 with water forms sulfurous acid. The equation is:

$$SO_2 + H_2O \rightarrow H_2SO_3$$

5. **(D)** Sulfuric acid is a strong dehydrating agent and draws water to itself so strongly that is can char sucrose by withdrawing the hydrogen and oxygen from $C_{12}H_{22}O_{11}$.

6. **(C)** The combining of hydrogen and nitrogen gases (Haber proess) is used to prepare ammonia commercially.

7. **(C)** The only nitrogen compound in the list that has a reddish brown color is nitrogen dioxide.

8. **(B)** The reaction of these two chemicals results in the production of ammonium hydroxide, which is unstable and forms ammonia gas and H_2O. The reaction is:

$$2NH_4Cl + Ca(OH)_2 \rightarrow 2NH_4OH + CaCl_2$$

Then, because the ammonium hydroxide is unstable at room temperatures, this reaction occurs: $NH_4OH \rightarrow NH_3\uparrow + H_2O$.

9. **(C)** The ammonia molecule is trigonal pyramidal with an unshared pair of electrons in one corner of the pyramid. This negative charge attracts a H^+ to form the ammonium ion, NH_4^+.

10. **(C)** An important ore of iron is hematite.

11. **(B)** CO, carbon monoxide, is used in the blast furnace as a reducing agent to react with oxide impurities.

12. **(A)** The metal with the given electron shell configuration is Cu. Cu, Ag, and Au are all in Group 11, but Ag and Au have higher atomic numbers and different electron shell configurations.

13. **(A)** Because the halogen family's physical state goes from a gas to a solid and they form ions by completing the outer p orbital, the only true statement is I, that fluorine is the most active.

14. **(C)** Not *all* metals are solids at room temperature. Most notable is mercury, which is a liquid at room temperature.

Carbon and Organic Chemistry

14

These skills are usually tested on the SAT Subject Test in Chemistry. You should be able to...

→ Describe the bonding patterns of carbon and its allotropic forms.

→ Explain the structural pattern and naming of the alkanes, alkenes, and alkynes, and their isomers.

→ Show graphically how hydrocarbons can be changed and the development of these functional groups, their structures, and their names: alcohols, aldehydes, ketones, esters, and amines.

This chapter will review and strengthen these skills. Be sure to do the Practice Exercises at the end of the chapter.

Carbon is unique. It forms inorganic substances such as carbon dioxide, graphite, and diamonds. It also forms organic substances without which life could not exist. It forms planar substances, tetrahedrons, and rings.

CARBON

Forms of Carbon

The element carbon occurs mainly in three allotropic forms: diamond, graphite, and amorphous (although some evidence shows the amorphous forms have some crystalline structure). In the mid-1980s, **fullerenes** were identified as a new allotropic form of carbon. They are found in soot that forms when carbon-containing materials are burned with limited oxygen. Their structure consists of near-spherical cages of carbon atoms resembling geodesic domes.

The **diamond** form has a close-packed crystal structure that gives it its property of extreme hardness. In it each carbon is bonded to four other carbons in a tetrahedron arrangement like this: These covalent solids form crystals that can be viewed as a single giant molecule made up of an almost endless number of covalent bonds. Because all of the bonds in this structure are equally strong, covalent solids are often very hard, and they are notoriously difficult to melt. Diamond is the hardest natural substance. At atmospheric pressure, it melts at 3,550°C.

TIP

<u>Allotropic forms of carbon:</u>

diamond

graphite

amorphous

fullerenes

Diamond uses
the *sp³* hybrid
orbitals to
explain its
tetrahedron
structure.

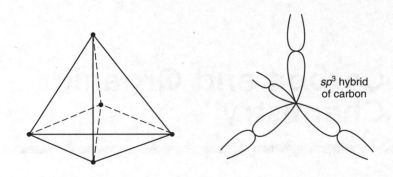

It has been possible to make synthetic diamonds in machines that subject carbon to extremely high pressures and temperatures. Most of these diamonds are used for industrial purposes, such as dies.

The graphite form is made up of planes of hexagonal structures that are weakly bonded to the planes above and below. This explains graphite's slippery feeling and makes it useful as a dry lubricant. Graphite is also mixed with clay to make "lead" for lead pencils. Its structure can be seen below. Graphite also has the property of being an electrical conductor.

TIP

Graphite uses the trigonal *sp²* hybrids

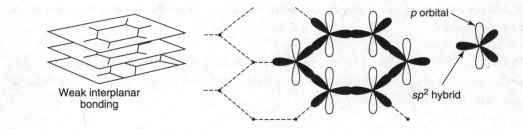

Some common amorphous forms of carbon are charcoal, coke, bone black, and lampblack.

Carbon Dioxide

Carbon dioxide (CO_2) is a widely distributed gas that makes up 0.04 percent of the air. There is a cycle that keeps this figure relatively stable. It is shown in Figure 37.

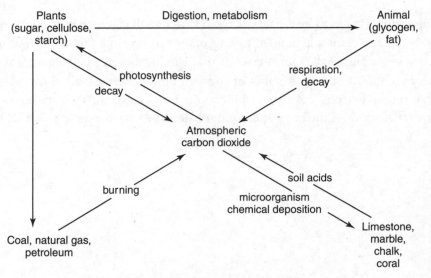

Figure 37. Carbon Dioxide Cycle

LABORATORY PREPARATION OF CO₂

The usual laboratory preparation consists of reacting calcium carbonate (marble chips) with hydrochloric acid, although any carbonate or bicarbonate and any common acid could be used. The gas is collected by water displacement or air displacement.

The test for carbon dioxide consists of passing it through limewater ($Ca(OH)_2$). If CO_2 is present the limewater turns cloudy because of the formation of a white precipitate of finely divided $CaCO_3$:

$$Ca(OH)_2(aq) + CO_2(g) \rightarrow CaCO_3(s) + H_2O(\ell)$$

Continued passing of CO_2 into the solution will eliminate the cloudy condition because the insoluble $CaCO_3$ becomes soluble calcium bicarbonate ($Ca(HCO_3)_2$):

$$CaCO_3(s) + H_2O(\ell) + CO_2(g) \rightarrow Ca^{2+}(HCO_3^-)^2(aq)$$

This reaction can easily be reversed with increased temperature or decreased pressure. This is the way stalagmites and stalactites form on the floors and roofs of caves, respectively. The ground water containing calcium bicarbonate is deposited on the roof and floor of the cave and decomposes into solid calcium carbonate formations.

IMPORTANT USES OF CO₂

1. Because CO_2 is the acid anhydride of carbonic acid, it forms the acid when reacted with soft drinks, thus making them "carbonated" beverages.

$$CO_2(g) + H_2O(\ell) \rightarrow H_2CO_3(aq)$$

2. Solid carbon dioxide (–78°C), or "dry ice," is used as a refrigerant because it has the advantage of not melting into a liquid; instead, it sublimes and in the process absorbs 3 times as much heat per gram as ice.

3. Fire extinguishers make use of CO_2 because of its properties of being $1\frac{1}{2}$ times heavier than air and not supporting ordinary combustion. It is used in the form of CO_2 extinguishers, which release CO_2 from a steel cylinder in the form of a gas to smother the fire.

4. Plants consume CO_2 in the **photosynthesis process**, in which chlorophyll (the catalyst) and sunlight (the energy source) must be present. The reactants and products of this reaction are:

$$6CO_2(g) + 6H_2O(\ell) \rightarrow \underset{\text{cellulose}}{C_6H_{12}O_6(s)} + 6O_2(g)$$

Know the photosynthesis process.

ORGANIC CHEMISTRY

Organic chemistry may be defined simply as the chemistry of the compounds of carbon. Since Friedrich Wöhler synthesized urea in 1828, chemists have synthesized thousands of carbon compounds in areas of dyes, plastic, textile fibers, medicines, and drugs. The number of organic compounds has been estimated to be in the neighborhood of a million and constantly increasing.

The carbon atom (atomic number 6) has four electrons in its outermost energy level, which show a tendency to be shared (electronegativity of 2.5) in covalent bonds. By this means, carbon bonds to other carbons, hydrogens, halogens, oxygen, and other elements to form the many compounds of organic chemistry.

HYDROCARBONS

Hydrocarbons, as the name implies, are compounds containing only carbon and hydrogen in their structures. The simplest hydrocarbon is methane, CH_4. As previously mentioned, this type of formula, which shows the kinds of atoms and their respective numbers, is called an **empirical** formula. In organic chemistry this is not sufficient to identify the compound it is used to represent. For example, the empirical formula C_2H_6O could denote either an ether or an ethyl alcohol. For this reason, a **structural** formula is used to indicate how the atoms are arranged in the molecule. The ether of C_2H_6O looks like this:

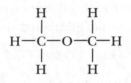

TIP

Organic chemistry makes use of structural formulas to show atomic arrangements.

whereas the ethyl alcohol is represented by this structural formula:

To avoid ambiguity, structural formulas are more often used than empirical formulas in organic chemistry. The structural formula of methane is

$$
\begin{array}{c}
\text{H} \\
| \\
\text{H}-\text{C}-\text{H} \\
| \\
\text{H}
\end{array}
$$

Alkane Series (Saturated)

TIP

Alkanes are C_nH_{2n+2}. They are homologous.

Methane is the first member of a hydrocarbon series called the **alkanes** (or paraffin series). The general formula for this series is C_nH_{2n+2}, where n is the number of carbons in the molecule. Table 14 provides some essential information about this series. Since many other organic structures use the stem of the alkane names, you should learn these names and structures well. Notice that, as the number of carbons in the chain increases, the boiling point also increases. The first four alkanes are gases at room temperature; the subsequent compounds are liquid, then become more viscous with increasing length of the chain.

Since the chain is increased by a carbon and two hydrogens in each subsequent molecule, the alkanes are referred to as a **homologous** series.

TIP

Learn the names of the first 10 alkanes.

The alkanes are found in petroleum and natural gas. They are usually extracted by fractional distillation, which separates the compounds by varying the temperature so that each vaporizes at its respective boiling point.

Table 14. The Alkanes

IUPAC Name	Molecular Formula	Number of Structural Isomers	Structure	State at Room Temperature	Boiling Point (°C)
Methane	CH_4	1	$H-\overset{\displaystyle H}{\underset{\displaystyle H}{C}}-H$	Gas	–162
Ethane	C_2H_6	1	$H-\overset{H}{\underset{H}{C}}-\overset{H}{\underset{H}{C}}-H$	Gas	–89
Propane	C_3H_8	1	(structural formula)	Gas	–42
n-Butane	C_4H_{10}	2	(structural formula)	Gas	0
n-Pentane	C_5H_{12}	3	(structural formula)	Liquid (*Note:* Solid at 17 carbons in the chain)	36
n-Hexane	C_6H_{14}	5	$CH_3–CH_2–CH_2–CH_2–CH_3$		69
n-Heptane	C_7H_{16}	7	$CH_3–CH_2–CH_2–CH_2–CH_2–CH_3$		98
n-Octane	C_8H_{18}	18	$CH_3–CH_2–CH_2–CH_2–CH_2–CH_2–CH_3$		126
n-Nonane	C_9H_{20}	35	$CH_3–CH_2–CH_2–CH_2–CH_2–CH_2–CH_2–CH_3$		151
n-Decane	$C_{10}H_{22}$	75	$CH_3–CH_2–CH_2–CH_2–CH_2–CH_2–CH_2–CH_2–CH_2–CH_3$		174

When the alkanes are burned with sufficient air, the compounds formed are CO_2 and H_2O. An example is:

$$2C_2H_6(g) + 7O_2(g) \rightarrow 4CO_2(g) + 6H_2O(g)$$

The alkanes can be reacted with halogens so that hydrogens are replaced by a halogen atom: These are called **alkyl halides**.

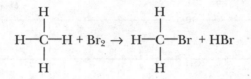

Some common substitution compounds of methane are:

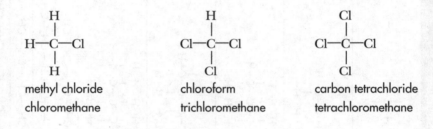

| methyl chloride | chloroform | carbon tetrachloride |
| chloromethane | trichloromethane | tetrachloromethane |

NAMING ALKANE SUBSTITUTIONS

When an alkane hydrocarbon has an end hydrogen removed, it is referred to as an alkyl substituent or group. The respective name of each is the alkane name with -*ane* replaced by -*yl*. These are called **alkyl groups**.

TIP

Replace -*ane* with -*yl* to form alkyl groups.

Alkane	Alkyl Group	Compounds
methane	methyl	bromomethane
butane	butyl	1-chlorobutane

One method of naming a substitution product is to use the alkyl name for the respective chain and the halide as shown above. The halogen takes the form of fluoro-, bromo-, iodo-, and so on, depending on the halogen, and is attached to an alkane name. It precedes the alkane name, as shown above in bromomethane and 1-chlorobutane.

The IUPAC system uses the name of the longest carbon chain as the parent chain. The carbon atoms are numbered in the parent chain to indicate where branching or substitution takes place. The direction of numbering is chosen so that the lowest numbers possible are given to the side chains. The complete name of the compound is arrived at by first naming the attached group, each of these being prefixed by the number of the carbon to which it is attached, and then the parent alkane. If a particular group appears more than once, the appropriate prefix (di, tri, and so on) is used to indicate how many times the group appears. A carbon atom number must be used to indicate the position of each such group. If two or more of the same group are attached to the same carbon atom, the number of the carbon atom is repeated. If two or more different substituted groups are in a name, they are arranged alphabetically.

➡ **Example 1** _____

2,2-dimethylbutane

TIP

Numbers have been added to the longest chain for identification only.

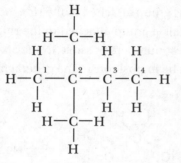

➡ **Example 2** _____

1,1-dichloro-3-ethyl-2,4-dimethylpentane

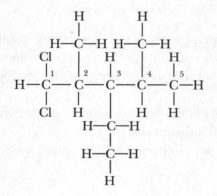

➡ **Example 3** _____

2-iodo-2-methylpropane

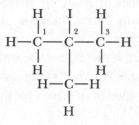

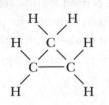

CYCLOALKANES

Starting with propane in the alkane series, it is possible to get a ring form by attaching the two chain ends. This reduces the number of hydrogens by two.

This hydrocarbon is called **cyclopropane**.

Cycloalkanes are named by adding the prefix cyclo- to the name of the straight-chain alkane with the same number of hydrocarbons, as shown above.

When there is only one alkyl group attached to the ring, no position number is necessary. When there is more than one alkyl group attached to the ring, the carbon atoms in the ring are numbered to give the lowest numbers possible to the alkyl groups. This means that one of the alkyl groups will always be in position 1. The general formula is C_nH_{2n}.

Here is an example:

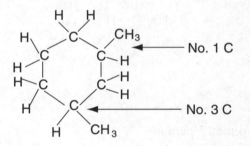

1,3-dimethylcyclohexane

If there are two or more alkyl groups attached to the ring, number the carbon atoms in the ring. Assign position number one to the alkyl group that comes first in alphabetical order, then number in the direction that gives the rest of the alkyl groups the lowest numbers possible.

Because all the members of the alkane series have single covalent bonds, this series and all such structures are said to be **saturated**.

If the hydrocarbon molecule contains double or triple covalent bonds, it is referred to as **unsaturated**.

PROPERTIES AND USES OF ALKANES

Properties for some straight-chain alkanes are indicated in Table 14. The trends in these properties can be explained by examining the structures of alkanes. The carbon-hydrogen bonds are nonpolar. The only forces of attraction between nonpolar molecules are weak intermolecular, or London dispersion, forces. These forces increase as the mass of a molecule increases.

The table also shows the physical states of alkanes. Smaller alkanes exist as gases at room temperature, while larger ones exist as liquids. Gasoline and kerosene consist mostly of liquid alkanes. Seventeen carbons are needed in the chain for the solid form to occur. Paraffin wax contains solid alkanes.

TIP

Cycloalkanes form single-bonded ring compounds.

The differences in the boiling points of mixtures of the liquid alkanes found in petroleum make it possible to separate the various components by **fractional distillation**. This is the major industrial process used in refining petroleum into gasoline, kerosene, lubricating oils, and several other minor components.

Alkene Series (Unsaturated)

The **alkene** series has a double covalent bond between two adjacent carbon atoms. The general formula of this series is C_nH_{2n}. In naming these compounds, the suffix of the alkane is replaced by *-ene*. Two examples:

Alkenes have the form C_nH_{2n}.

$$
\begin{array}{ccc}
& H & H \\
& | & | \\
H - & C = C & - H \\
\end{array}
\qquad \text{ethene (common name: ethylene)}
$$

$$
\begin{array}{cccc}
& H & H & H \\
& | & | & | \\
H - & C = C & - C & - H \\
& & & | \\
& & & H \\
\end{array}
\qquad \text{1-propene (common name: propylene)}
$$

Naming a more complex example is:

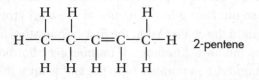

$$CH_2{-}CH_3$$
$$|$$
$$CH_2{=}C{-}CH_2{-}CH_2{-}CH_3$$

The position number and name of the alkyl group are in front of the double-bond position number. The alkyl group above is an ethyl group. It is on the second carbon atom of the parent hydrocarbon.

The name is 2-ethyl-1-pentene.

If the double bond occurs on an interior carbon, the chain is numbered so that the position of the double bond is designated by the lowest possible number assigned to the first doubly bonded carbon. For example:

$$
\begin{array}{ccccc}
H & H & & H & \\
| & | & & | & \\
H - C - C - C = C - C - H & & \text{2-pentene} \\
| & | & | & | & | \\
H & H & H & H & H \\
\end{array}
$$

The bonding is more complex in the double covalent bond than in the single bonds in the molecule. Using the orbital pictures of the atom, we can show this as follows:

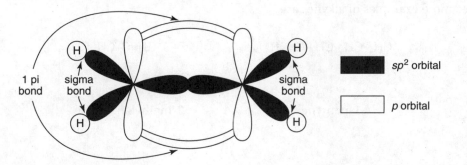

The two *p* lobes attached above and below constitute *one* bond called a pi (π) bond.

The *sp²* orbital bonds between the carbons and with each hydrogen are referred to as sigma (σ) bonds.

Alkyne Series (Unsaturated)

TIP

Alkynes have the form C_nH_{2n-2}.

The **alkyne** series has a triple covalent bond between two adjacent carbons. The general formula of this series is C_nH_{2n-2}. In naming these compounds, the alkane suffix is replaced by *-yne*. Two examples:

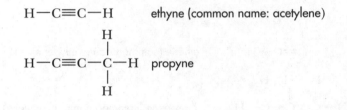

The orbital structure of ethyne can be shown as follows:

TIP

Notice that *pi bonds* are between *p* orbitals and that *sigma bonds* are between *s* and *p* orbitals.

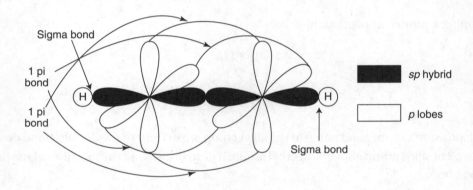

The bonds formed by the *p* orbitals and the one bond between the *sp* orbitals make up the triple bond.

The preceding examples show only one triple bond. If there is more than one triple bond, modify the suffix to indicate the number of triple bonds. For example, 2 would be a diyne, 3 would be a triyne, and so on. Next add the names of the alkyl groups if they are attached. Number the carbon atoms in the chain so that the first carbon atom in the triple bond nearest the end of the chain has the lowest number. If numbering from both ends gives the same positions for two triple bonds, then number from the end nearest the first alkyl group. Then, place the position numbers of the triple bonds immediately before the name of the parent hydrocarbon alkyne and place the alkyl group position numbers immediately before the name of the corresponding alkyl group.

Two more examples of alkynes are:

$$CH_3-CH_2-CH_2-C\equiv CH \qquad\qquad CH\equiv C-CH-CH_3$$

$$\qquad\qquad\qquad\qquad\qquad\qquad\qquad\qquad\qquad\qquad\qquad CH_3$$

1-pentyne   2-methyl-1-butyne

Naming a more complex example is:

$$CH_2-CH_3$$
$$|$$
$$CH\equiv C-CH_2-CH_2-CH_3$$

The position number and the name of the alkyl group are placed in front of the double-bond position number. The alkyl group above is an ethyl group. It is on the second carbon atom of the parent hydrocarbon.

The name is 2-ethyl-1-pentyne.

Aromatics

The aromatic compounds are unsaturated ring structures. The basic formula of this series is C_nH_{2n-6}, and the simplest compound is benzene (C_6H_6). The benzene structure is a resonance structure that is represented like this:

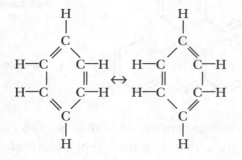

Note: The carbon-to-carbon bonds are neither single nor double bonds but hybrid bonds. This structural representation is called **resonance structures.**

The benzene resonance structure can also be shown like this:

The orbital structure can be represented like this:

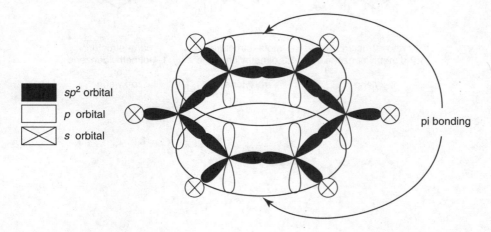

TIP

C_6H_6, benzene is the simplest aromatic compound.

Most of the aromatics have an aroma, thus the name "aromatic."

The C_6H_5 group is a substituent called phenyl. This is the benzene structure with one hydrogen missing. If the phenyl substituent adds a methyl group, the compound is called toluene or methyl benzene.

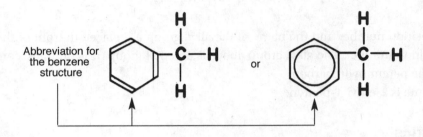

Abbreviation for the benzene structure

or

Two other members of the benzene series and their structures:

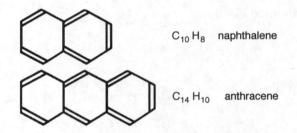

$C_{10}H_8$ naphthalene

$C_{14}H_{10}$ anthracene

The IUPAC system of naming benzene derivatives, as with chain compounds, involves numbering the carbon atoms in the ring in order to pinpoint the locations of the side chains. However, if only two groups are substituted in the benzene ring, the compound formed will be a benzene derivative having three possible isomeric forms. In such cases, the prefixes **ortho-**, **meta-**, and **para-**, abbreviated as o-, m-, and p-, may be used to name the isomers. In the ortho- structure, the two substituted groups are located on adjacent carbon atoms. In the meta- structure, they are separated by one carbon atom. In the para- structure, they are separated by two carbon atoms.

ortho- structure
1,2-dimethylbenzene
or
o-xylene

meta- structure
1,3-dimethylbenzene
or
m-xylene

para- structure
1,4-dimethylbenzene
or
p-xylene

Isomers

Many of the chain hydrocarbons can have the same formula, but their structures may differ. For example, butane is the first compound that can have two different structures or **isomers** for the same formula.

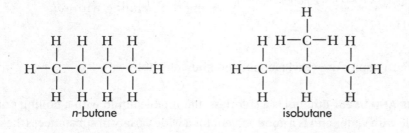

n-butane   isobutane

This isomerization can be shown by the following equation:

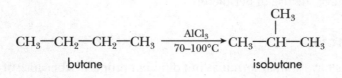

butane   isobutane

The isomers have different properties, both physical and chemical, from those of hydrocarbons with the normal structure.

HYDROCARBON DERIVATIVES

Alcohols—Methanol and Ethanol

The simplest alcohols are alkanes that have one or more hydrogen atoms replaced by the hydroxyl group, –OH. This is called its **functional group**.

- ■ **Methanol**

Methanol is the simplest alcohol. Its structure is

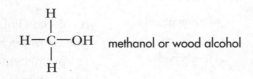

methanol or wood alcohol

PROPERTIES AND USES. Methanol is a colorless, flammable liquid with a boiling point of 65°C. It is miscible with water, is exceedingly poisonous, and can cause blindness if taken internally. It can be used as a fuel, as a solvent, and as a denaturant to make ethyl alcohol, unsuitable for drinking.

■ Ethanol

Ethanol is the best known and most used alcohol. Its structure is

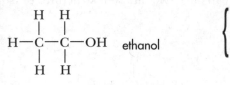 ethanol

> Notice that the alcohol names are derived from the alkanes by replacing the *e* ending with *-ol.*

Its common names are ethyl alcohol and grain alcohol.

PROPERTIES AND USES. Ethanol is a colorless, flammable liquid with a boiling point of 78°C. It is miscible with water and is a good solvent for a wide variety of substances (these solutions are often referred to as "tinctures"). It can be used as an antifreeze because of its low freezing point, –115°C, and for making acetaldehyde and ether. It is presently used in gasoline as an alternative to reduce the use of petroleum.

Other Alcohols

Isomeric alcohols have similar formulas but different properties because of their differences in structure. If the —OH is attached to an end carbon, the alcohol is called a primary alcohol. If attached to a "middle" carbon, it is called a secondary alcohol. Some examples:

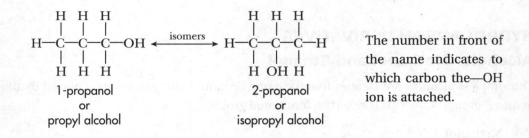

The number in front of the name indicates to which carbon the—OH ion is attached.

Other alcohols with more than one —OH group:

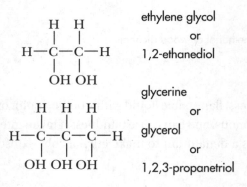

A colorless liquid, high boiling point, low freezing point. Used as permanent antifreeze in automobiles.

Colorless liquid, odorless, viscous, sweet taste. Used to make nitroglycerine, resins for paint, and cellophane.

Aldehydes

The functional group of an aldehyde is the $-C\!\!\diagup_{\diagdown H}^{O}$, formyl group. The general formula is RCHO, where R represents a hydrocarbon radical.

PREPARATION FROM AN ALCOHOL. Aldehydes can be prepared by the oxidation of an alcohol. This can be done by inserting a hot copper wire into the alcohol. A typical reaction is:

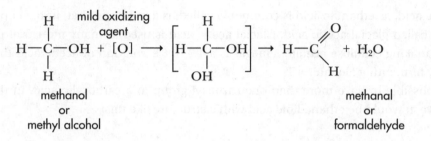

The middle structure is an intermediate structure; since two hydroxyl groups do not stay attached to the same carbon, it changes to the aldehyde by a water molecule "breaking away."

The aldehyde name is derived from the alcohol name by dropping the -*ol* and adding -*al*. Ethanol forms ethanal (acetaldehyde) in the same manner.

Organic Acids or Carboxylic Acids

The functional group of an organic acid is the $-C\!\!\diagup_{\diagdown OH}^{O}$, carboxyl group. The general formula is R—COOH.

PREPARATION FROM AN ALDEHYDE. Organic acids can be prepared by the mild oxidation of an aldehyde. The simplest acid is methanoic acid, which is present in ants, bees, and other insects. A typical reaction is:

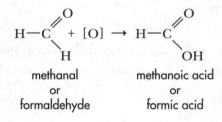

Notice that in the IUPAC system the name is derived from the alkane stem by adding -*oic*.

Ethanal can be oxidized to ethanoic acid:

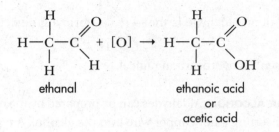

ethanal   ethanoic acid
                                 or
                           acetic acid

Acetic acid, as ethanoic acid is commonly called, is a mild acid that, in the concentrated form, is called glacial acetic acid. Glacial acetic acid is used in many industrial processes, such as making cellulose acetate. Vinegar is a 4% to 8% solution of acetic acid that can be made by fermenting alcohol.

It is possible to have more than one carboxyl group in a carboxylic acid. In the ethane derivative, it would be ethanedioic acid with a structure like this:

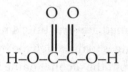

ethanedioic acid

Summary of Oxygen Derivatives

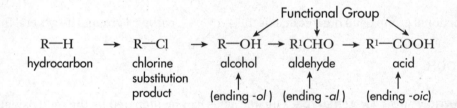

Functional Group

| R—H | → | R—Cl | → | R—OH | → | R¹CHO | → | R¹—COOH |

R—H   →   R—Cl   →   R—OH → R¹CHO → R¹—COOH

hydrocarbon   chlorine   alcohol   aldehyde   acid
                  substitution
                  product   (ending -*ol*)   (ending -*al*)   (ending -*oic*)

Note: R¹ indicates a hydrocarbon chain different from R by having one less carbon in the chain.

An actual example using ethane:

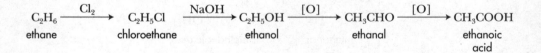

C_2H_6 $\xrightarrow{Cl_2}$ C_2H_5Cl $\xrightarrow{NaOH}$ C_2H_5OH $\xrightarrow{[O]}$ CH_3CHO $\xrightarrow{[O]}$ CH_3COOH

ethane   chloroethane   ethanol   ethanal   ethanoic
                                                                          acid

Ketones

When a secondary alcohol is slightly oxidized, it forms a compound having the functional group , and called a ketone. The R¹ indicates that this group need not be the same as R. An example is:

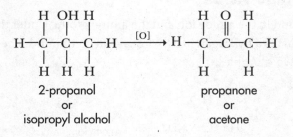

2-propanol
or
isopropyl alcohol

propanone
or
acetone

Example in a longer chain:

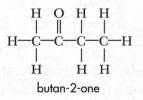

butan-2-one

In the IUPAC method the name of the ketone has the ending -*one* with a digit indicating the carbon that has the double-bonded oxygen preceding the ending in larger chains, as shown in butan-2-one. Another method of designating a ketone is to name the radicals on either side of the ketone structure and use the word **ketone**. In the preceding reaction, the product would be dimethyl ketone.

Note that both aldehydes and ketones contain the carbonyl group in their structures. In the aldehydes, it is at the end of the chain, and, in acids, it is the interior of the chain.

Ethers

When a primary alcohol, such as ethanol, is dehydrated with sulfuric acid, an ether forms. The functional group is R—O—R¹, in which R¹ may be the same hydrocarbon group, as shown in example 1 below, or a different hydrocarbon group, as shown in example 2.

1.

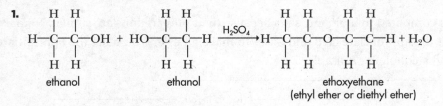

ethanol

ethanol

ethoxyethane
(ethyl ether or diethyl ether)

2. Another ether with unlike groups, R—O—R¹:

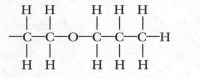

ethoxypropane
(ethyl propyl ether)

In the IUPAC method, the ether name, as shown in the examples, is made up of two attached alkyl chains to the oxygen. The shorter of the two chains becomes the first part of the name, with the *–ane* suffix changed to *–oxy* and the name of the longer alkane chain as the suffix. Examples are ethoxyethane and ethoxypropane.

Diethyl ether is commonly referred to as ether and is used as an anesthetic.

Amines and Amino Acids

The group NH_2^- is found in the amide ion and the amino group. Under the proper conditions, the amide ion can replace a hydrogen in a hydrocarbon compound. The resulting compound is called an **amine**. Two examples:

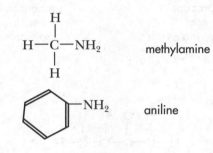

TIP

The amide functional group is

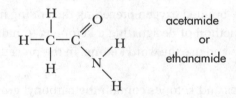

In *amides*, the NH_2^- group replaces a hydrogen in the carboxyl group. When naming amides, the *-ic* of the common name or the *-oic* of the IUPAC name of the parent acid is replaced by *-amide*. For example:

acetamide
or
ethanamide

TIP

Esters can be compared to inorganic salts.

Amino acids are organic acids that contain one or more amino groups. The simplest uncombined amino acid is glycine, or amino acetic acid, NH_2CH_2—COOH. More than 20 amino acids are known, about half of which are essential in the human diet because they are needed to make up the body proteins.

Esters

TIP

Note the functional group

$$R-\overset{\overset{\displaystyle O}{\|}}{C}-O-R^1$$

Esters are often compared to inorganic salts because their preparations are similar. To make a salt, you react the appropriate acid and base. To make an ester, you react the appropriate organic acid and alcohol. For example:

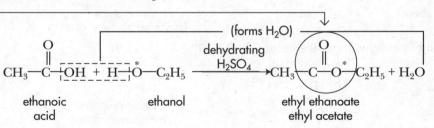

ethanoic acid   ethanol   ethyl ethanoate
                                          ethyl acetate

The name is made up of the alkyl substituent of the alcohol and the acid name, in which -*ic* is replaced with -*ate*.

The general equation is:

**Note the
functional group.**

$$\underset{\text{alcohol}}{R\overset{*}{O}-H} + \underset{\text{acid}}{R^1CO-OH} \rightarrow \underset{\text{ester}}{R^1CO\overset{*}{O}-R} + HOH$$

Esters usually have sweet smells and are used in perfumes and flavor extracts.

The following chart summarizes the organic structures and formulas discussed in this section.

Classes of Organic Compounds

Class	Functional Group	General Formula
Alcohol	— OH	R — OH
Alkyl halides	— X X = F, Cl, Br, or I	R — X
Ether	— O —	R — O — R′
Aldehyde	$-\overset{\displaystyle O}{\underset{\displaystyle H}{C}}$	$R-\overset{\displaystyle O}{\underset{\displaystyle H}{C}}$
Ketone	$-\overset{\displaystyle O}{C}-$	$R-\overset{\displaystyle O}{C}-R'$
Carboxylic acid	$-\overset{\displaystyle O}{\underset{\displaystyle OH}{C}}$	$R-\overset{\displaystyle O}{\underset{\displaystyle OH}{C}}$
Ester	$-\overset{\displaystyle O}{C}-O-$	$R-\overset{\displaystyle O}{C}-O-R'$
Amine	$-N{\overset{\displaystyle H}{\underset{\displaystyle H}{}}}$	$R-N{\overset{\displaystyle H}{\underset{\displaystyle H}{}}}$

The following terms summarize all the concepts and ideas that were introduced in this chapter. You should be able to explain their meaning and how you would use them in chemistry. They appear in boldface type in this chapter to draw your attention to them. The boldface type also makes it easier for you to look them up if you need to. You could also use the Internet search engine *google.com* on your computer to get a quick and expanded explanation of these terms, laws, and formulas.

alcohol	aromatics	isomer
aldehyde	diamond	ketone
alkane	ester	ortho-, meta-, para positions
alkene	ether	photosynthesis
alkyl groups	fullerene	saturated hydrocarbon
alkyne	functional groups	unsaturated hydrocarbon
amine	hydrocarbon	
amino acid	hydrogenation	

INTERNET RESOURCES

Online content that reinforces major concepts discussed in this chapter can be found at the following Internet addresses if they are still available. *Some may have been changed or deleted.*

The Chemistry of Carbon

http://www.nyu.edu/pages/mathmol/modules/carbon/carbon1.html
This site offers an interesting tutorial on the chemistry of carbon.

Organic Nomenclature

http://www.chem.ucalgary.ca/courses/351/WebContent/orgnom/index.html
This website offers a tutorial on naming organic compounds.

1. Carbon atoms usually

    (A) lose 4 electrons
    (B) gain 4 electrons
    (C) form 4 covalent bonds
    (D) share the 2 electrons in the first principal energy level

2. Coke is produced from bituminous coal by

    (A) cracking
    (B) synthesis
    (C) substitution
    (D) destructive distillation

3. The usual method for preparing carbon dioxide in the laboratory is

    (A) heating a carbonate
    (B) fermentation
    (C) reacting an acid and a carbonate
    (D) burning carbonaceous materials

4. The precipitate formed when carbon dioxide is bubbled into limewater is

    (A) $CaCl_2$
    (B) H_2CO_3
    (C) CaO
    (D) $CaCO_3$

5. The "lead" in a lead pencil is

    (A) bone black
    (B) graphite and clay
    (C) lead oxide
    (D) lead peroxide

6. The first and simplest alkane is

    (A) ethane
    (B) methane
    (C) C_2H_2
    (D) methene
    (E) CCl_4

7. Slight oxidation of a primary alcohol gives

    (A) a ketone
    (B) an organic acid
    (C) an ether
    (D) an aldehyde
    (E) an ester

8. The characteristic group of an organic ester is

    (A) —CO—
    (B) —COOH
    (C) —CHO
    (D) —O—
    (E) —COO—

9. The organic acid that can be made from ethanol is

    (A) acetic acid
    (B) formic acid
    (C) C_3H_7OH
    (D) found in bees and ants
    (E) butanoic acid

10. An ester can be prepared by the reaction of

    (A) two alcohols
    (B) an alcohol and an aldehyde
    (C) an alcohol and an organic acid
    (D) an organic acid and an aldehyde
    (E) an acid and a ketone

11. Compounds that have the same composition but differ in their structural formulas

    (A) are used for substitution products
    (B) are called isomers
    (C) are called polymers
    (D) have the same properties
    (E) are usually alkanes

12. Ethene is the first member of the

    (A) alkane series
    (B) saturated hydrocarbons
    (C) alkyne series
    (D) unsaturated hydrocarbons
    (E) aromatic hydrocarbons

The following questions are in the format that is used on the **SAT Subject Test in Chemistry**. If you are not familiar with these types of questions, study pages xiii–xviii before doing the remainder of the review questions.

> **Directions:** Each of the following sets of lettered choices refers to the numbered questions immediately below it. For each numbered item, choose the one lettered choice that fits it best. Every choice in a set may be used once, more than once, or not at all.

Questions 13–20

(A) $CH_3-CH_2-CH_3$

(B) $CH_3-C\overset{\displaystyle O}{\underset{OH}{\diagdown}}$

(C) $CH_3-O-C_3H_7$

(D) $CH_3-\overset{\displaystyle O}{\overset{\|}{C}}-CH_3$

(E) $CH_3-CH_2-N\overset{H}{\underset{H}{\diagup}}$

13. Which organic structure is ethylamine?

14. Which organic structure is methyl propyl ether (methoxypropane)?

15. Which organic structure is propane?

16. Which organic structure is ethanoic acid?

17. Which organic structure is propanone?

Questions 18–20

Using the same choices, match the functional groups named to the structure that contains it.

18. Which structure contains an organic acid functional group?

19. Which structure contains a ketone grouping?

20. Which structure contains an amine group?

Answers and Explanations

1. **(C)** Because carbon has 4 electrons in its outer energy level, it usually forms four covalent bonds to fill each of four sp^3 orbitals.

2. **(D)** Bituminous coal has too much gaseous impurities to burn at a high temperature needed to refine iron ore. It is heated in coke ovens to form the hotter and cleaner burning coke.

3. **(C)** The reaction of an acid and a carbonate is the usual way to prepare CO_2.

4. **(D)** The reaction is: $CO_2 + Ca(OH)_2 \rightarrow CaCO_3\downarrow + H_2O$.

5. **(B)** The lead in a lead pencil is a mixture of graphite and clay.

6. **(B)** The first alkane is methane, CH_4.

7. **(D)** The slight oxidation of a primary alcohol produced an aldehyde.

An example is:

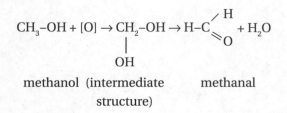

methanol (intermediate   methanal
structure)

8. **(E)** The functional group for an ester is shown as $-COO-$.

9. **(A)** Ethanol can be oxidized into the organic acid enthanoic acid, which has the common name of acetic acid.

10. **(C)** The formation of an ester is from the reaction of an organic acid and an alcohol. The general equation is:

$$\underset{\text{alcohol}}{RO-H} + \underset{\text{acid}}{R^1CO-OH} \rightarrow \underset{\text{ester}}{R^1COO-R} + HOH$$

11. **(B)** Isomers are compounds that have the same composition but differ in their structural formulas.

12. **(D)** Ethene is the first member of the alkene series that has one double bond. Because it has a double bond, it is said to be unsaturated. The alkane series, which has all single bonds between the carbon atoms in the chain, is called a saturated series.

13. **(E)** The NH_2^- group, called an amine group, makes this structure ethylamine. This type of organic structure is called an amide.

14. **(C)** The methyl (CH_3-) group and the propyl ($-C_3H_7$) group attached to a center oxygen ($-O-$) makes this methyl propyl ether (methoxypropane).

15. **(A)** Propane is the third member of the alkane series, which is made up of a chain of single-bonded carbons and hydrogens with the general formula of C_nH_{2n+2}.

16. **(B)** Ethanoic acid is composed of a methyl group attached to the carboxyl group (–COOH). The latter is the functional group for an organic acid.

17. **(D)** The propan- part of the name tells you its basic structure is from propane, which is a three-carbon alkane. The –one part tells you it is a ketone that has a double-bonded oxygen attached to the second carbon in the chain.

18. **(B)** The ethanoic acid contains the carboxyl group (–COOH) that is shown in (B). This is the identifying functional group for organic acids.

19. **(D)** The functional group for ketones is a carbon in the chain double bonded to an oxygen atom ($-\overset{\overset{\displaystyle O}{\|}}{C}-$). The number in the front tells you which carbon has the double-bonded oxygen attached to it.

20. **(E)** The amine group is the nitrogen with two hydrogens, (–NH$_2$), attached to a chain carbon. These are basic to the amino acid structures in the body.

The Laboratory

These skills are usually tested on the SAT Subject Test in Chemistry. You should be able to...

> → **Name, identify, and explain proper laboratory rules and procedures.**
> → **Identify and explain the proper use of laboratory equipment.**
> → **Use laboratory data and observations to make proper interpretations and conclusions.**

This chapter will review and strengthen these skills. Be sure to do the Practice Exercises at the end of the chapter.

Laboratory setups vary from school to school depending on whether the lab is equipped with macro- or microscale equipment. Microlabs use specialized equipment that allows lab work to be done on a much smaller scale. The basic principles are the same as when using full-sized equipment, but microscale equipment lowers the cost of materials, results in less waste, and poses less danger. The examples in this book are of macroscale experiments.

Along with learning to use microscale equipment, most labs require a student to learn how to use technological tools to assist in experiments. The most common are:

Gravimetric balance with direct readings to thousandths of a gram instead of a triple-beam balance

pH meters that give pH readings directly instead of using indicators

Spectrophotometer, which measures the percentage of light transmitted at specific frequencies so that the molarity of a sample can be determined without doing a titration

Computer-assisted labs that use probes to take readings, e.g., temperature and pressure, so that programs available for computers can print out a graph of the relationship of readings taken over time

LABORATORY SAFETY RULES

The Ten Commandments of Lab Safety

The following is a summary of rules you should be well aware of in your own chemistry lab.

1. Dress appropriately for the lab. Wear safety goggles and a lab apron or coat. Tie back long hair. Do not wear open-toed shoes.
2. Know what safety equipment is available and how to use it. This includes the eyewash fountain, fire blanket, fire extinguisher, and emergency shower.

3. Know the dangers of the chemicals in use, and read labels carefully. Do not taste or sniff chemicals.

4. Dispose of chemicals according to instructions. Use designated disposal sites, and follow the rules. Never return unneeded chemicals to the original containers.

5. Always add acids and bases to water slowly to avoid splattering. This is especially important when using strong acids and bases that can generate significant heat, form steam, and splash out of the container.

6. Never point heating test tubes at yourself or others. Be aware of reactions that are occurring so that you can remove them from the heat if necessary before they "shoot" out of the test tube.

7. Do not pipette anything by mouth! Never use your mouth as a suction pump, not even at home with toxic or flammable liquids.

8. Use the fume hood when dealing with toxic fumes! If you can smell them, you are exposing yourself to a dose that can harm you.

9. Do not eat or drink in the lab! It is too easy to take in some dangerous substance accidentally.

10. Follow all directions. Never haphazardly mix chemicals. Pay attention to the order in which chemicals are to be added to each other, and do not deviate!

SOME BASIC SETUPS

Throughout this book, drawings of laboratory setups that serve specific needs have been presented. You should be familiar with the assembly and use of each of these setups. The following list, with page references, will enable you to review them in context with their uses:

The following are additional laboratory setups with which you should be familiar:

1. PREPARATION OF A GASEOUS PRODUCT, SOLUBLE IN WATER AND LIGHTER THAN AIR, BY THE DOWNWARD DISPLACEMENT OF AIR. SEE FIGURE 38.

➡ **Example** _____

Preparation of ammonia (NH_3).

$$2NH_4Cl(s) + Ca(OH)_2(s) \rightarrow CaCl_2(s) + 2H_2O(g) + 2NH_3(g)$$

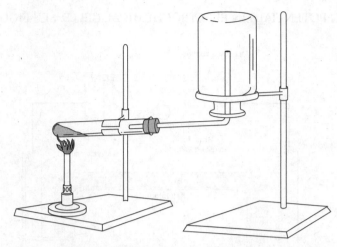

Figure 38. Preparation of Ammonia

2. SEPARATION OF A MIXTURE BY CHROMATOGRAPHY. SEE FIGURE 39.

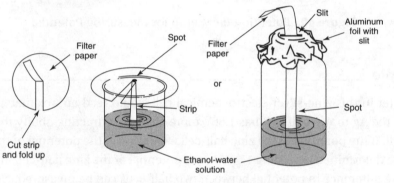

Figure 39. Chromatography Setup

➡ **Example** _____

Chromatography is a process used to separate parts of a mixture. The component parts separate as the solvent carrier moves past the spot of material to be separated by capillary action. Because of variations in solubility, attraction to the filter paper, and density, each fraction moves at a different rate. Once separation occurs, the fractions are either identified by color or removed for other tests. A usual example is the use of Shaeffer Skrip Ink No. 32, which separates into yellow, red, and blue streaks of dyes.

3. MEASURING POTENTIALS IN ELECTROCHEMICAL CELLS. SEE FIGURE 40.

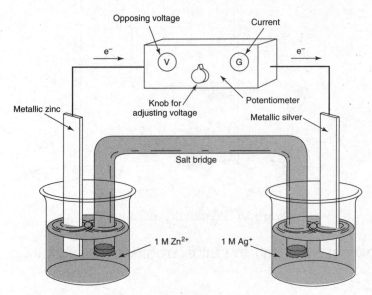

Figure 40. Potentiometer Setup for Measuring Potential

➥ Example _____

The voltmeter in this zinc-silver electrochemical cell would read approximately 1.56V. This means that the Ag to Ag^+ half-cell has 1.56V more electron-attracting ability than the Zn to Zn^{2+} half-cell. If the potential of the zinc half-cell were known, the potential of the silver half-cell could be determined by adding 1.56V to the potential of the zinc half-cell. In a setup like this, only the difference in potential between two half-cells can be measured. Notice the use of the **salt bridge** instead of a porous barrier.

4. REPLACEMENT OF HYDROGEN BY A METAL. SEE FIGURE 41.

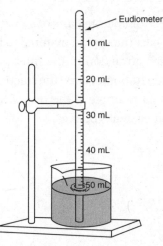

Figure 41. Eudiometer Apparatus

Example

Measure the mass of a strip of magnesium with an analytical balance to the nearest 0.001g. Using a coiled strip with a mass of about 0.040g produces about 40mL of H_2. Pour 5mL of concentrated HCl into a eudiometer, and slowly fill the remainder with water. Try to minimize mixing. Lower the coil of Mg strip into the tube, invert it, and lower it to the bottom of the beaker. After the reaction is complete, you can measure the volume of the gas released and calculate the mass of hydrogen replaced by the magnesium. (Refer to Chapter 5 for a discussion of gas laws.)

SUMMARY OF QUALITATIVE TESTS

I. Identification of Some Common Gases

Gas	Test	Result
Ammonia NH_3	1. Smell cautiously.	1. Sharp odor.
	2. Test with litmus.	2. Red litmus turns blue.
	3. Expose to HCl fumes.	3. White fumes form, NH_4Cl.
Carbon dioxide CO_2	1. Pass through limewater, $Ca(OH)_2$.	1. White precipitate forms, $CaCO_3$.
Carbon monoxide CO	1. Burn it and pass product through limewater, $Ca(OH)_2$.	1. White precipitate forms, $CaCO_3$.
Hydrogen H_2	1. Allow it to mix with some air, then ignite.	1. Gas explodes.
	2. Burn it—trap product.	2. Burns with blue flame—product H_2O turns cobalt chloride paper from blue to pink.
Hydrogen chloride HCl	1. Smell cautiously.	1. Choking odor.
	2. Exhale over the gas.	2. Vapor fumes form.
	3. Dissolve in water and test with litmus.	3. Blue litmus turns red.
	4. Add $AgNO_3$ to the solution.	4. White precipitate forms.
Hydrogen sulfide H_2S	1. Smell cautiously.	1. Rotten egg odor.
	2. Test with moist lead acetate paper.	2. Turns brown-black (PbS).
Oxygen O_2	1. Insert glowing splint.	1. Bursts into flame.
	2. Add nitric oxide gas.	2. Turns reddish brown.

II. Identification of Some Negative Ions

Ion	Test	Result
Acetate $C_2H_3O_2^-$	Add concentrated H_2SO_4 and warm gently.	Odor of vinegar released.
Carbonate CO_3^-	Add HCl acid; pass released gas through limewater.	White, cloudy precipitate forms.
Chloride Cl^-	1. Add silver nitrate solution.	1. White precipitate forms.
	2. Then add nitric acid, later followed by ammonium hydroxide.	2. Precipitate insoluble in HNO_3 but dissolves in NH_4OH.
Hydroxide OH^-	Test with red litmus paper.	Turns blue.
Sulfate SO_4^-	Add solution of $BaCl_2$, then HCl.	White precipitate forms; insoluble in HCl.
Sulfide S^{2-}	Add HCl and test gas released with lead acetate paper.	Gas, with rotten egg odor, turns paper brown-black.

III. Identification of Some Positive Ions

Ion	Test	Result
Ammonium NH_4^+	Add strong base (NaOH); heat gently.	Odor of ammonia.
Ferrous Fe^{2+}	Add solution of potassium ferricyanide, $K_3Fe(CN)_6$.	Dark blue precipitate forms (Turnball's blue).
Ferric Fe^{3+}	Add solution of potassium ferrocyanide, $K_4Fe(CN)_6$.	Dark blue precipitate forms (Prussian blue).
Hydrogen H^+	Test with blue litmus paper.	Turns red.

IV. Qualitative Tests of Some Metals

FLAME TESTS. Carefully clean a platinum wire by dipping it into dilute HNO_3 and heating in the Bunsen flame. Repeat until the flame is colorless. Dip heated wire into the substance being tested (either solid or solution), and then hold it in the hot outer part of the Bunsen flame.

Compound of	Color of Flame
Sodium (Na)	Yellow
Potassium (K)	Violet (use cobalt-blue glass to screen out Na impurities)
Lithium (Li)	Crimson
Calcium (Ca)	Orange-red
Barium (Ba)	Green
Strontium (Sr)	Bright red

HYDROGEN SULFIDE TESTS. Bubble hydrogen sulfide gas through the solution of a salt of the metal being tested. Check color of the precipitate formed.

Compound of	Color of Sulfide Precipitate
Lead (Pb)	Brown-black (PbS)
Copper (Cu)	Black (CuS)
Silver (Ag)	Black (Ag_2S)
Mercury (Hg)	Black (HgS)
Nickel (Ni)	Black (NiS)
Iron (Fe)	Black (FeS)
Cadmium (Cd)	Yellow (CdS)
Arsenic (As)	Light yellow (As_2S_3)
Antimony (Sb)	Orange (Sb_2S_3)
Zinc (Zn)	White (ZnS)
Bismuth (Bi)	Brown (Bi_2S_3)

CHAPTER SUMMARY

The following terms summarize all the concepts and ideas that were introduced in this chapter. You should be able to explain their meaning and how you would use them in chemistry. They appear in boldface type in this chapter to draw your attention to them. The boldface type also makes it easier for you to look them up if you need to. You could also use the Internet search engine *goole.com* on your computer to get a quick and expanded explanation of these terms, laws, and formulas.

gravimetric balance with direct readings

pH meters

salt bridge

spectrophotometer

computer-assisted labs

INTERNET RESOURCES

Online content that reinforces the major concepts discussed in this chapter can be found at the following Internet addresses if they are still available. *Some may have been changed or deleted.*

Chemistry Safety Rules

chemistry.about.com/od/labtechniques/

This website provides additional information on common lab techniques of which students of chemistry should be aware.

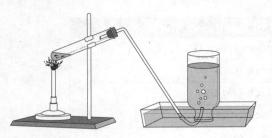

1. In the reaction setup shown above, which of the following are true?

  I. This setup can be used to prepare a soluble gas by water displacement.
  II. This setup involves a decomposition reaction if the substance heated is potassium chlorate.
  III. This setup can be used to prepare an insoluble gas by water displacement.

  (A) I only
  (B) II only
  (C) I and III
  (D) II and III
  (E) I, II, and III

Questions 2–4 refer to the following diagram:

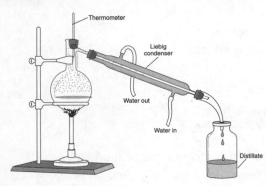

  (A) Around the thermometer
  (B) In the condenser
  (C) In the circulating water
  (D) In the heated flask
  (E) In the distillate

2. In this laboratory setup for distillation, where does the vaporization take place?

3. If the liquid being distilled contains dissolved magnesium chloride, where will it be found after distillation is completed?

4. If the liquid being distilled contains dissolved ammonia gas, where will it be found after distillation is completed?

_____

5. If the flame used to heat a flask is an orange color and blackens the bottom of the flask, what correction should you make to solve this problem?

  (A) Move the flask farther from the flame.
  (B) Move the flask closer to the flame.
  (C) Allow less air into the collar of the burner.
  (D) Allow more air into the collar of the burner.
  (E) The problem is in the supply of the gas, and you cannot fix it.

Questions 6–8 refer to the following diagram:

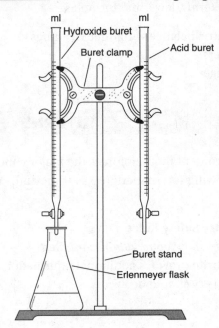

6. In the above titration setup, if you introduce 15 mL of the NaOH with an unknown molarity into the flask and then add 5 drops of phenolphthalein indicator, what will you observe?

(A) A pinkish color will appear throughout the solution.
(B) A blue color will appear throughout the solution.
(C) There will be a temporary pinkish color that will dissipate.
(D) There will be a temporary blue color that will dissipate.
(E) There will not be a color change.

7. If the HCl is 0.1 M standard solution and you must add 30 mL to reach the end point, what is the molarity of the NaOH?

(A) 0.1 M
(B) 0.2 M
(C) 0.3 M
(D) 1 M
(E) 2 M

8. When is the end point reached and the volume of the HCl recorded in this reaction?

I. When the color first disappears and returns in the flask.
II. When equal amounts of HCl and NaOH are in the flask.
III. When the color disappears and does not return in the flask.

(A) I only
(B) III only
(C) I and III
(D) II and III
(E) I, II, and III

Questions 9–11 refer to the following diagram:

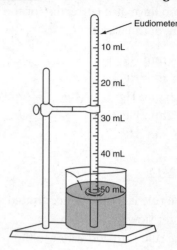

In this setup, a clean strip of magnesium with a mass of 0.040 g was introduced into the bottom of the tube, which contained a dilute solution of HCl, and allowed to react completely. The hydrogen gas formed was collected and the following data recorded:

Air pressure in the room = 730 mm Hg
Temperature of the water solution = 302 K
Vapor pressure of water at 302 K = 30.0 mm Hg

The gas collected did not fill the eudiometer. The height of the meniscus above the level of the water was 40.8 mm.

9. What is the theoretical yield (in mL) at STP of hydrogen gas produced when the 0.040 g of Mg reacted completely?

(A) 10 mL
(B) 25 mL
(C) 37 mL
(D) 46 mL
(E) 51 mL

10. What is the correction to the atmospheric pressure due to the 40.8 mm height of the solution in the tube and above the level in the beaker?

(A) 3.0 mm Hg
(B) 6.0 mm Hg
(C) 13.6 mm Hg
(D) 27.2 mm Hg
(E) 40.8 mm Hg

11. What is the pressure of the collected gas once you have also corrected for the vapor pressure of the water?

(A) 730 mm Hg
(B) 727 mm Hg
(C) 30.0 mm Hg
(D) 697 mm Hg
(E) 760 mm Hg

Questions 12–14

(A) The rule is to add concentrated acid to water slowly.
(B) The rule is to add water to the concentrated acid slowly.
(C) Carefully replace unused or excess chemicals into their properly labeled containers from which they came.
(D) Flush eyes with water at the eyewash fountain for at least 15 minutes, and then report the accident for further help.
(E) Dispose of chemicals in the proper places and following posted procedures. Do not return them to their original containers.

12. Which of the above choices is the proper way to dilute a concentrated acid?

13. How do you properly dispose of chemicals not needed in the experiment?

14. What should you do if a chemical splatters into your eye?

———————————————

15. What instrument is used in chemistry labs to measure the molarity of a colored solution by measuring the light transmitted through it?

(A) Electronic gravimetric balance
(B) pH meter
(C) Spectrophotometer
(D) Computer assisted probes
(E) Galvanometer

Answers and Explanations

1. **(D)** This setup can be used to prepare an insoluble gas but not a soluble one. If the substance is potassium chlorate, if does decompose into potassium chloride and oxygen.

2. **(D)** Vaporization occurs in the heated flask.

3. **(D)** The magnesium chloride will be left behind in the heated flask since it is not volatile as the liquid boils off.

4. **(E)** Since the dissolved ammonia is volatile below the boiling point of water, it will be found in the distillate. Some will also escape as a gas.

5. **(D)** One of the basic adjustments to a burner is to assure enough air is mixing with the gas to form a blue cone-shaped flame. With insufficient air, carbon deposits will form on the flask and the flame will be orange.

6. **(A)** Phenolphthalein indicator is colorless below a pH of 8.3 but is red to pink in basic solutions above this pH. In this NaOH solution, it will be red to pink.

7. **(B)** Calculate the molarity by using the formula:

$$M_{acid} \times V_{acid} = M_{base} \times V_{base}$$
$$0.1\,M \times 30\,mL = M_{base} \times 15\,mL$$
$$M_{base} = 0.2\,M$$

Notice that as long as the units of volume are the same, they cancel out of the equation.

8. **(B)** The end point is reached when the color of the indicator disappears and does not return. The color will first disappear temporarily before the end point but finally will not return.

9. **(C)** The theoretical yield at STP can be found from the chemical equation of the reaction:

$$
\begin{array}{cccccc}
0.040\text{ g Mg} & & & & & x\text{L} \\
\text{Mg} & + & 2\text{HCl} & \rightarrow & \text{MgCl}_2 & + & \text{H}_2(g) \\
24.0\text{ g/mol} & & & & 22.4\text{ L}
\end{array}
$$

Solving for xL = .037 L or 37 mL

10. **(A)** The 40.8 mm of water being held up in the tube by atmospheric pressure can be changed to its equivalent height of mercury by dividing by 13.6, since 1 mm of Hg = 13.6 mm of water.

$$40.8\,mm\,H_2O \times \frac{1\,mm\,Hg}{13.6\,mm\,H_2O} = 3\,mm\,Hg$$

By correcting the atmospheric pressure, we get: 730 mm Hg – 3 mm Hg = 727 mm Hg

11. **(D)** Correcting the previous pressure by subtracting the given amount for the vapor pressure of water at 302 K gives you 727 mm Hg – 30 mm Hg vapor pressure = 697 mm Hg as the final pressure.

12. **(A)** The correct and safe way to dilute concentrated acids is to add water down the side of the beaker slowly and be aware of any heat buildup.

13. **(E)** You never return chemicals or solutions to their original containers for fear of contaminating the original source.

14. **(D)** It is essential to get your eyes washed of any chemicals. Know where the eyewash fountains are, and know how to use them.

15. **(C)** One of the new technological additions to chemistry labs is the spectron 20 that uses the absorption of light waves to do qualitative and quantitative investigations in the lab.

PART 3
Practice Tests

Additional tests are available online at
http://barronsbooks.com/tp/sat/chemistry/

Practice Subject Tests in Chemistry

The Subject Tests in Chemistry are planned to test the principles and concepts drawn from the factual material found largely in inorganic chemistry and, to a much lesser extent, in organic chemistry. Only a few questions are asked concerning industrial or analytical chemistry.

A detailed description of every aspect of the test is given in the introduction of this book. Read it carefully. Study the types of questions asked on the test. Carefully read the instructions given for each type of question at the beginning of each section.

You will be provided with a Periodic Table to use during the test. All necessary information regarding atomic numbers and atomic masses is given on the chart. Before you attempt any of the practice tests, read the information in the Introduction and the material that precedes the Diagnostic Test in the front of this book. When you understand the information there, are aware of the types of questions and their respective directions, and know how the test will be scored, you are ready to take Practice Test 1.

Remember that you have **1 hour** and that you **may not use a calculator. Use the Periodic Table** provided for the Diagnostic Test in the front of this book, and record your answers in the appropriate spaces on the answer sheet. After you have taken each test, follow the instructions for diagnosing your strengths and weaknesses and for how to improve in those areas.

ANSWER SHEET
Practice Test 1

Determine the correct answer for each question. Then, using a No. 2 pencil, blacken completely the oval containing the letter of your choice.

1. Ⓐ Ⓑ Ⓒ Ⓓ Ⓔ
2. Ⓐ Ⓑ Ⓒ Ⓓ Ⓔ
3. Ⓐ Ⓑ Ⓒ Ⓓ Ⓔ
4. Ⓐ Ⓑ Ⓒ Ⓓ Ⓔ
5. Ⓐ Ⓑ Ⓒ Ⓓ Ⓔ
6. Ⓐ Ⓑ Ⓒ Ⓓ Ⓔ
7. Ⓐ Ⓑ Ⓒ Ⓓ Ⓔ
8. Ⓐ Ⓑ Ⓒ Ⓓ Ⓔ
9. Ⓐ Ⓑ Ⓒ Ⓓ Ⓔ
10. Ⓐ Ⓑ Ⓒ Ⓓ Ⓔ
11. Ⓐ Ⓑ Ⓒ Ⓓ Ⓔ
12. Ⓐ Ⓑ Ⓒ Ⓓ Ⓔ
13. Ⓐ Ⓑ Ⓒ Ⓓ Ⓔ
14. Ⓐ Ⓑ Ⓒ Ⓓ Ⓔ
15. Ⓐ Ⓑ Ⓒ Ⓓ Ⓔ
16. Ⓐ Ⓑ Ⓒ Ⓓ Ⓔ

17. Ⓐ Ⓑ Ⓒ Ⓓ Ⓔ
18. Ⓐ Ⓑ Ⓒ Ⓓ Ⓔ
19. Ⓐ Ⓑ Ⓒ Ⓓ Ⓔ
20. Ⓐ Ⓑ Ⓒ Ⓓ Ⓔ
21. Ⓐ Ⓑ Ⓒ Ⓓ Ⓔ
22. Ⓐ Ⓑ Ⓒ Ⓓ Ⓔ
23. Ⓐ Ⓑ Ⓒ Ⓓ Ⓔ

ON THE ACTUAL CHEMISTRY TEST. THE FOLLOWING TYPE OF QUESTION MUST BE ANSWERED ON A SPECIAL SECTION (LABELED "CHEMISTRY") AT THE LOWER LEFT-HAND CORNER OF PAGE 2 OF YOUR ANSWER SHEET. THESE QUESTIONS WILL BE NUMBERED BEGINING WITH 101 AND MUST BE ANSWERED ACCORDING TO THE DIRECTIONS.

CHEMISTRY* Fill in oval CE only if II is a correct explanation of 1.

	I	II	CE*
101.	Ⓣ Ⓕ	Ⓣ Ⓕ	◯
102.	Ⓣ Ⓕ	Ⓣ Ⓕ	◯
103.	Ⓣ Ⓕ	Ⓣ Ⓕ	◯
104.	Ⓣ Ⓕ	Ⓣ Ⓕ	◯
105.	Ⓣ Ⓕ	Ⓣ Ⓕ	◯
106.	Ⓣ Ⓕ	Ⓣ Ⓕ	◯
107.	Ⓣ Ⓕ	Ⓣ Ⓕ	◯
108.	Ⓣ Ⓕ	Ⓣ Ⓕ	◯
109.	Ⓣ Ⓕ	Ⓣ Ⓕ	◯
110.	Ⓣ Ⓕ	Ⓣ Ⓕ	◯
111.	Ⓣ Ⓕ	Ⓣ Ⓕ	◯
112.	Ⓣ Ⓕ	Ⓣ Ⓕ	◯
113.	Ⓣ Ⓕ	Ⓣ Ⓕ	◯
114.	Ⓣ Ⓕ	Ⓣ Ⓕ	◯
115.	Ⓣ Ⓕ	Ⓣ Ⓕ	◯

ANSWER SHEET
Practice Test 1

ON THE ACTUAL CHEMISTRY TEST, THE REMAINING QUESTIONS MUST BE ANSWERED BY RETURNING TO THE SECTION OF YOUR ANSWER SHEET YOU STARTED FOR CHEMISTRY.

24. Ⓐ Ⓑ Ⓒ Ⓓ Ⓔ
25. Ⓐ Ⓑ Ⓒ Ⓓ Ⓔ
26. Ⓐ Ⓑ Ⓒ Ⓓ Ⓔ
27. Ⓐ Ⓑ Ⓒ Ⓓ Ⓔ
28. Ⓐ Ⓑ Ⓒ Ⓓ Ⓔ
29. Ⓐ Ⓑ Ⓒ Ⓓ Ⓔ
30. Ⓐ Ⓑ Ⓒ Ⓓ Ⓔ
31. Ⓐ Ⓑ Ⓒ Ⓓ Ⓔ
32. Ⓐ Ⓑ Ⓒ Ⓓ Ⓔ
33. Ⓐ Ⓑ Ⓒ Ⓓ Ⓔ
34. Ⓐ Ⓑ Ⓒ Ⓓ Ⓔ
35. Ⓐ Ⓑ Ⓒ Ⓓ Ⓔ
36. Ⓐ Ⓑ Ⓒ Ⓓ Ⓔ
37. Ⓐ Ⓑ Ⓒ Ⓓ Ⓔ
38. Ⓐ Ⓑ Ⓒ Ⓓ Ⓔ
39. Ⓐ Ⓑ Ⓒ Ⓓ Ⓔ

40. Ⓐ Ⓑ Ⓒ Ⓓ Ⓔ
41. Ⓐ Ⓑ Ⓒ Ⓓ Ⓔ
42. Ⓐ Ⓑ Ⓒ Ⓓ Ⓔ
43. Ⓐ Ⓑ Ⓒ Ⓓ Ⓔ
44. Ⓐ Ⓑ Ⓒ Ⓓ Ⓔ
45. Ⓐ Ⓑ Ⓒ Ⓓ Ⓔ
46. Ⓐ Ⓑ Ⓒ Ⓓ Ⓔ
47. Ⓐ Ⓑ Ⓒ Ⓓ Ⓔ
48. Ⓐ Ⓑ Ⓒ Ⓓ Ⓔ
49. Ⓐ Ⓑ Ⓒ Ⓓ Ⓔ
50. Ⓐ Ⓑ Ⓒ Ⓓ Ⓔ
51. Ⓐ Ⓑ Ⓒ Ⓓ Ⓔ
52. Ⓐ Ⓑ Ⓒ Ⓓ Ⓔ
53. Ⓐ Ⓑ Ⓒ Ⓓ Ⓔ
54. Ⓐ Ⓑ Ⓒ Ⓓ Ⓔ
55. Ⓐ Ⓑ Ⓒ Ⓓ Ⓔ

56. Ⓐ Ⓑ Ⓒ Ⓓ Ⓔ
57. Ⓐ Ⓑ Ⓒ Ⓓ Ⓔ
58. Ⓐ Ⓑ Ⓒ Ⓓ Ⓔ
59. Ⓐ Ⓑ Ⓒ Ⓓ Ⓔ
60. Ⓐ Ⓑ Ⓒ Ⓓ Ⓔ
61. Ⓐ Ⓑ Ⓒ Ⓓ Ⓔ
62. Ⓐ Ⓑ Ⓒ Ⓓ Ⓔ
63. Ⓐ Ⓑ Ⓒ Ⓓ Ⓔ
64. Ⓐ Ⓑ Ⓒ Ⓓ Ⓔ
65. Ⓐ Ⓑ Ⓒ Ⓓ Ⓔ
66. Ⓐ Ⓑ Ⓒ Ⓓ Ⓔ
67. Ⓐ Ⓑ Ⓒ Ⓓ Ⓔ
68. Ⓐ Ⓑ Ⓒ Ⓓ Ⓔ
69. Ⓐ Ⓑ Ⓒ Ⓓ Ⓔ
70. Ⓐ Ⓑ Ⓒ Ⓓ Ⓔ

PRACTICE TEST 1

Note: For all questions involving solutions and/or chemical equations, assume that the system is in water unless otherwise stated.

Reminder: You may *not* use a calculator on these tests.

The following symbols have the meanings listed unless otherwise noted.

H	= enthalpy	g	= gram(s)
M	= molar	J	= joule(s)
n	= number of moles	kJ	= kilojoule(s)
P	= pressure	L	= liter(s)
R	= molar gas constant	mL	= milliliter(s)
S	= entropy	mm	= millimeter(s)
T	= temperature	mol	= mole(s)
V	= volume	V	= volt(s)
atm	= atmosphere(s)		

PART A

Directions: Every set of the given lettered choices below refers to the numbered statements or formulas immediately following it. Choose the one lettered choice that best fits each statement or formula and then fill in the corresponding oval on the answer sheet. Each choice may be used once, more than once, or not at all in each set.

Questions 1–6 refer to the choices in the following table:

Periodic Table (abbreviated)

						(D)	¹⁰Ne
³Li							
	(A)	...			(C)		
(B)	²⁰Ca	...					(E)

1. The most electronegative element

2. The element with a possible oxidation number of –2

3. The element that would react in a 1:1 ratio with (D)

4. The element with the smallest ionic radius

5. The element with the smallest first ionization potential

6. The element with a complete p orbital as its outermost energy level

Questions 7–9 refer to the following terms.

(A) Reduction potential
(B) Ionization energy
(C) Electronegativity
(D) Heat of formation
(E) Activation energy

7. This is the energy change that accompanies the combining of elements in their natural states to form one mole of a compound.

8. This is the energy needed to remove an electron from a gaseous atom in its ground state.

9. This is the minimum energy needed for molecules to react.

GO ON TO THE NEXT PAGE

Questions 10–12 refer to the following heating curve for water:

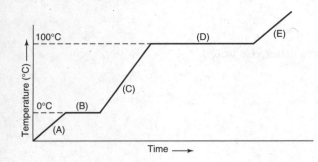

10. In which part of the curve is the state of H_2O only a solid?

11. Which part of the graph shows a phase change requiring the greatest amount of energy?

12. Where is the temperature of H_2O changing at 4.18 J/g°C (or 1 cal/g°C)?

Questions 13–15 refer to the following diagram:

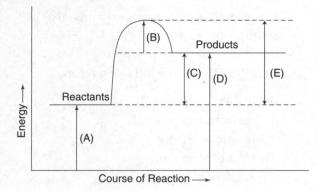

13. Indicates the activation energy of the forward reaction

14. Indicates the activation energy of the reverse reaction

15. Indicates the difference between the activation energies for the reverse and forward reactions and equals the energy change in the reaction

Questions 16–18 refer to the following elements in the ground state:

(A) Fe
(B) Au
(C) Na
(D) Ar
(E) U

16. A common metal element that resists reaction with most acids

17. A monatomic element that exists in the gaseous state at STP

18. A transition element described as having its inner $3d$ orbital partially filled

Questions 19 and 20 refer to the following:

(A) Radioactive isotope
(B) Monoclinic crystal
(C) Sulfur trioxide
(D) Sulfate salt
(E) Allotropic form

19. A substance that exhibits a resonance structure

20. A product formed from a base reacting with H_2SO_4.

Questions 21–23 refer to the following terms:

(A) Dilute
(B) Concentrated
(C) Unsaturated
(D) Saturated
(E) Supersaturated

21. The condition, unrelated to quantities, that indicates that the rate going into solution is equal to the rate coming out of solution

GO ON TO THE NEXT PAGE

22. The condition that exists when a water solution that has been at equilibrium and saturated is heated to a higher temperature with a higher solubility, but no additional solute is added

23. The descriptive term that indicates there is a large quantity of solute, compared with the amount of solvent, in a solution

PART B

ON THE ACTUAL CHEMISTRY TEST, THE FOLLOWING TYPE OF QUESTION MUST BE ANSWERED ON A SPECIAL SECTION (LABELED "CHEMISTRY") AT THE LOWER LEFT-HAND CORNER OF PAGE 2 OF YOUR ANSWER SHEET. THESE QUESTIONS WILL BE NUMBERED BEGINNING WITH 101 AND MUST BE ANSWERED ACCORDING TO THE FOLLOWING DIRECTIONS.

Directions: Every question below contains two statements, I in the left-hand column and II in the right-hand column. For each question, decide if statement I is true or false **and** if statement II is true or false and fill in the corresponding T or F ovals on your answer sheet. *Fill in oval CE only if statement II is a correct explanation of statement I.

Sample Answer Grid:

CHEMISTRY * Fill in oval CE only if II is a correct explanation of I.

	I	II	CE*
101.	T F	T F	◯

I

101. Nonmetallic oxides are usually acid anhydrides

102. When HCl gas and NH_3 gas come into contact, a white smoke forms

103. The reaction of barium chloride and sodium sulfate does not go to completion

104. When two elements react exothermically to form a compound, the compound should be relatively stable

105. The ion of a nonmetallic atom is larger in radius than the atom

II

101. BECAUSE nonmetallic oxides form acids when placed in water.

102. BECAUSE NH_3 and HCl react to form a white solid, ammonium chlorate.

103. BECAUSE the compound barium sulfate is formed as an insoluble precipitate.

104. BECAUSE the release of energy from a combination reaction indicates that the compound formed is at a lower energy level than the reactants and thus relatively stable.

105. BECAUSE when a nonmetallic ion is formed, it gains electrons in the outer orbital and thus increases the size of the electron cloud around the nucleus.

GO ON TO THE NEXT PAGE

	I		**II**

106. Oxidation and reduction occur together   BECAUSE   in redox reactions, electrons must be gained and lost.

107. Decreasing the atmospheric pressure on a pot of boiling water causes the water to stop boiling   BECAUSE   changes in pressure are directly related to the boiling point of water.

108. The reaction of hydrogen with oxygen to form water is an exothermic reaction   BECAUSE   water molecules have polar covalent bonds.

109. Atoms of different elements can have the same mass number   BECAUSE   the atoms of each element have a characteristic number of protons in the nucleus.

110. The proton and the neutron have essentially the same mass   BECAUSE   the proton and the neutron have essentially the same charge.

111. $^{13}_{6}C$ and $^{14}_{6}C$ are isotopes of the element carbon   BECAUSE   isotopes of an element have the same number of protons in the nucleus but have a different number of neutrons.

112. The Cu^{2+} ion needs to be oxidized to form Cu metal   BECAUSE   oxidation is a gain of electrons.

113. The volume of a gas at 373 K and a pressure of 600 millimeters of mercury will be decreased at STP   BECAUSE   decreasing the temperature and increasing the pressure will cause the volume to decrease because

$$V_2 = V_1 \times \frac{P_1}{P_2} \times \frac{T_2}{T_1}.$$

114. The pH of a 0.01 molar solution of HCl is 2   BECAUSE   dilute HCl dissociates into two essentially ionic particles.

115. Nuclear fusion on the sun converts hydrogen to helium with a release of energy   BECAUSE   some mass is converted to energy in a solar fusion.

GO ON TO THE NEXT PAGE

Directions: Every question or incomplete statement below is followed by five suggested answers or completions. Choose the one that is best in each case and then fill in the corresponding oval on the answer sheet.

24. What is the approximate molar mass of $Ca(NO_3)_2$?

    (A) 70 g
    (B) 82 g
    (C) 102 g
    (D) 150 g
    (E) 164 g

25. In the reaction

    $$2KClO_3 + MnO_2 \rightarrow 2KCl + 3O_2 \text{ (g)} + MnO_2,$$

    which substance is the catalyst?

    (A) O_2
    (B) $KClO_3$
    (C) MnO_2
    (D) KCl
    (E) O_2 in the MnO_2

26. The normal configuration for ethyne (acetylene) is

    (A) H—C=C—H
    (B) H—C—C—H
    (C) H—CH$_2$—CH$_2$—H
    (D) H—C≡C—H
    (E) H—CH—CH—H

27. According to the Kinetic-Molecular Theory, molecules increase in kinetic energy when they

    (A) are mixed with other molecules at lower temperature
    (B) are frozen into a solid
    (C) are condensed into a liquid
    (D) are heated to a higher temperature
    (E) collide with each other in a container at a lower temperature

28. How many atoms are represented in the formula $Ca_3(PO_4)_2$?

    (A) 5
    (B) 8
    (C) 9
    (D) 12
    (E) 13

29. All of the following have covalent bonds EXCEPT

    (A) HCl
    (B) CCl_4
    (C) H_2O
    (D) CsF
    (E) CO_2

30. Which of the following is (are) generally the WEAKEST attractive force?

    (A) Dipole–dipole forces
    (B) Coordinate covalent bonding
    (C) Covalent bonding
    (D) Polar covalent bonding
    (E) Ionic bonding

GO ON TO THE NEXT PAGE

31. Which of these resembles the molecular structure of the water molecule?

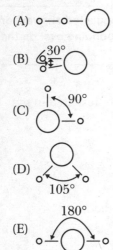

(A)
(B)
(C)
(D)
(E)

32. The two most important considerations in deciding whether a reaction will occur spontaneously are

(A) the stability and state of the reactants
(B) the energy gained and the heat evolved
(C) a negative value for ΔH and a positive value for ΔS
(D) a positive value for ΔH and a negative value for ΔS
(E) the endothermic energy and the structure of the products

33. The reaction of an acid such as HCl and a base such as NaOH always

(A) forms a precipitate
(B) forms a volatile product
(C) forms an insoluble salt and water
(D) forms a sulfate salt and water
(E) forms a salt and water

34. The oxidation number of sulfur in H_2SO_4 is

(A) +2
(B) +3
(C) +4
(D) +6
(E) +8

35. Which of the substances in the following reaction is being reduced?

$$FeO + CO \rightarrow Fe + CO_2$$

(A) Fe and C
(B) Fe
(C) CO_2
(D) C
(E) CO

36. Which of the following when placed into water will test as an acid solution?

I. $HCl(g) + H_2O$
II. Excess $H_3O^+ + H_2O$
III. $CuSO_4(s) + H_2O$

(A) I only
(B) III only
(C) I and II only
(D) II and III only
(E) I, II, and III

37. The property of matter that is independent of its surrounding conditions and position is

(A) volume
(B) density
(C) mass
(D) weight
(E) state

38. Where are the highest ionization energies found in the Periodic Table?

(A) Upper left corner
(B) Lower left corner
(C) Upper right corner
(D) Lower right corner
(E) Middle of transition elements

GO ON TO THE NEXT PAGE

39. Which of the following pairs of compounds can be used to illustrate the Law of Multiple Proportions?

   (A) NO and NO_2
   (B) CH_4 and CO_2
   (C) ZnO_2 and $ZnCl_2$
   (D) NH_3 and NH_4Cl
   (E) H_2O and HCl

40. In this equilibrium reaction: $A + B \rightleftharpoons AB +$ heat (in a closed container), how could the forward reaction rate be increased?

   I. By increasing the concentration of AB
   II. By increasing the concentration of A
   III. By removing some of product AB

   (A) I only
   (B) III only
   (C) I and III only
   (D) II and III only
   (E) I, II, and III

41. For the reaction of sodium with water, the balanced equation using the smallest whole numbers has which of the following coefficients?

   I. 1
   II. 2
   III. 3

   (A) I only
   (B) III only
   (C) I and II only
   (D) II and III only
   (E) I, II, and III

42. If 10 liters of CO gas react with sufficient oxygen for a complete reaction, how many liters of CO_2 gas are formed?

   (A) 5
   (B) 10
   (C) 15
   (D) 20
   (E) 40

43. If 49 grams of H_2SO_4 react with 80.0 grams of NaOH, how much reactant will be left over after the reaction is complete?

   (A) 24.5 g H_2SO_4
   (B) none of either compound
   (C) 20. g NaOH
   (D) 40. g NaOH
   (E) 60. g NaOH

44. If the molar concentration of Ag^+ ions in 1 liter of a saturated water solution of silver chloride is 1.4×10^{-5} mole/liter, what is the K_{sp} of silver chloride?

   (A) 0.34×10^{-10}
   (B) 0.69×10^{-5}
   (C) 2.0×10^{-10}
   (D) 2.0×10^{-5}
   (E) 3.0×10^{-10}

45. If the density of a diatomic gas at STP is 1.43 grams/liter, what is the molar mass of the gas?

   (A) 14.3 g
   (B) 32.0 g
   (C) 48.0 g
   (D) 64.3 g
   (E) 224 g

46. From 2 moles of $KClO_3$ how many liters of O_2 can be produced at STP by decomposition of all the $KClO_3$?

   (A) 11.2
   (B) 22.4
   (C) 33.6
   (D) 44.8
   (E) 67.2

GO ON TO THE NEXT PAGE

47. Which value best determines whether a reaction is spontaneous?

(A) change in Gibbs free energy, ΔG
(B) change in entropy, ΔS
(C) change in kinetic energy, ΔKE
(D) change in enthalpy, ΔH
(E) change in potential energy, ΔPE

Questions 48–52 refer to the following experimental scenario and data:

Silver oxide is placed into a crucible and heated strongly for 15 minutes over a flame. After cooling, the mass of the crucible and contents is determined. The whole system is heated strongly again for 5 minutes and cooled. Then the mass is measured again.

Recorded Data:

Mass of crucible   = 14.03 g

Mass of crucible and silver oxide = 18.67 g

Mass of crucible and product   = 18.36 g
(after 1st heating)

Mass of crucible and product   = 18.35 g
(after 2nd heating)

48. The reaction that occurs by heating the silver oxide produces both silver metal and oxygen gas. The type of reaction that occurs is called

(A) neutralization
(B) combustion
(C) redox
(D) synthesis
(E) double replacement

49. The data from this experiment can be used

(A) to find the activity of silver in silver oxide
(B) to find the percent composition of silver oxide
(C) to verify the Law of Multiple Proportions
(D) to verify the Second Law of Thermodynamics
(E) to prove that silver is a malleable metal

50. The number of moles of oxygen atoms released from the silver oxide after two rounds of heating is

(A) 0.020 mol
(B) 0.10 mol
(C) 0.32 mol
(D) 1.0 mol
(E) 3.2 mol

51. The purpose of the second heating is to

(A) ensure that all of the silver is removed from the crucible
(B) melt the silver
(C) ensure that all of the oxygen reacts with the silver
(D) ensure that all of the silver reacts with the oxygen in the air
(E) ensure that all of the oxygen is removed from the silver oxide

52. To prove that oxygen is a product in this reaction, which common laboratory test could be used?

(A) The "pop" test
(B) The "limewater" test
(C) The "cobalt chloride" test
(D) The "glowing splint" test
(E) The "oil drop" test

53. When HCl fumes and NH_3 fumes are introduced into opposite ends of a long, dry glass tube, a white ring forms in the tube. Which statement explains this phenomenon?

(A) NH_4Cl forms.
(B) HCl diffuses faster.
(C) NH_3 diffuses faster.
(D) The ring occurs closer to the end into which HCl was introduced.
(E) The ring occurs in the very middle of the tube.

GO ON TO THE NEXT PAGE

54. The correct formula for calcium hydrogen sulfate is

(A) CaH_2SO_4
(B) $CaHSO_4$
(C) $Ca(HSO_4)_2$
(D) Ca_2HSO_4
(E) $Ca_2H_2SO_4$

55. Forty grams of sodium hydroxide is dissolved in enough water to make 1 liter of solution. What is the molarity of the solution?

(A) 0.25 M
(B) 0.5 M
(C) 1 M
(D) 1.5 M
(E) 4 M

56. For a saturated solution of salt in water, which statement is true?

(A) All dissolving has stopped.
(B) Crystals begin to grow.
(C) An equilibrium has been established.
(D) Crystals of the solute will visibly continue to dissolve.
(E) The solute is exceeding its solubility.

57. In which of the following series is the pi bond present in the bonding structure?

  I. Alkane
 II. Alkene
III. Alkyne

(A) I only
(B) III only
(C) I and III only
(D) II and III only
(E) I, II, and III

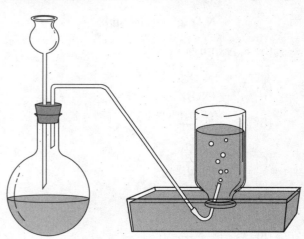

58. Why could you NOT use this setup for preparing H_2 if the generator contained Zn + vinegar?

(A) Hydrogen would not be produced.
(B) The setup of the generator is improper.
(C) The generator must be heated with a burner.
(D) The delivery tube setup is wrong.
(E) The gas cannot be collected over water.

59. In a proper laboratory setup for collecting a gas by water displacement, which of these gases could NOT be collected over H_2O because of its solubility?

(A) CO_2
(B) NO
(C) O_2
(D) NH_3
(E) CH_4

60. What is the approximate percentage of oxygen in the formula mass of $Ca(NO_3)_2$?

(A) 28
(B) 42
(C) 58
(D) 68
(E) 84

GO ON TO THE NEXT PAGE

61. For the following reaction:

$$N_2O_4(g) \rightleftharpoons 2NO_2(g),$$

the K_{eq} expression is

(A) $K_{eq} = \dfrac{[N_2O_4]}{[NO_2]}$

(B) $K_{eq} = \dfrac{[N_2O_4]}{[NO_2]^2}$

(C) $K_{eq} = \dfrac{[NO_2]}{[N_2O_4]}$

(D) $K_{eq} = \dfrac{[NO_2]^2}{[N_2O_4]}$

(E) $K_{eq} = \dfrac{[N_2O_4]^2}{[NO_2]}$

62. What is the K_{eq} for the reaction in question 61 if at equilibrium the concentration of N_2O_4 is 4×10^{-2} mole/liter and that of NO_2 is 2×10^{-2} mole/liter?

(A) 1×10^{-2}
(B) 2×10^{-2}
(C) 4×10^{-2}
(D) 4×10^{-4}
(E) 8×10^{-2}

63. How much water, in liters, must be added to 0.50 liter of 6.0 M HCl to make the solution 2.0 M?

(A) 0.33
(B) 0.50
(C) 1.0
(D) 1.5
(E) 2.0

64. 4.0 grams of hydrogen are ignited with 4.0 grams of oxygen. How many grams of water can be formed?

(A) 0.50
(B) 2.5
(C) 4.5
(D) 8.0
(E) 36

65. Which structure is an ester?

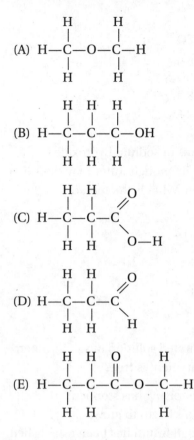

66. What piece of apparatus can be used to introduce more liquid into a reaction and also serve as a pressure valve?

(A) Stopcock
(B) Pinchcock
(C) Thistle tube
(D) Flask
(E) Condenser

67. Which formulas could represent the empirical formula and the molecular formula of a given compound?

(A) CH_2O and $C_4H_6O_4$
(B) CHO and $C_6H_{12}O_6$
(C) CH_4 and C_5H_{12}
(D) CH_2 and C_3H_6
(E) CO and CO_2

GO ON TO THE NEXT PAGE

68. The reaction

$$CH_4 + O_2 \rightarrow CO_2 + H_2O$$

could be classified as

I. Synthesis
II. Combustion
III. Redox

(A) I only
(B) I and II
(C) I and III
(D) II only
(E) II and III

69. $2Na(s) + Cl_2(g) \rightarrow 2NaCl(s) + 822$ kJ

How much heat is released by the above reaction if 0.5 mole of sodium reacts completely with chlorine?

(A) 205 kJ
(B) 411 kJ
(C) 822 kJ
(D) 1,640 kJ
(E) 3,290 kJ

70. The water molecule

I. contains polar bonds.
II. is a polar molecule.
III. exhibits hydrogen bonding between hydrogen and oxygen atoms within the molecule.

(A) I only
(B) I and II
(C) I and III
(D) II only
(E) I and III

If you finish before one hour is up, you may go back to check your work or complete unanswered questions.

1. **D**	7. **D**	13. **E**	19. **C**
2. **C**	8. **B**	14. **B**	20. **D**
3. **B**	9. **E**	15. **C**	21. **D**
4. **A**	10. **A**	16. **B**	22. **C**
5. **B**	11. **D**	17. **D**	23. **B**
6. **E**	12. **C**	18. **A**	

101. **T, T, CE**	105. **T, T, CE**	109. **T, T**	113. **T, T, CE**
102. **T, F**	106. **T, T, CE**	110. **T, F**	114. **T, T**
103. **F, T**	107. **F, T**	111. **T, T, CE**	115. **T, T, CE**
104. **T, T, CE**	108. **T, T**	112. **F, F**	

24. **E**	36. **E**	48. **C**	60. **C**
25. **C**	37. **C**	49. **B**	61. **D**
26. **D**	38. **C**	50. **A**	62. **A**
27. **D**	39. **A**	51. **E**	63. **C**
28. **E**	40. **D**	52. **D**	64. **C**
29. **D**	41. **C**	53. **A**	65. **E**
30. **A**	42. **B**	54. **C**	66. **C**
31. **D**	43. **D**	55. **C**	67. **D**
32. **C**	44. **C**	56. **C**	68. **E**
33. **E**	45. **B**	57. **D**	69. **A**
34. **D**	46. **E**	58. **B**	70. **B**
35. **B**	47. **A**	59. **D**	

ANSWERS EXPLAINED

1. **(D)** The most electronegative element, F, would be found in the upper right corner of the table; the noble gases are exceptions at the far right.

2. **(C)** Elements in the group with (C) have a possible oxidation number of –2.

3. **(B)** Elements in the group with (B) react in a 1:1 ratio with elements in the group with (D), since one has an electron to lose and the other one needs an electron to complete its outer energy level.

4. **(A)** (A) loses 2 electrons to form an ion whose remaining electrons, being close to the nucleus, are pulled in closer than what occurs in (D) because of a greater number of protons.

5. **(B)** Since (B) has only one electron in the outer $4s$ orbital, it can more easily be removed than can an electron from the $3s$ orbital of (A), which is closer to the positive nucleus.

6. **(E)** All Group VIII elements have a complete p orbital as the outer energy level. This explains why these elements are "inert."

7. **(D)** The heat of formation is the energy change caused by the difference in the initial energy and final energy of the system when elements in their standard state react to form a compound.

8. **(B)** The ionization energy is defined as the energy needed to remove an electron from the ground state of the isolated gaseous atom (or ion).

9. **(E)** The activation energy is defined as the minimum energy required for molecules to react. This is true for both exothermic and endothermic reactions.

10. **(A)** H_2O is ice in Part A.

11. **(D)** H_2O changes state at Parts B and D. The heat of vaporization at D (540 cal/g or 2.26 $\times 10^3$ J/g) is greater than the heat of fusion (80 cal/g or 3.34×10^2 J/g) at B.

12. **(C)** Water is heating at 1°C/cal/g or 4.18 J/g/1°C in Part C. This is liquid water's specific heat.

13. **(E)** This is the energy needed to start the forward reaction.

14. **(B)** The part indicated by B represents the activation energy for the reverse reaction.

15. **(C)** The net energy change is the endothermic quantity indicated by C.

16. **(B)** Gold is known as a common noble metal because of its resistance to acids. Aqua regia, a mixture of HNO_3 and HCl, will react with gold. It is very low in the activity series.

17. **(D)** Argon is the only element among the choices that is monoatomic in the ground state. It exists as individual atoms and is a gas at STP.

18. **(A)** Iron has 5 electrons in the d orbitals, which are partially filled.

19. **(C)** Only sulfur trioxide has a resonance structure (as shown here):

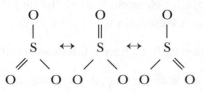

20. **(D)** Sulfuric acid reacts with a base to form a sulfate salt.

21. **(D)** The condition described is the equilibrium that exists at saturation.

22. **(C)** With the increased temperature more solute may go into solution; therefore, the solution is now unsaturated.

23. **(B)** The term "concentrated" means that there is a large amount of solute in the solvent.

101. **(T, T, CE)** Nonmetallic oxides are usually acid anhydrides, and they form acids in water.

102. **(T, F)** The white smoke formed is ammonium chloride, not ammonium chlorate.

103. **(F, T)** The reaction of barium chloride and sodium sulfate *does*, essentially, go to completion since barium sulfate is a precipitate.

104. **(T, T, CE)** The product of an exothermic reaction is relatively stable because it is at a lower energy level than the reactants.

105. **(T, T, CE)** Both statements are true, and the reason explains the assertion.

106. **(T, T, CE)** Both statements are true, and the reason explains the assertion.

107. **(F, T)** The statement is false while the reason is true. Decreasing the pressure on a boiling pot will only cause the water to boil more vigorously.

108. **(T, T)** The statements are true, but the reason doesn't explain the assertion.

109. **(T, T)** The statements are true, but the reason doesn't explain the assertion.

110. **(T, F)** The mass of the proton and the neutron are both 1.7×10^{-24} g. The charge on a proton, however, is positive while the charge on a neutron is neutral.

111. **(T, T, CE)** The two configurations of carbon are isotopes because they have the same atomic number but different mass numbers because $^{13}_{6}C$ has 6 protons and 7 neutrons while $^{14}_{6}C$ has 6 protons and 8 neutrons.

112. **(F, F)** Both the statement and the reason are false.

113. **(T, T, CE)** In going from 100°C to 0°C, the volume decreases as the gas gets colder; therefore, the temperature fraction expressed in kelvins must be $\dfrac{273}{373}$ to decrease the volume. To go from 600 mm to 760 mm of pressure increases the pressure, thus causing the gas to contract. The fraction must then be $\dfrac{600}{760}$ to cause the volume to decrease. You could use the formula

$$V_2 = V_1 \times \frac{T_2}{T_1} \times \frac{P_1}{P_2}$$

114. **(T, T)** Since HCl is a strong acid and ionizes completely in dilute solution of water (the $[H^+]$ and $[H_3O^+]$ are the same thing), the molar concentration of a 0.01 molar solution is 1×10^{-2} mol/L.

$$pH = -\log[H^+]$$
$$pH = -\log[1 \times 10^{-2}]$$
$$pH = -(-2) = 2$$

The pH is 2, but the reason, although true, does not explain the statement.

115. **(T, T, CE)** Both statements are true, and the reason explains the assertion.

24. **(E)** The total molar mass is:

$$
\begin{aligned}
Ca &= 40 \text{ g/mol} \\
N &= 28 \text{ g/2 mol} \\
O &= 96 \text{ g/6 mol} \\
\hline
\text{Total } &164 \text{ g/mol}
\end{aligned}
$$

25. **(C)** The catalyst, by definition, is not consumed in the reaction and ends up in its original form as one of the products. In this reaction, the catalyst is the MnO_2.

26. **(D)** The ethyne molecule is the first member of the alkyne series with a general formula of C_nH_{2n-2}. It contains a triple bond between the two C atoms: H—C≡C—H.

27. **(D)** Heating molecules increases their kinetic energy.

28. **(E)** $3Ca + 2P + 8O = 13$ atoms.

29. **(D)** Cesium and fluorine are from the most electropositive and electronegative portions, respectively, of the periodic chart, and thus form an ionic bond by cesium giving an electron to fluorine to form the respective ions.

30. **(A)** Dipole-dipole forces are due to the weak attraction between permanently polar molecules. They are much weaker than the others named.

31. **(D)** The molecular structure of water is that of a polar covalent compound with the hydrogens 105° apart.

32. **(C)** The most important considerations for a spontaneous reaction are (1) that the reaction is exothermic with a negative ΔH, so that once started it tends to continue on its own because of the energy released, and (2) that the reaction tends to go to the highest state of randomness, shown by a positive value for ΔS.

33. **(E)** Normal H^+ acids and OH^- bases form water and a salt (not necessarily a soluble salt).

34. **(D)** Every compound has a charge of 0. H usually has +1, and O usually has –2, so

$$\begin{array}{rl} H_2 = & +2 \\ O_4 = & -8 \\ S = & x \\ \hline \text{Total} = & 0 \\ +2 - 8 + x = & 0 \\ x = -2 + 8 = & +6 \end{array}$$

35. **(B)** Iron's oxidation state is changing from +2 to 0. It is the substance being reduced.

36. **(E)** HCl and H_3O^+ give acid solutions, as does $CuSO_4$ (the salt of a weak base and a strong acid) when it hydrolyzes in water. I, II, and III are correct.

37. **(C)** Mass is a constant and is not dependent on position or surrounding conditions.

38. **(C)** The complete outer energy levels of electrons of the smallest noble gases have the highest ionization potential.

39. **(A)** Only NO and NO_2 fit the definition of the Law of Multiple Proportions, in which one substance stays the same and the other varies in units of whole integers.

40. **(D)** Increasing the concentration of one or both of the reactants and removing some of the product formed would cause the forward reaction to increase in rate to try to regain the equilibrium condition. II and III are correct.

41. **(C)** I and II are correct. The equation is $2Na + 2H_2O \rightarrow 2NaOH + H_2(g)$. The coefficients include a 1 and a 2.

42. **(B)** $2CO + O_2 \rightarrow 2CO_2$ indicates 2 volumes of CO react with 1 volume of O_2 to form 2 volumes of CO_2. Therefore, 10 L of CO form 10 L of CO_2.

43. **(D)** $H_2SO_4 + 2NaOH \rightarrow 2H_2O + Na_2SO_4$ is the equation for this reaction. 1 mol $H_2SO_4 =$ 98 g. 1 mol NaOH = 40 g. Then 49 g of $H_2SO_4 = \frac{1}{2}$ mol H_2SO_4. The equation shows that

1 mol of H_2SO_4 reacts with 2 mol of NaOH for a ratio of 1:2. Therefore, $\frac{1}{2}$ mol of H_2SO_4

reacts with 1 mol of NaOH in this reaction. Since 1 mol of NaOH equals 40. g, and 80. g of NaOH is given, 40. g of it will remain after the reaction has gone to completion.

44. **(C)** $K_{sp} = [Ag^+][Cl^-] = 2.0 \times 10^{-10}$. The molar concentrations of Ag^+ and Cl^- are 1.4×10^{-5} mol/L, so

$$[Ag^+][Cl^-] = K_{sp}$$
$$[1.4 \times 10^{-5}][1.4 \times 10^{-5}] = 2.0 \times 10^{-10}$$

45. **(B)** If 1.43 g is the mass of 1 L, then the mass of 22.4 L, which is the molar volume of a gas at STP, will give the molar mass. Then, $\frac{22.4\,L}{\text{mol}} \times 1.43\,g/L = 32.0$, the molar mass.

46. **(E)** The equation is:

$$2KClO_3(s) \rightarrow 2KCl(s) + 3O_2(g)$$

This shows that 2 mol of $KClO_3$ yields 3 mol of O_2. Three moles of $O_2 = 3 \times 22.4$ L $= 67.2$ L of O_2.

47. **(A)** Gibbs free energy combines the overall energy changes and the entropy change. The formula is $\Delta G = \Delta H - T\Delta S$. Only if ΔG is negative will the reaction be spontaneous in the forward direction.

48. **(C)** When silver oxide decomposes, the oxidation states of the silver and the oxygen change from +1 to 0 and from −2 to 0, respectively. Therefore, the decomposition of silver oxide is a redox reaction.

49. **(B)** Since the mass of the silver oxide (4.64 g), the mass of the oxygen released upon heating (0.32 g), and the mass of the silver remaining in the crucible (4.32 g) can all be calculated, the percent composition of the silver oxide can be found. In other words, (0.32 g/4.64 g) × 100% = 6.9% oxygen in silver oxide. Similarly, (4.32 g/4.64 g) × 100% = 93.1% silver in silver oxide.

50. **(A)** The mass of oxygen released from the silver oxide can be converted into moles by using dimensional analysis.

$$\frac{0.32\,\text{g O}}{1} \times \frac{1\,\text{mol O}}{16\,\text{g O}} = 0.020\,\text{mol O}$$

Because the numbers in the problem are easily relatable to each other, this problem could be solved without a calculator!

51. **(E)** The purpose of the second heating is to ensure that all of the oxygen is removed from the silver oxide. Since the mass after the second heating is only 0.01 g different from that of the first heating, the experimenter could feel confident that all of the oxygen has been practically removed because the difference is so small.

52. **(D)** A splint of wood with an ember placed into a container where oxygen may have been produced offers evidence of the production of oxygen if the splint starts burning. This is called the "glowing splint" test.

53. **(A)** The phenomenon is the formation of the white ring, which is NH_4Cl. Although the distance traveled by each gas could be measured to verify Graham's Law of Gaseous Diffusion, this was not asked. The relationship is that the diffusion rate is inversely proportional to the square root of the gas's molar mass.

54. **(C)** Since Ca^{2+} and HSO_4^- combine, the formula is $Ca(HSO_4)_2$.

55. **(C)** NaOH is 40 g/mol. 40 g in 1 L makes a 1 M solution.

56. **(C)** A saturated solution represents a condition where the solute is going into solution as rapidly as some solute is coming out of the solution.

57. **(D)** II and III are correct since the double-bonded carbons in the alkene and the triple-bonded carbons in the alkyne series have pi bonds.

58. **(B)** The thistle tube is not below the level of the liquid in the generator and the gas would escape into the air. (Vinegar is an acid and would produce hydrogen.)

59. **(D)** NH_3 is very soluble and could not be collected in this manner. All others are not sufficiently soluble to hamper this method of collection.

60. **(C)** The formula (mass) is the total of (Ca =) 40 + (2N =) 28 + (6O =) 96, or 164. Since oxygen is 96 amu of 164 total, the percentage is $96/164 \times 100 = 58.5\%$.

61. **(D)** The K_{eq} expression consists of the concentration(s) of the products over those of the reactants, with the coefficients becoming exponents:

$$K_{eq} = \frac{[NO_2]^2}{[N_2O_4]}$$

62. **(A)** $K_{eq} = \dfrac{[2 \times 10^{-2}]^2}{[4 \times 10^{-2}]} = \dfrac{4 \times 10^{-4}}{4 \times 10^{-2}}$

$= 1 \times 10^{-2}$

63. **(C)** In dilution problems, the expressions $M_1 \times V_1 = M_2 \times V_2$ can be used. Substituting (6.0 M) (0.50 L) = (2.0 M) (x L) gives x = 1.5 L total volume. Since there was 0.50 L to begin with, an additional 1L must be added.

64. **(C)** The reaction equation and information given can be set up like this:

Given	Given	
4.0 g	4.0 g	x g
$2H_2$	+ O_2	→ $2H_2O$
4.0 g	32 g	36 g

Studying this shows that the limiting element will be the 4 g of oxygen since 4 g of H_2 would require 32 g of O_2. The solution setup is

$$\frac{4.0\,g\,O_2}{32\,g\,O_2} = \frac{x\,g\,H_2O}{36\,g\,H_2O}$$

$$x = 4.5\,g\,H_2O$$

65. **(E)** The functional group of an ester is

$$\overset{\displaystyle O}{\underset{\displaystyle \|}{R-C-O-R_1}}.$$ This appears only in (E).

66. **(C)** The thistle tube serves both these purposes.

67. **(D)** The empirical formula is a representation of the elements in their simplest ratio. Therefore, CH_2 is the simplest ratio of the molecular formula C_3H_6.

68. **(E)** Synthesis reactions contain only one product. This reaction contains two products. Therefore, it can't be described as a synthesis reaction. It is, however, both combustion and redox. All combustion reactions are redox reactions because the elemental oxygen (with an oxidation state of 0) will enter into compounds and change its oxidation state.

69. **(A)** In the equation, 2 mol of Na release 822 kJ of heat. If only 0.5 mol of Na is consumed, only one-fourth as much heat will be released: $\frac{1}{4} \times 822 = 205$ kJ.

70. **(B)** The water molecule contains atoms of hydrogen and oxygen covalently bonded to each other. Since there is a large electronegativity difference between these atoms, the bond is very polar. The polar bonds as well as the two lone pair electrons that are not symmetrically balanced about the oxygen atom make the overall molecule polar. Hydrogen bonding is not a type of intramolecular bond and doesn't apply when atoms held together within a molecule are being considered.

CALCULATING YOUR SCORE

Your score on practice Test 1 can now be computed manually. The actual test is scored by machine, but the same method is used to arrive at the raw score. You get one point for each correct answer. For each wrong answer, you lose one-fourth of a point. Questions that you omit or that have more than one answer are not counted. On your answer sheet mark all correct answers with a "C" and all incorrect answers with an "X."

Determining Your Raw Test Score

Total the number of correct answers you have recorded on your answer sheet. It should be the same as the total of all the numbers you place in the block in the lower left corner of each area of the Subject Area summary in the next section.

  A. Enter the total number of correct answers here: _____
    Now count the number of wrong answers you recorded on your answer sheet.
  B. Enter the total number of wrong answers here: _____
    Multiply the number of wrong answers in B by 0.25.
  C. Enter that product here: _____
    Subtract the result in C from the total number of right answers in A.
  D. Enter the result of your subtraction here: _____
  E. Round the result in D to the nearest whole number: _____.
    This is your raw test score.

Conversion of Raw Scores to Scaled Scores

Your raw score is converted by the College Board into a scaled score. The College Board scores range from 200 to 800. This conversion is done to ensure that a score earned on any edition of a particular SAT Subject Test in Chemistry is comparable to the same scaled score earned on any other edition of the same test. Because some editions of the tests may be slightly easier or more difficult than others, scaled scores are adjusted so that they indicate the same level of performance regardless of the edition of the test taken and the ability of the group that takes it. Consequently, a specific raw score on one edition of a particular test will not necessarily translate to the same scaled score on another edition of the same test.

Because the practice tests in this book have no large population of scores with which they can be scaled, scaled scores cannot be determined.

Results from previous SAT Chemistry tests appear to indicate that the conversion of raw scores to scaled scores GENERALLY follows this pattern:

Raw Score	Scaled Score	Raw Score	Scaled Score
85–82	800–800	30–25	540–520
81–75	790–760	25–20	520–490
75–70	760–740	20–15	490–460
70–65	740–710	15–10	460–430
65–60	710–690	10–5	430–400
60–55	690–670	5–0	400–370
55–50	670–640	0 to –5	370–340
50–45	640–620	–5 to –10	340–310
45–40	620–590	–10 to –15	310–290
40–35	590–570	–15 to –20	290–270
35–30	570–540	–20 or lower	270–200

Note that this scale provides only a *general idea* of what a raw score may translate into on a scaled score range of 800–200. Scaling on every test is usually slightly different. Some students who had taken the SAT Subject Test in Chemistry after using this book had reported that they have scored slightly higher on the SAT test than on the practice tests in this book. They *all* reported that preparing well for the test paid off in a better score!

DIAGNOSING YOUR NEEDS

After taking Practice Test 1, check your answers against the correct ones. Then fill in the chart below.

In the space under each question number, place a check if you answered that question correctly.

➥ Example _____

If your answer to question 5 was correct, place a check in the appropriate box.

Next, total the check marks for each section and insert the number in the designated block. Now do the arithmetic indicated, and insert your percent for each area.

Subject Area*	(✔) Questions Answered Correctly								
I. Atomic Theory and Structure, including periodic relationships	6	8	105	109	111	31	38	110	115
☐ No. of checks ÷ 9 × 100 = _____%									

II. Chemical Bonding and Molecular Structure	3	17	18	19	29	30	39	54	57	70
☐ No. of checks ÷ 10 × 100 = _____%										

III. States of Matter and Kinetic Molecular Theory of Gases	10	11	12	107	108	113	27
☐ No. of checks ÷ 7 × 100 = _____%							

IV. Solutions, including concentration units, solubility, and colligative properties	21	103	44	55	63
☐ No. of checks ÷ 5 × 100 = _____%					

V. Acids and Bases	20	101	102	114	33	36	53
☐ No. of checks ÷ 7 × 100 = _____%						

VI. Oxidation-Reduction	106	112	34	35	48	68
☐ No. of checks ÷ 6 × 100 = _____%					

VII. Stoichiometry	24	28	41	42	43	45	46	50	60	64
☐ No. of checks ÷ 10 × 100 = _____%										

VIII. Reaction Rates	25	32
☐ No. of checks ÷ 2 × 100 = _____%		

IX. Equilibrium	22	40	56	61	62
☐ No. of checks ÷ 5 × 100 = _____%					

X. Thermodynamics: energy changes in chemical reactions, randomness, and criteria for spontaneity	13	14	15	104	47	64
☐ No. of checks ÷ 6 × 100 = _____%					

The subject areas have been expanded to identify specific areas in the text.

Subject Area*	(✔) Questions Answered Correctly					
XI. Descriptive Chemistry: physical and chemical properties of elements and their familiar compounds; organic chemistry; periodic properties	1	2	4	5	7	9
	16	23	26	37	65	67
☐ No. of checks ÷ 12 × 100 = _____%						
XII. Laboratory: equipment, procedures, observations, safety, calculations, and interpretation of results	49	51	52	58	59	66
☐ No. of checks ÷ 6 × 100 = _____%						

The subject areas have been expanded to identify specific areas in the text.

To develop a study plan, refer to pages 31–33.

ANSWER SHEET
Practice Test 2

Determine the correct answer for each question. Then, using a No. 2 pencil, blacken completely the oval containing the letter of your choice.

1. Ⓐ Ⓑ Ⓒ Ⓓ Ⓔ
2. Ⓐ Ⓑ Ⓒ Ⓓ Ⓔ
3. Ⓐ Ⓑ Ⓒ Ⓓ Ⓔ
4. Ⓐ Ⓑ Ⓒ Ⓓ Ⓔ
5. Ⓐ Ⓑ Ⓒ Ⓓ Ⓔ
6. Ⓐ Ⓑ Ⓒ Ⓓ Ⓔ
7. Ⓐ Ⓑ Ⓒ Ⓓ Ⓔ
8. Ⓐ Ⓑ Ⓒ Ⓓ Ⓔ
9. Ⓐ Ⓑ Ⓒ Ⓓ Ⓔ
10. Ⓐ Ⓑ Ⓒ Ⓓ Ⓔ
11. Ⓐ Ⓑ Ⓒ Ⓓ Ⓔ
12. Ⓐ Ⓑ Ⓒ Ⓓ Ⓔ
13. Ⓐ Ⓑ Ⓒ Ⓓ Ⓔ
14. Ⓐ Ⓑ Ⓒ Ⓓ Ⓔ
15. Ⓐ Ⓑ Ⓒ Ⓓ Ⓔ
16. Ⓐ Ⓑ Ⓒ Ⓓ Ⓔ

17. Ⓐ Ⓑ Ⓒ Ⓓ Ⓔ
18. Ⓐ Ⓑ Ⓒ Ⓓ Ⓔ
19. Ⓐ Ⓑ Ⓒ Ⓓ Ⓔ
20. Ⓐ Ⓑ Ⓒ Ⓓ Ⓔ
21. Ⓐ Ⓑ Ⓒ Ⓓ Ⓔ
22. Ⓐ Ⓑ Ⓒ Ⓓ Ⓔ
23. Ⓐ Ⓑ Ⓒ Ⓓ Ⓔ

ON THE ACTUAL CHEMISTRY TEST. THE FOLLOWING TYPE OF QUESTION MUST BE ANSWERED ON A SPECIAL SECTION (LABELED "CHEMISTRY") AT THE LOWER LEFT-HAND CORNER OF PAGE 2 OF YOUR ANSWER SHEET. THESE QUESTIONS WILL BE NUMBERED BEGINING WITH 101 AND MUST BE ANSWERED ACCORDING TO THE DIRECTIONS.

CHEMISTRY* Fill in oval CE only if II is a correct explanation of 1.

	I	II	CE*
101.	Ⓣ Ⓕ	Ⓣ Ⓕ	◯
102.	Ⓣ Ⓕ	Ⓣ Ⓕ	◯
103.	Ⓣ Ⓕ	Ⓣ Ⓕ	◯
104.	Ⓣ Ⓕ	Ⓣ Ⓕ	◯
105.	Ⓣ Ⓕ	Ⓣ Ⓕ	◯
106.	Ⓣ Ⓕ	Ⓣ Ⓕ	◯
107.	Ⓣ Ⓕ	Ⓣ Ⓕ	◯
108.	Ⓣ Ⓕ	Ⓣ Ⓕ	◯
109.	Ⓣ Ⓕ	Ⓣ Ⓕ	◯
110.	Ⓣ Ⓕ	Ⓣ Ⓕ	◯
111.	Ⓣ Ⓕ	Ⓣ Ⓕ	◯
112.	Ⓣ Ⓕ	Ⓣ Ⓕ	◯
113.	Ⓣ Ⓕ	Ⓣ Ⓕ	◯
114.	Ⓣ Ⓕ	Ⓣ Ⓕ	◯
115.	Ⓣ Ⓕ	Ⓣ Ⓕ	◯

ANSWER SHEET
Practice Test 2

ON THE ACTUAL CHEMISTRY TEST, THE REMAINING QUESTIONS MUST BE ANSWERED BY RETURNING TO THE SECTION OF YOUR ANSWER SHEET YOU STARTED FOR CHEMISTRY.

24. Ⓐ Ⓑ Ⓒ Ⓓ Ⓔ   40. Ⓐ Ⓑ Ⓒ Ⓓ Ⓔ   56. Ⓐ Ⓑ Ⓒ Ⓓ Ⓔ
25. Ⓐ Ⓑ Ⓒ Ⓓ Ⓔ   41. Ⓐ Ⓑ Ⓒ Ⓓ Ⓔ   57. Ⓐ Ⓑ Ⓒ Ⓓ Ⓔ
26. Ⓐ Ⓑ Ⓒ Ⓓ Ⓔ   42. Ⓐ Ⓑ Ⓒ Ⓓ Ⓔ   58. Ⓐ Ⓑ Ⓒ Ⓓ Ⓔ
27. Ⓐ Ⓑ Ⓒ Ⓓ Ⓔ   43. Ⓐ Ⓑ Ⓒ Ⓓ Ⓔ   59. Ⓐ Ⓑ Ⓒ Ⓓ Ⓔ
28. Ⓐ Ⓑ Ⓒ Ⓓ Ⓔ   44. Ⓐ Ⓑ Ⓒ Ⓓ Ⓔ   60. Ⓐ Ⓑ Ⓒ Ⓓ Ⓔ
29. Ⓐ Ⓑ Ⓒ Ⓓ Ⓔ   45. Ⓐ Ⓑ Ⓒ Ⓓ Ⓔ   61. Ⓐ Ⓑ Ⓒ Ⓓ Ⓔ
30. Ⓐ Ⓑ Ⓒ Ⓓ Ⓔ   46. Ⓐ Ⓑ Ⓒ Ⓓ Ⓔ   62. Ⓐ Ⓑ Ⓒ Ⓓ Ⓔ
31. Ⓐ Ⓑ Ⓒ Ⓓ Ⓔ   47. Ⓐ Ⓑ Ⓒ Ⓓ Ⓔ   63. Ⓐ Ⓑ Ⓒ Ⓓ Ⓔ
32. Ⓐ Ⓑ Ⓒ Ⓓ Ⓔ   48. Ⓐ Ⓑ Ⓒ Ⓓ Ⓔ   64. Ⓐ Ⓑ Ⓒ Ⓓ Ⓔ
33. Ⓐ Ⓑ Ⓒ Ⓓ Ⓔ   49. Ⓐ Ⓑ Ⓒ Ⓓ Ⓔ   65. Ⓐ Ⓑ Ⓒ Ⓓ Ⓔ
34. Ⓐ Ⓑ Ⓒ Ⓓ Ⓔ   50. Ⓐ Ⓑ Ⓒ Ⓓ Ⓔ   66. Ⓐ Ⓑ Ⓒ Ⓓ Ⓔ
35. Ⓐ Ⓑ Ⓒ Ⓓ Ⓔ   51. Ⓐ Ⓑ Ⓒ Ⓓ Ⓔ   67. Ⓐ Ⓑ Ⓒ Ⓓ Ⓔ
36. Ⓐ Ⓑ Ⓒ Ⓓ Ⓔ   52. Ⓐ Ⓑ Ⓒ Ⓓ Ⓔ   68. Ⓐ Ⓑ Ⓒ Ⓓ Ⓔ
37. Ⓐ Ⓑ Ⓒ Ⓓ Ⓔ   53. Ⓐ Ⓑ Ⓒ Ⓓ Ⓔ   69. Ⓐ Ⓑ Ⓒ Ⓓ Ⓔ
38. Ⓐ Ⓑ Ⓒ Ⓓ Ⓔ   54. Ⓐ Ⓑ Ⓒ Ⓓ Ⓔ   70. Ⓐ Ⓑ Ⓒ Ⓓ Ⓔ
39. Ⓐ Ⓑ Ⓒ Ⓓ Ⓔ   55. Ⓐ Ⓑ Ⓒ Ⓓ Ⓔ

PRACTICE TEST 2

Note: For all questions involving solutions and/or chemical equations, assume that the system is in water unless otherwise stated.

Reminder: You may *not* use a calculator on these tests.

The following symbols have the meanings listed unless otherwise noted.

H = enthalpy	T = temperature	L = liter(s)
M = molar	V = volume	mL = milliliter(s)
n = number of moles	atm = atmosphere(s)	mm = millimeter(s)
P = pressure	g = gram(s)	mol = mole(s)
R = molar gas constant	J = joule(s)	V = volt(s)
S = entropy	kJ = kilojoule(s)

PART A

Directions: Every set of the given lettered choices below refers to the numbered statements or formulas immediately following it. Choose the one lettered choice that best fits each statement or formula and then fill in the corresponding oval on the answer sheet. Each choice may be used once, more than once, or not at all in each set.

Questions 1–4 refer to the following terms:

    (A) Boiling point
    (B) Melting point
    (C) Critical point
    (D) Freezing point
    (E) Triple point

1. The temperature and pressure at which three states of a substance may coexist

2. The temperature at which a solid becomes a liquid

3. The temperature of 373 K for H_2O at standard pressure

4. The temperature at which the vapor pressure of a liquid equals the atmospheric pressure

Questions 5–7 refer to the following diagram:

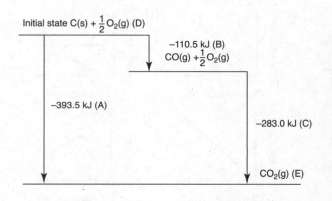

5. The ΔH of the reaction to form CO from $C + O_2$

6. The ΔH of the reaction to form CO_2 from $CO + O_2$

7. The ΔH of the reaction to form CO_2 from $C + O_2$

GO ON TO THE NEXT PAGE

Questions 8–11

(A) Hydrogen bond
(B) Ionic bond
(C) Polar covalent bond
(D) Nonpolar covalent bond
(E) Metallic bond

8. The type of bond between atoms of potassium and chloride when they form a crystal of potassium chloride

9. The type of bond between the atoms in a nitrogen molecule

10. The type of bond between the atoms in a molecule of CO_2 (electronegativity difference = 1)

11. The type of bond between the atoms of calcium in a crystal of calcium

Questions 12–14 refer to the following graphs:

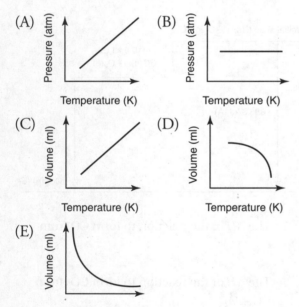

12. The graph of volume vs. pressure for a gas at constant temperature

13. The graph of pressure vs. temperature for a gas at constant volume

14. The graph of volume vs. temperature for a gas at constant pressure

Questions 15–18

(A) Least-reactive family of elements
(B) Alkali metals
(C) Halogen family of elements
(D) Noble gases
(E) Family whose oxides form acids in water

15. The elements that most actively react with water to release hydrogen

16. The elements least likely to become involved in chemical reactions

17. Family that contains elements in the colored gaseous state, in the liquid state, and with metallic properties

18. Group of nonmetallic elements containing N and P

Questions 19–23

(A) $1s$
(B) $2s$
(C) $3s$
(D) $3p$
(E) $3d$

19. Electron energy sublevel filled by the first period of transition metals

20. The lowest energy orbital of those shown

21. Of the electron energy sublevels shown, the one that holds a maximum of 6 electrons

22. Largest of the orbitals with a spherical probability distribution

23. Orbital that describes the probability distribution for sodium's outermost electron in the ground state

GO ON TO THE NEXT PAGE

ON THE ACTUAL CHEMISTRY TEST, THE FOLLOWING TYPE OF QUESTION MUST BE ANSWERED ON A SPECIAL SECTION (LABELED "CHEMISTRY") AT THE LOWER LEFT-HAND CORNER OF PAGE 2 OF YOUR ANSWER SHEET. THESE QUESTIONS WILL BE NUMBERED BEGINNING WITH 101 AND MUST BE ANSWERED ACCORDING TO THE FOLLOWING DIRECTIONS.

Directions: Every question below contains two statements, I in the left-hand column and II in the right-hand column. For each question, decide if statement I is true or false <u>and</u> if statement II is true or false and fill in the corresponding T or F ovals on your answer sheet. *<u>Fill in oval CE only if statement II is a correct explanation of statement I.</u>

Sample Answer Grid:

CHEMISTRY * Fill in oval CE only if II is a correct explanation of I.

	I	II	CE*
101.	Ⓣ Ⓕ	Ⓣ Ⓕ	◯

	I		II
101.	The structure of SO_3 is shown by using more than one structural formula	BECAUSE	SO_3 is very unstable and resonates between these possible structures.
102.	When the ΔG of a reaction at a given temperature is negative, the reaction occurs spontaneously	BECAUSE	when ΔG is negative, ΔH is also negative.
103.	One mole of CO_2 has a greater mass than 1 mole of H_2O	BECAUSE	the molecular mass of CO_2 is greater than the molecular mass of H_2O.
104.	Hydrosulfuric acid is often used in qualitative tests	BECAUSE	$H_2S(aq)$ reacts with many metallic ions to give colored precipitates.
105.	Crystals of sodium chloride go into solution in water as ions	BECAUSE	the sodium ion has a 1+ charge and the chloride ion has a 1− charge and they are hydrated by the water molecules.
106.	If some phosphoric acid, H_3PO_4, is added to the equilibrium mixture represented by the equation $H_3PO_4 + H_2O \leftrightarrow PO_4^{3-} + H_3O^+$, the concentration of H_3O^+ decreases	BECAUSE	the equilibrium constant of a reaction changes as the concentration of the reactants changes.

GO ON TO THE NEXT PAGE

PRACTICE TEST 2

	I		**II**

107. The $\Delta H_{reaction}$ of a particular reaction can be arrived at by the summation of the $\Delta H_{reaction}$ values of two or more reactions that, added together, give the $\Delta H_{reaction}$ of the particular reaction

BECAUSE Hess's Law conforms to the First Law of Thermodynamics, which states that the total energy of the universe is a constant.

108. In a reaction that has both a forward and a reverse reaction, A + B ⇌ AB, when only A and B are introduced into a reacting vessel, the forward reaction rate is the highest at the beginning and begins to decrease from that point until equilibrium is reached

BECAUSE the reverse reaction does not begin until equilibrium is reached.

109. At equilibrium, the forward reaction and reverse reaction stop

BECAUSE at equilibrium, the reactants and products have reached the equilibrium concentrations.

110. The hydrid orbital form of carbon in acetylene is believed to be the *sp* form

BECAUSE C_2H_2 is a linear molecule with a triple bond between the carbons.

111. The weakest of the bonds between molecules are coordinate covalent bonds

BECAUSE coordinate covalent bonds represent the weak attractive force of the electrons of one molecule for the positively charged nucleus of another.

112. A saturated solution is not necessarily concentrated

BECAUSE *dilute* and *concentrated* are terms that relate only to the relative amount of solute dissolved in the solvent.

113. Lithium is the most active metal in the first group of the Periodic Table

BECAUSE lithium has only one electron in the outer energy level.

114. The oxidation state of carbon is always +4

BECAUSE carbon has 4 valence electrons.

115. The atomic number of a neutral atom that has a mass of 39 and has 19 electrons is 19

BECAUSE the number of protons in a neutral atom is equal to the number of electrons.

GO ON TO THE NEXT PAGE

PRACTICE TEST 2

Directions: Each of the questions or incomplete statements below is followed by five suggested answers or completions. Select the one that is best in each case and then fill in the corresponding oval on the answer sheet.

24. All of the following involve a chemical change EXCEPT

   (A) the formation of HCl from H_2 and Cl_2
   (B) the color change when NO is exposed to air
   (C) the formation of steam from burning H_2 and O_2
   (D) the solidification of vegetable oil at low temperatures
   (E) the odor of NH_3 when NH_4Cl is rubbed together with $Ca(OH)_2$ powder

25. When most fuels burn, the products include carbon dioxide and

   (A) hydrocarbons
   (B) hydrogen
   (C) water
   (D) hydroxide
   (E) hydrogen peroxide

26. In the metric system, the prefix *kilo*- means

   (A) 10^0
   (B) 10^{-1}
   (C) 10^{-2}
   (D) 10^2
   (E) 10^3

27. How many atoms are in 1 mole of water?

   (A) 3
   (B) 54
   (C) 6.02×10^{23}
   (D) $2(6.02 \times 10^{23})$
   (E) $3(6.02 \times 10^{23})$

28. Which of the following elements normally exist as monoatomic molecules?

   (A) Cl
   (B) H
   (C) O
   (D) N
   (E) He

29. The shape of a PCl_3 molecule is described as

   (A) bent
   (B) trigonal planar
   (C) linear
   (D) trigonal pyramidal
   (E) tetrahedral

30. The complete loss of an electron of one atom to another atom with the consequent formation of electrostatic charges is referred to as

   (A) a covalent bond
   (B) a polar covalent bond
   (C) an ionic bond
   (D) a coordinate covalent bond
   (E) a pi bond between *p* orbitals

31. In the decomposition of water with electricity (electrolysis), the following reaction occurs.

$$2H_2O(\ell) \rightarrow 2\,H_2(g) + O_2(g)$$

   The hydrogen is

   (A) oxidized from +1 to 0
   (B) oxidized from 0 to +1
   (C) reduced from 0 to +1
   (D) reduced from +1 to 0
   (E) not changing oxidation states

GO ON TO THE NEXT PAGE

32. Which of the following radiation emissions has no mass?

   (A) Alpha particle
   (B) Beta particle
   (C) Proton
   (D) Neutron
   (E) Gamma ray

33. If a radioactive element with a half-life of 100 years is found to have transmutated so that only 25% of the original sample remains, what is the age, in years, of the sample?

   (A) 25
   (B) 50
   (C) 100
   (D) 200
   (E) 400

34. What is the pH of an acetic acid solution if the $[H_3O^+] = 1 \times 10^{-4}$ mole/liter?

   (A) 1
   (B) 2
   (C) 3
   (D) 4
   (E) 5

35. The polarity of water is useful in explaining which of the following?

   I. The solution process
   II. The ionization process
   III. The high conductivity of distilled water

   (A) I only
   (B) II only
   (C) I and II only
   (D) II and III only
   (E) I, II, and III

36. When sulfur dioxide is bubbled through water, the solution will contain

   (A) sulfurous acid
   (B) sulfuric acid
   (C) hyposulfuric acid
   (D) persulfuric acid
   (E) anhydrous sulfuric acid

37. Four grams of hydrogen gas at STP contain

   (A) 6.02×10^{23} atoms
   (B) 12.04×10^{23} atoms
   (C) 12.04×10^{46} atoms
   (D) 1.2×10^{23} molecules
   (E) 12.04×10^{23} molecules

38. Analysis of a gas gave: C = 85.7% and H = 14.3%. If the formula mass of this gas is 42 atomic mass units, what are the empirical formula and the true formula?

   (A) CH; C_4H_4
   (B) CH_2; C_3H_6
   (C) CH_3; C_3H_9
   (D) C_2H_2; C_3H_6
   (E) C_2H_4; C_3H_6

39. Which fraction would be used to correct a given volume of gas at 303 K to its new volume when it is heated to 333 K and the pressure is kept constant?

   (A) $\dfrac{303 - 273}{60 + 273}$

   (B) $\dfrac{60}{30}$

   (C) $\dfrac{273}{333}$

   (D) $\dfrac{303}{333}$

   (E) $\dfrac{333}{303}$

40. Which of the substances listed decreases the freezing point of benzene (C_6H_6) more than the others if a lab tech tries to dissolve 5.00 grams in 500.0 g of benzene?

   (A) paradichlorobenzene, $C_6H_4Cl_2$
   (B) sodium chloride, NaCl
   (C) aluminum chloride, $AlCl_3$
   (D) ethanol, C_2H_5OH
   (E) sucrose, $C_{12}H_{22}O_{11}$

GO ON TO THE NEXT PAGE

41. What is the approximate pH of a 0.005 M solution of H_2SO_4?

(A) 1
(B) 2
(C) 5
(D) 9
(E) 13

42. How many grams of NaOH are needed to make 100 grams of a 5% solution?

(A) 2
(B) 5
(C) 20
(D) 40
(E) 95

43. For the Haber process: $N_2 + 3H_2 \rightleftharpoons 2NH_3 +$ heat (at equilibrium), which of the following statements concerning the reaction rate is/are true?

   I. The reaction to the right will increase when pressure is increased.
   II. The reaction to the right will decrease when the temperature is increased.
   III. The reaction to the right will decrease when NH_3 is removed from the chamber.

(A) I only
(B) II only
(C) I and II only
(D) II and III only
(E) I, II, and III

44. If you titrate 1.0 M H_2SO_4 solution against 50. milliliters of 1.0 M NaOH solution, what volume of H_2SO_4, in milliliters, will be needed for neutralization?

(A) 10.
(B) 25.
(C) 40.
(D) 50.
(E) 100

45. How many grams of CO_2 can be prepared from 150 grams of calcium carbonate reacting with an excess of hydrochloric acid solution?

(A) 11
(B) 22
(C) 33
(D) 44
(E) 66

Question 46 refers to the following diagram:

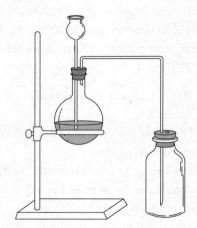

46. The diagram represents a setup that may be used to prepare and collect

(A) NH_3
(B) NO
(C) H_2
(D) SO_3
(E) CO_2

GO ON TO THE NEXT PAGE

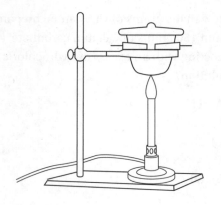

Recorded Data:

Mass of evaporating dish   = 14.30 g

Mass of dish and hydrated BaCl$_2$ = 26.50 g

Mass of dish and anhydrous BaCl$_2$ = 24.77 g
(after 1st heating)

Mass of dish and anhydrous BaCl$_2$ = 24.70 g
(after 2nd heating)

Mass of dish and anhydrous BaCl$_2$ = 24.70 g
(after 3rd heating)

47. The lab setup shown above was used for the gravimetric analysis of the empirical formula of MgO. In synthesizing MgO from a Mg strip in the crucible, which of the following is NOT true?

   (A) The initial strip of Mg should be cleaned.
   (B) The lid of the crucible should fit tightly to exclude oxygen.
   (C) The heating of the covered crucible should continue until the Mg is fully reacted.
   (D) The crucible, lid, and the contents should be cooled to room temperature before measuring their mass.
   (E) When the Mg appears to be fully reacted, the crucible lid should be partially removed and heating continued.

Questions 48–50 refer to the following experimental scenario and data:

A sample of the hydrated form of barium chloride is placed into an evaporating dish and heated strongly for 20 minutes on a hot plate. After cooling in a desiccator, the mass of the dish and contents is determined. The whole system is then heated strongly again for 5 minutes and cooled. Then the mass is measured again. The heating, cooling, and measuring of the mass is repeated a third time.

48. The purpose of cooling the anhydrous BaCl$_2$ in the desiccator is to

   (A) protect the experimenter from the heat of the dish
   (B) cool the dish slowly
   (C) keep the contents of the dish from rehydrating
   (D) cool the dish quickly
   (E) rehydrate the contents of the dish

49. If the person running the experiment failed to notice a small amount of the solid substance jump out of the evaporating dish during heating, the resulting data would suggest the amount of water in the hydrate is

   (A) less than what it should be
   (B) the same as if nothing jumped out of the dish
   (C) more than what it should be
   (D) equal to the mass of the BaCl$_2$ that also left the dish
   (E) more than the mass of the BaCl$_2$ that also left the dish

50. The number of moles of water released from the hydrated barium chloride after three rounds of heating is

   (A) 0.0200 mol
   (B) 0.0400 mol
   (C) 0.0800 mol
   (D) 0.100 mol
   (E) 1.20 mol

GO ON TO THE NEXT PAGE

51. What is the mass, in grams, of 1 mole of $KAl(SO_4)_2 \cdot 12H_2O$?

(A) 132
(B) 180
(C) 394
(D) 474
(E) 516

52. What mass of aluminum will be completely oxidized by 2 moles of oxygen at STP?

(A) 18 g
(B) 37.8 g
(C) 50.4 g
(D) 72.0 g
(E) 100.8 g

53. In general, when metal oxides react with water, they form solutions that are

(A) acidic
(B) basic
(C) neutral
(D) unstable
(E) colored

Questions 54–56 refer to the following spontane-ous reactions:

$$Zn(s) + SnCl_2(aq) \rightarrow Sn(s) + ZnCl_2(aq)$$

and

$$Sn(s) + CuCl_2(aq) \rightarrow Cu(s) + SnCl_2(aq)$$

54. The significant driving force associated with both of these reactions is

(A) the transfer of electrons
(B) the transfer of protons
(C) the transfer of mass
(D) the transfer of ions
(E) the transfer of atoms

55. The most active of the metals shown is

(A) Sn
(B) Cu
(C) Zn
(D) Cl
(E) $ZnCl_2$

56. Substances acting as oxidizing agents include

(A) tin(II) chloride
(B) tin
(C) zinc chloride
(D) both tin *and* copper(II) chloride
(E) both zinc chloride *and* copper(II) chloride

57. How many liters of oxygen (STP) can be prepared from the decomposition of 212 grams of sodium chlorate (1 mol = 106 g)?

(A) 11.2
(B) 22.4
(C) 44.8
(D) 67.2
(E) 78.4

58. In this equation: $Al(OH)_3 + H_2SO_4 \rightarrow Al_2(SO_4)_3 + H_2O$, the whole-number coefficients of the balanced equation are

(A) 1, 3, 1, 2
(B) 2, 3, 2, 6
(C) 2, 3, 1, 6
(D) 2, 6, 1, 3
(E) 1, 3, 1, 6

GO ON TO THE NEXT PAGE

59. What is $\Delta H_{reaction}$ for the decomposition of 1 mole of sodium chlorate? (ΔH_f^0 values: $NaClO_3(s) = -85.7$ kcal/mol, $NaCl(s) = -98.2$ kcal/mol, $O_2(g) = 0$ kcal/mol)

(A) −183.9 kcal
(B) −91.9 kcal
(C) +45.3 kcal
(D) +22.5 kcal
(E) −12.5 kcal

60. Isotopes of an element are related because which of the following is (are) the same in these isotopes?

I. Atomic mass
II. Atomic number
III. Arrangement of orbital electrons

(A) I only
(B) II only
(C) I and II only
(D) II and III only
(E) I, II, and III

61. In the reaction of zinc with dilute HCl to form H_2, which of the following will increase the reaction rate?

I. Increasing the temperature
II. Increasing the exposed surface of zinc
III. Using a more concentrated solution of HCl

(A) I only
(B) II only
(C) I and III only
(D) II and III only
(E) I, II, and III

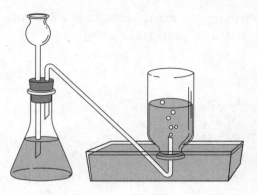

62. The laboratory setup shown above can be used to prepare a

(A) gas less dense than air and soluble in water
(B) gas more dense than air and soluble in water
(C) gas soluble in water that reacts with water
(D) gas insoluble in water
(E) gas that reacts with water

63. In this reaction:

$$CaCO_3 + 2HCl \rightarrow CaCl_2 + H_2O + CO_2.$$

If 4.0 moles of HCl are available to the reaction with an unlimited supply of $CaCO_3$, how many moles of CO_2 can be produced at STP?

(A) 1.0
(B) 1.5
(C) 2.0
(D) 2.5
(E) 3.0

64. A saturated solution of PbS at 25°C contains 5.0×10^{-14} mole/liter of Pb^{2+} ions. What is the K_{sp} value of this salt?

(A) 5.0×10^{-14}
(B) 5.0×10^{-15}
(C) 2.5×10^{-14}
(D) 2.5×10^{-28}
(E) 2.5×10^{-27}

GO ON TO THE NEXT PAGE

65. If 0.1 mole of K_2S was added to the solution in question 64, what would happen to the Pb^{2+} concentration?

(A) It would increase.
(B) It would decrease.
(C) It would remain the same.
(D) It would first increase, then decrease.
(E) It would first decrease, then increase.

66. Which of the following will definitely cause the volume of a gas to increase?

 I. Decreasing the pressure with the temperature held constant.
 II. Increasing the pressure with a temperature decrease.
 III. Increasing the temperature with a pressure increase.

(A) I only
(B) II only
(C) I and III only
(D) II and III only
(E) I, II, and III

67. The number of oxygen atoms in 0.50 mole of $Al_2(CO_3)_3$ is

(A) 4.5×10^{23}
(B) 9.0×10^{23}
(C) 3.6×10^{24}
(D) 2.7×10^{24}
(E) 5.4×10^{24}

Question 68 refers to a solution of 1 M acid, HA, with $K_a = 1 \times 10^{-6}$.

68. What is the H_3O^+ concentration? (Assume $[HA] = 1$, $[H_3O^+] = x$, $[A^-] = x$.)

(A) 1×10^{-5}
(B) 1×10^{-4}
(C) 1×10^{-2}
(D) 1×10^{-3}
(E) 0.9×10^{-3}

69. What is the percent dissociation of acetic acid in a 0.1 M solution if the $[H_3O^+]$ is 1×10^{-3} mole/liter?

(A) 0.01%
(B) 0.1%
(C) 1.0%
(D) 1.5%
(E) 2.0%

70. The oxidation state of chromium in the polyatomic ion $Cr_2O_7^{2-}$ is

(A) –3
(B) –2
(C) +2
(D) +3
(E) +6

STOP

If you finish before one hour is up, you may go back to check your work or complete unanswered questions.

1. **E**	7. **A**	13. **A**	19. **E**
2. **B**	8. **B**	14. **C**	20. **A**
3. **A**	9. **D**	15. **B**	21. **D**
4. **A**	10. **C**	16. **D**	22. **C**
5. **B**	11. **E**	17. **C**	23. **C**
6. **C**	12. **E**	18. **E**	

101. **T, F**	105. **T, T, CE**	109. **F, T**	113. **T, T**
102. **T, F**	106. **F, F**	110. **T, T, CE**	114. **F, T**
103. **T, T, CE**	107. **T, T, CE**	111. **F, F**	115. **T, T, CE**
104. **T, T, CE**	108. **T, F**	112. **T, T, CE**	

24. **D**	36. **A**	48. **C**	60. **D**
25. **C**	37. **E**	49. **C**	61. **E**
26. **E**	38. **B**	50. **D**	62. **D**
27. **E**	39. **E**	51. **D**	63. **C**
28. **E**	40. **A**	52. **D**	64. **E**
29. **D**	41. **B**	53. **B**	65. **B**
30. **C**	42. **B**	54. **A**	66. **A**
31. **D**	43. **C**	55. **C**	67. **D**
32. **E**	44. **B**	56. **A**	68. **D**
33. **D**	45. **E**	57. **D**	69. **C**
34. **D**	46. **E**	58. **C**	70. **E**
35. **C**	47. **B**	59. **E**	

ANSWERS EXPLAINED

1. **(E)** A phase diagram shows that all three states can exist at the triple point.

2. **(B)** When a solid turns into a liquid, the process is called melting. This happens at a particular temperature when the pressure on the solid is constant. It's called the melting point.

3. **(A)** Water boils at 100.0°C under standard pressure (1.00 atmosphere). On the absolute temperature scale, that's 373 K.

4. **(A)** When the vapor pressure of a liquid is the same as the atmospheric pressure, bubbles can be created in the body of the liquid and boiling can take place.

5. **(B)** The first step is the ΔH for $C + \frac{1}{2}O_2 \rightarrow CO$. It releases –110.5kJ of heat. This is written as –110.5 kJ because it is exothermic.

6. **(C)** This is the second step on the diagram. It releases –283.0 kJ of heat.

7. **(A)** To arrive at the ΔH, take the total drop (–393.5 kJ) or add these reactions:

$$
\begin{array}{lll}
C + \frac{1}{2}O_2 & \rightarrow CO & -110.5 \text{ kJ} \\
CO + \frac{1}{2}O_2 & \rightarrow CO_2 & -283.0 \text{ kJ} \\
\hline
C + O_2 & \rightarrow CO_2 & -393.5 \text{ kJ}
\end{array}
$$

8. **(B)** Potassium and chlorine have a large enough difference in their electronegativities to form ionic bonds. The respective positions of these two elements in the periodic chart also are indicative of the large difference in their electronegativity values.

9. **(D)** Two atoms of an element that forms a diatomic molecule always have a nonpolar covalent bond between them since the electron attraction or electronegativity of the two atoms is the same.

10. **(C)** Electronegativity differences between 0.5 and 1.7 are usually indicative of polar covalent bonding. CO_2 is an interesting example of a *nonpolar molecule* with polar covalent bonds since the bonds are symmetrical in the molecule.

11. **(E)** Calcium is a metal and forms a metallic bond between atoms.

12. **(E)** This graph shows the volume decreasing as the pressure is increased and the temperature is held constant. It is an example of Boyle's Law ($PV = k$).

13. **(A)** This graph shows the pressure increasing as the temperature is increased and the volume is held constant. It is an example of Gay-Lussac's Law ($P/T = k$).

14. **(C)** This graph shows the volume increasing as the temperature is increased and the pressure is held constant. It is an example of Charles's Law ($V/T = k$).

15. **(B)** The alkali metals react with water to form hydroxides and release hydrogen. A typical reaction is:

$$2Na(s) + 2H_2O(\ell) \rightarrow 2NaOH(aq) + H_2(g)$$

16. **(D)** The noble gases are the least reactive because of their completed outer orbital.

17. **(C)** The halogen family contains the colored gases fluorine and chlorine at room temperatures, the reddish liquid bromine, and metallic-like purple iodine.

18. **(E)** These nonmetals, when they are oxides, react as acidic anhydrides with water to form acid solutions.

19. **(E)** The first transition metal on the Periodic Table is scandium. It is in the third period. Scandium's electron notation shows that this element is the first to have the $3d$ sublevel filled. The nine elements that follow scandium fill up the $3d$ sublevel.

20. **(A)** The lowest-energy orbital in any atom is the $1s$. The $1s$ orbital represents the probability distribution for the electron when it is closest to the nucleus.

21. **(D)** The type of sublevel that can hold 6 electrons is called p type. A p sublevel contains 3 degenerate orbitals, each with a capacity of 2 electrons.

22. **(C)** An s-type orbital has a spherical probability distribution for the electrons in that energy state. Energy level 3 represents a higher energy than energy levels 1 or 2; therefore, the $3s$ orbital is larger than the $1s$ or $2s$ orbitals. The higher the energy of the electrons, the farther away the electrons will likely be found from the nucleus.

23. **(C)** The electron notation for sodium is $1s^2\,2s^2\,2p^6\,3s^1$. Hence, the outermost electron in the sodium atom is in the $3s$ orbital.

101. **(T, F)** Sulfur trioxide is shown by three structural formulas because each bond is "hybrid" of a single and double bond. Resonance in chemistry does not mean that the bonds resonate between the structures shown in the structural drawing.

102. **(T, F)** When ΔG is negative in the Gibbs equation, the reaction is spontaneous. However, the total equation determines this, not just the ΔH. The Gibbs equation is:

$$\Delta G = \Delta H - T\Delta S$$

103. **(T, T, CE)** One mole of each of the substances contains 6.02×10^{23} molecules, but their molecular masses are different. CO_2 is found by adding one C = 12 and two O = 32, or a total of 44 amu. The H_2O, however, adds up to two H = 2 plus one O = 16, or a total of 18 amu. Thus, it is true that 1 mol of CO_2 at 44 g/mol is heavier than 1 mol of H_2O at 18 g/mol.

104. **(T, T, CE)** Hydrosulfuric acid is a weak acid but is used in qualitative tests because of the distinctly colored precipitates of sulfides that it forms with many metallic ions.

105. **(T, T, CE)** Sodium chloride is an ionic crystal, not a molecule, and its ions are hydrated by the polar water molecules.

106. **(F, F)** The addition of more H_3PO_4 causes the equilibrium to shift to the right and increase the concentration of H_3O^+ ions until equilibrium is restored. Therefore, the first statement is false. The second is also false since the equilibrium constant remains the same at a given temperature.

107. **(T, T, CE)** The statement is true, and the reason is also true and explains the statement.

108. **(T, F)** The statement is true, but not the reason. In an equilibrium reaction, concentrations can be shown to progress like this:

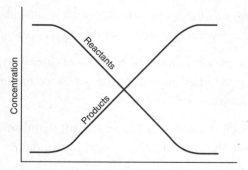

until equilibrium is reached. Then the concentrations stabilize.

109. **(F, T)** The forward and reverse reactions are occurring at equal rates when equilibrium is reached. The reactions do not stop. The concentrations remain the same at this point.

110. **(T, T, CE)** Since acetylene is known to be a linear molecule with a triple bond between the two carbons, the *sp* orbitals along the central axis with the hydrogens bonded on either end fit the experimental evidence.

111. **(F, F)** The weakest bonds between molecules are van der Waals forces, not coordinate covalent bonds.

112. **(T, T, CE)** The terms *dilute* and *concentrated* merely indicate a relatively small amount of solvent and a large amount of solvent, respectively. You can have a dilute saturated solution if the solute is only slightly soluble.

113. **(T, T)** Lithium replaces any other metal in a single replacement reaction; therefore, it is the most active metal. It's also true that lithium has only one electron in its outer energy level. Although lithium's outer configuration of only one electron aids in its high activity, it is not the only reason for its high activity. Other metals (like sodium and potassium) have only one electron in their outer energy levels and are fairly active, but they aren't as active as lithium.

114. **(F, T)** The oxidation state of carbon is often +4 but not always. Carbon does have 4 valence electrons that it often relinquishes responsibility for in terms of oxidation numbers (which is why carbon can have an oxidation state of +4); however, it doesn't have to do so, and in many compounds it exhibits a different oxidation state.

115. **(T, T, CE)** There are as many electrons as there are protons in a neutral atom, and the atomic number represents the number of protons.

24. **(D)** The solidification of vegetable oil is merely a physical change, like the formation of ice from liquid water at lower temperatures. All the other choices involve actual recombinations of atoms and thus are chemical changes.

25. **(C)** Water is formed because most common fuels contain hydrogen in their structures.

26. **(E)** The other choices, in order, represent 1, $\frac{1}{10}$ or deci-, $\frac{1}{100}$ or centi-, and 100 or hecto-.

27. **(E)** One mole of any substance contains 6.02×10^{23} molecules. Since each water molecule is triatomic, there would be $3(6.02 \times 10^{23})$ atoms present.

28. **(E)** The noble gases are all monatomic because of their complete outer energy levels. A rule to help you remember diatomic gases is: Gases ending in *-gen* or *-ine* usually form diatomic molecules.

29. **(D)** By both the VSEPR (Valence Shell Electron Pair Repulsion) method and the orbital structure method, the PCl_3 molecule is trigonal pyramidal:

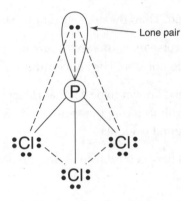

30. **(C)** The complete loss and gain of electrons is an ionic bond. All other bonds indicated are "sharing of electrons" type bonds or some form of covalent bonding.

31. **(D)** The cathode reaction releases only H_2 gas. This half-reaction is as given in (D).

32. **(E)** The beta particle is a high-speed electron and has the smallest mass of the first four choices. However, gamma rays are electromagnetic waves. They have no mass.

33. **(D)** If 25% of the sample now remains, then 100 years ago 50% would be present. If you go back another 100 years, the sample would contain 100% of the radioactive element. Therefore, the sample is 100 + 100 = 200 years old.

34. **(D)** pH = $-\log[H_3O^+]$ = $-\log[1 \times 10^{-4}]$ = $-(-4)$ = 4.

35. **(C)** Only I and II are true. Distilled water does not significantly conduct an electric current. The polarity of the water molecule is helpful in ionization and in causing substances to go into solution.

36. **(A)** SO_2 is the acid anhydride of H_2SO_3 or sulfurous acid. $H_2O + SO_2 \rightarrow H_2SO_3$.

37. **(E)** Four grams of hydrogen gas at STP represent 2 mol of hydrogen since 2 g is the gram-molecular mass of hydrogen. Each mole of a gas contains 6.02×10^{23} molecules, so 2 mol contains $2 \times 6.02 \times 10^{23}$ or 12.04×10^{23} molecules.

38. **(B)** To solve percent composition problems, first, using estimation, divide the % given by the atomic mass:

$$\frac{85.7}{12} \approx 7 \qquad \frac{14.3}{1} \approx 14 \text{ (This shows a 1 : 2 ratio)}$$

The empirical formula is CH_2. Since the molecular mass is 42 and the empirical formula has a molecular mass of 14, the true formula must be 3 times the empirical formula, or C_3H_6.

39. **(E)** Because the temperature (in kelvins) increases from 303 K to 333 K, the volume of the gas should increase with pressure held constant. The correct fraction is $\dfrac{333}{303}$.

40. **(A)** Paradichlorobenzene, a nonpolar solid, is the only substance that will appreciably dissolve in benzene. Therefore, it will be the only substance capable of depressing the freezing point of benzene.

41. **(B)** The pH is $-\log[H^+]$. A 0.005 molar solution of H_2SO_4 ionizes in a dilute solution to release two H^+ ions per molecule of H_2SO_4. Therefore, the molar concentration of H^+ ion is 2×0.005 mol/L or 0.010 mol/L. Substituting this in the formula, you have:

$$pH = -\log [0.01] = -\log [1 \times 10^{-2}]$$

The log of a number is the exponent to which the base 10 is raised to express that number in exponential form:

$$-\log [1 \times 10^{-2}] = -[-2] = 2$$

42. **(B)** If the solution is to be 5% sodium hydroxide, then 5% of 100 g is 5 g. Percent is always by mass unless otherwise specified.

43. **(C)** Because this equation is exothermic, higher temperatures will decrease the reaction to the right and increase the reaction to the left, so II is true.

Also, I is true because with an increase of pressure the reaction will try to relieve that pressure by going in the direction that has the least volume: in this reaction, to the right. Statement III is false because removing product in this reaction would increase the forward reaction. Statements I and II are true.

44. **(B)** The reaction is:

$$H_2SO_4(aq) + 2NaOH(aq) \rightarrow Na_2SO_4(aq) + H_2O(\ell)$$
$$\begin{array}{cc} 1\,\text{mol} & 2\,\text{mol} \\ \text{acid} & \text{base} \end{array}$$

$$1 \text{ mol acid} = \frac{1}{2} \text{ (mol) of base}$$

Since

molarity $\times$ volume (L) = moles,

then

$$M_a V_a = \frac{1}{2} M_b V_b$$

$$(1.0\,\text{M})(x\,\text{L}) = \frac{1}{2}(1\,\text{M})(0.05\,\text{L})$$

$$x\,\text{L} = \frac{1}{2}(0.05\,\text{L})$$

$$x\,\text{L} = 0.025\,\text{L or } 25.\,\text{mL}$$

45. **(E)** The reaction is:

$$150 \text{ g}$$
$$CaCO_3(s) + 2HCl(aq) \rightarrow CaCl_2(aq) + H_2O(\ell) + CO_2(g)$$

The gram-molecular mass of calcium carbonate is 100 g. Then $150 \text{ g} = \dfrac{150}{100}$ or 1.5 mol of calcium carbonate. According to the equation, 1 mol of $CaCO_3$ yields 1 mol of carbon dioxide, so 1.5 mol of $CaCO_3$ yields 1.5 mol of CO_2.

The gram-molecular mass of CO_2 = 44 g:

$$1.5 \text{ mol of } CO_2 = 1.5 \times 44 \text{ g} = 66 \text{ g } CO_2$$

46. **(E)** The other choices are wrong because:

(A) is less dense than air and will escape through hole in stopper
(B) reacts with air
(C) is less dense than air and will escape through hole in stopper
(D) needs heat to be evolved

47. **(B)** The Mg needs oxygen to form MgO so the lid cannot be tightly sealed. Oxygen is needed for the Mg to oxidize to MgO. All other choices are true.

48. **(C)** A desiccator is a device used to keep things dry. After the hydrated $BaCl_2$ has had the water removed, the desiccator keeps the water from returning to the substance.

49. **(C)** If material jumped out of the evaporating dish while heating, barium chloride and water would both be removed. The purpose of the heating is to remove only the water. Therefore, it would seem like more water had been removed than was actually present, and the mass lost from the barium chloride would inaccurately be counted as mass lost from water.

50. **(D)** As a result of the complete heating, 1.80 grams of water is removed from the hydrate. The mass of water released from the barium chloride can be converted into moles by using dimensional analysis.

$$\frac{1.80 \text{ g } H_2O}{1} \times \frac{1 \text{ mol } H_2O}{18.0 \text{ g } H_2O} = 0.100 \text{ mol } H_2O$$

51. **(D)** $1K = 39$, $1Al = 27$, $2(SO_4) = 2(32 + 16 \times 4) = 192$, and $12H_2O = 12(2 + 16) = 216$. This totals 474 g.

52. **(D)**

$$\overset{x \text{ g}}{4\underset{108 \text{ g}}{Al}} + \overset{2 \text{ mol}}{3\underset{3 \text{ mol}}{O_2}} \rightarrow 2Al_2O_3$$

$$\frac{x \text{ g}}{108 \text{ g}} = \frac{2 \text{ mol}}{3 \text{ mol}}, \text{ so } x = 72 \text{ g}$$

Or, using the mole method: 44.8 L = 2 mol

$$\overset{x \text{ mol}}{4\underset{4 \text{ mol}}{Al}} + \overset{2 \text{ mol}}{3\underset{3 \text{ mol}}{O_2}} \rightarrow 2Al_2O_3$$

This shows that:

$$\frac{x \, \text{mol Al}}{4 \, \text{mol Al}} = \frac{2 \, \text{mol O}_2}{3 \, \text{mol O}_2}$$

$$x = \frac{2 \, \text{mol O}_2 \times 4 \, \text{mol Al}}{3 \, \text{mol O}_2} = \frac{8}{3} \, \text{mol Al}$$

Since molar mass of Al = 27 g/mol

$$x = \frac{8}{3} \, \text{mol Al} \times 27 \, \text{g/mol} = 72 \, \text{g Al}$$

53. **(B)** Metal oxides are generally basic anhydrides.

54. **(A)** Both of the reactions shown are single replacement reactions. Single replacement reactions are also redox processes in which electrons are transferred from one substance to another. They can be viewed as a driving force for reactions to occur.

55. **(C)** Tin replaces copper in reaction #2 and is therefore more active than copper. Zinc replaces tin in reaction #1, however, and is therefore more active than both tin and copper. This is reflected in the activity series where zinc is at a higher level.

56. **(A)** Zinc is being oxidized (from 0 to +2) by the tin(II) chloride in the first reaction. Therefore, tin(II) chloride is the oxidizing agent in that reaction. Copper(II) chloride is the oxidizing agent in the second reaction, but is not matched up with another oxidizing agent in any of the other answers.

57. **(D)** $\overset{212\,\text{g}}{2\text{NaClO}_3} \rightarrow 2\text{NaCl} + \overset{x\,\text{L}}{3\text{O}_2}$

$$\frac{212 \, \text{g}}{106 \, \text{g/mol}} = 2 \, \text{mol}$$

Equation shows

$$2 \, \text{mol} \rightarrow 3 \, \text{mol O}_2$$

$$3 \, \text{mol} \times 22.4 \, \text{L/mol} = 67.2 \, \text{L}$$

58. **(C)** The balanced equation has the coefficients 2, 3, 1, and 6: $2\text{Al(OH)}_3 + 3\text{H}_2\text{SO}_4 \rightarrow \text{Al}_2(\text{SO}_4)_3 + 6\text{H}_2\text{O}$.

59. **(E)** The reaction is $\text{NaClO}_3(s) \rightarrow \text{NaCl}(s) + \frac{3}{2}\text{O}_2(g)$.

$$\Delta H_{\text{reaction}} = \Delta H_{f(\text{products})} - \Delta H_{f(\text{reactants})}$$

$$\Delta H_{\text{reaction}} = (-98.2 + 0) - (-85.7)$$

$$\Delta H_{\text{reaction}} = -12.5 \, \text{kcal}$$

60. **(D)** II and III are identical; isotopes differ only in the number of neutrons in the nucleus and this affects the atomic mass only.

61. **(E)** I, II, and III will increase the rate of this reaction. Each of them causes the rate of this reaction to increase.

62. **(D)** This setup depends on water displacement of an insoluble gas.

63. **(C)** The coefficients give the molar relations, so 2.0 mol of HCl give off 1.0 mol of CO_2. Given 4.0 mol of HCl, you have

$$\frac{4.0}{2.0} = \frac{x}{1.0}, \text{ then } x = \frac{4.0}{2.0} = 2.0 \text{ mol}$$

64. **(E)** $K_{sp} = [Pb^{2+}][S^{2-}]$. Since $[Pb^{2+}]$ is given at 5.0×10^{-14} and because the formula is a 1:1 ratio, the $[S^{2-}]$ must also be that value, the K_{sp} can be calculated by

$$K_{sp} = [5.0 \times 10^{-14}][5.0 \times 10^{-14}] = 25 \times 10^{-28}$$

Putting the answer in scientific notation, i.e., one digit to the left of the decimal point in the first factor, the answer is 2.5×10^{-27}.

65. **(B)** The introduction of the "common ion" S^{2-} at 0.1 molar forces the equilibrium to shift to the left and reduce the Pb^{2+} concentration.

66. **(A)** According to the gas laws, only I will definitely cause an increase in the volume of a confined gas.

67. **(D)** In 1 mol of $Al_2(CO_3)_3$, nine oxygens (three carbonates with three oxygen atoms each) are in each formula unit, or 9 mol of O atoms are in 1 mol of $Al_2(CO_3)_3$. Because only 0.50 mol is given, there are $\frac{1}{2}(9)$ or 4.5 mol of O atoms. In 4.5 mol of oxygen, there are

$$4.5 \text{ mol O atoms} \times \frac{6.0 \times 10^{23} \text{ atoms}}{1 \text{ mol O atoms}}$$

27.0×10^{23} atoms or 2.7×10^{24} atoms.

68. **(D)** When HA ionizes, it forms equal amounts of H^+ and A^- ions, but these amounts are very small because the K_a is very small. K_a can be expressed as $[H^+][A^-]/[HA]$. Because you are told to assume $[HA] = 1$, you have:

$$\frac{[x][x]}{[1]} = 1 \times 10^{-6}$$
$$x^2 = 1 \times 10^{-6}$$
$$x = 1 \times 10^{-3}$$

69. **(C)** Percent dissociation =

$$\frac{\text{moles/liter of } H_3O^+}{\text{original concentration}} \times 100$$

$$\% \text{ dissociation} = \frac{1 \times 10^{-3}}{1 \times 10^{-1}} \times 100 = 1\%$$

70. **(E)** Since the oxidation state of oxygen is generally –2 and the overall charge of the ion is 2–, the oxidation state of the chromium in the dichromate ion must be +6. This is because the oxidation state of the 7 oxygen atoms totals to –14 and the 2 chromium at +6 each would total to +12. Comparing the oxygen to the chromium leaves of net of 2– charge—the overall charge of the ion.

CALCULATING YOUR SCORE

Your score on Practice Test 2 can now be computed manually. The actual test is scored by machine, but the same method is used to arrive at the raw score. You get one point for each correct answer. For each wrong answer, you lose one-fourth of a point. Questions that you omit or that have more than one answer are not counted. On your answer sheet mark all correct answers with a "C" and all incorrect answers with an "X."

Determining Your Raw Test Score

Total the number of correct answers you have recorded on your answer sheet. It should be the same as the total of all the numbers you place in the block in the lower left corner of each area of the Subject Area summary in the next section.

A. Enter the total number of correct answers here:
   Now count the number of wrong answers you recorded on your answer sheet.
B. Enter the total number of wrong answers here:
   Multiply the number of wrong answers in B by 0.25.
C. Enter that product here:
   Subtract the result in C from the total number of right answers in A.
D. Enter the result of your subtraction here:
E. Round the result in D to the nearest whole number:
   This is your raw test score.

Conversion of Raw Scores to Scaled Scores

Your raw score is converted by the College Board into a scaled score. The College Board scores range from 200 to 800. This conversion is done to ensure that a score earned on any edition of a particular SAT Subject Test in Chemistry is comparable to the same scaled score earned on any other edition of the same test. Because some editions of the tests may be slightly easier or more difficult than others, scaled scores are adjusted so that they indicate the same level of performance regardless of the edition of the test taken and the ability of the group that takes it. Consequently, a specific raw score on one edition of a particular test will not necessarily translate to the same scaled score on another edition of the same test.

Because the practice tests in this book have no large population of scores with which they can be scaled, scaled scores cannot be determined.

Results from previous SAT Chemistry tests appear to indicate that the conversion of raw scores to scaled scores GENERALLY follows this pattern:

Raw Score	Scaled Score	Raw Score	Scaled Score
85–82	800–800	30–25	540–520
81–75	790–760	25–20	520–490
75–70	760–740	20–15	490–460
70–65	740–710	15–10	460–430
65–60	710–690	10–5	430–400
60–55	690–670	5–0	400–370
55–50	670–640	0 to –5	370–340
50–45	640–620	–5 to –10	340–310
45–40	620–590	–10 to –15	310–290
40–35	590–570	–15 to –20	290–270
35–30	570–540	–20 or lower	270–200

Note that this scale provides only a *general idea* of what a raw score may translate into on a scaled score range of 800–200. Scaling on every test is usually slightly different. Some students who had taken the SAT Subject Test in Chemistry after using this book had reported that they have scored slightly higher on the SAT test than on the practice tests in this book. They *all* reported that preparing well for the test paid off in a better score!

DIAGNOSING YOUR NEEDS

After taking Practice Test 2, check your answers against the correct ones. Then fill in the chart below.

In the space under each question number, place a check if you answered that question correctly.

➡ Example _____

If your answer to question 5 was correct, place a check in the appropriate box.

Next, total the check marks for each section and insert the number in the designated block. Now do the arithmetic indicated and insert your percent for each area.

Subject Area*	(✔) Questions Answered Correctly											
I. Atomic Theory and Structure, including periodic relationships			19	20	21	23	101	115	28	32	33	60
☐ No. of checks ÷ 10 × 100 = _____%												
II. Chemical Bonding and Molecular Structure					8	9	10	11	110	111	29	30
☐ No. of checks ÷ 8 × 100 = _____%												
III. States of Matter and Kinetic Molecular Theory of Gases				2	3	4	12	13	14	27	39	65
☐ No. of checks ÷ 9 × 100 = _____%												
IV. Solutions, including concentration units, solubility, and colligative properties									105	112	40	44
☐ No. of checks ÷ 4 × 100 = _____%												
V. Acids and Bases							34	36	41	53	68	69
☐ No. of checks ÷ 6 × 100 = _____%												
VI. Oxidation-Reduction							114	70	31	35	54	55
☐ No. of checks ÷ 6 × 100 = _____%												
VII. Stoichiometry	103	37	38	42	45	50	51	52	57	58	63	67
☐ No. of checks ÷ 12 × 100 = _____%												
VIII. Reaction Rates											43	61
☐ No. of checks ÷ 2 × 100 = _____%												
IX. Equilibrium								106	108	109	64	66
☐ No. of checks ÷ 5 × 100 = _____%												
X. Thermodynamics: energy changes in chemical reactions, randomness, and criteria for spontaneity							5	6	7	102	107	59
☐ No. of checks ÷ 6 × 100 = _____%												

The subject areas have been expanded to identify specific areas in the text.

Subject Area*	(✔) Questions Answered Correctly					
XI. **Descriptive Chemistry:** physical and chemical properties of elements and their familiar compounds; organic chemistry; periodic properties	1	15	16	17	18	22
	104	113	24	25	26	56
☐ No. of checks ÷ 12 × 100 = _____%						
XII. **Laboratory:** equipment, procedures, observations, safety, calculations, and interpretation of results	46	47	48	49	62	
☐ No. of checks ÷ 5 × 100 = _____%						

The subject areas have been expanded to identify specific areas in the text.

To develop a study plan, refer to pages 31–33.

PRACTICE TEST 2

ANSWER SHEET
Practice Test 3

Determine the correct answer for each question. Then, using a No. 2 pencil, blacken completely the oval containing the letter of your choice.

1. Ⓐ Ⓑ Ⓒ Ⓓ Ⓔ
2. Ⓐ Ⓑ Ⓒ Ⓓ Ⓔ
3. Ⓐ Ⓑ Ⓒ Ⓓ Ⓔ
4. Ⓐ Ⓑ Ⓒ Ⓓ Ⓔ
5. Ⓐ Ⓑ Ⓒ Ⓓ Ⓔ
6. Ⓐ Ⓑ Ⓒ Ⓓ Ⓔ
7. Ⓐ Ⓑ Ⓒ Ⓓ Ⓔ
8. Ⓐ Ⓑ Ⓒ Ⓓ Ⓔ
9. Ⓐ Ⓑ Ⓒ Ⓓ Ⓔ
10. Ⓐ Ⓑ Ⓒ Ⓓ Ⓔ
11. Ⓐ Ⓑ Ⓒ Ⓓ Ⓔ
12. Ⓐ Ⓑ Ⓒ Ⓓ Ⓔ
13. Ⓐ Ⓑ Ⓒ Ⓓ Ⓔ
14. Ⓐ Ⓑ Ⓒ Ⓓ Ⓔ
15. Ⓐ Ⓑ Ⓒ Ⓓ Ⓔ
16. Ⓐ Ⓑ Ⓒ Ⓓ Ⓔ

17. Ⓐ Ⓑ Ⓒ Ⓓ Ⓔ
18. Ⓐ Ⓑ Ⓒ Ⓓ Ⓔ
19. Ⓐ Ⓑ Ⓒ Ⓓ Ⓔ
20. Ⓐ Ⓑ Ⓒ Ⓓ Ⓔ
21. Ⓐ Ⓑ Ⓒ Ⓓ Ⓔ
22. Ⓐ Ⓑ Ⓒ Ⓓ Ⓔ
23. Ⓐ Ⓑ Ⓒ Ⓓ Ⓔ

ON THE ACTUAL CHEMISTRY TEST. THE FOLLOWING TYPE OF QUESTION MUST BE ANSWERED ON A SPECIAL SECTION (LABELED "CHEMISTRY") AT THE LOWER LEFT-HAND CORNER OF PAGE 2 OF YOUR ANSWER SHEET. THESE QUESTIONS WILL BE NUMBERED BEGINING WITH 101 AND MUST BE ANSWERED ACCORDING TO THE DIRECTIONS.

CHEMISTRY* Fill in oval CE only if II is a correct explanation of I.

	I	II	CE*
101.	Ⓣ Ⓕ	Ⓣ Ⓕ	◯
102.	Ⓣ Ⓕ	Ⓣ Ⓕ	◯
103.	Ⓣ Ⓕ	Ⓣ Ⓕ	◯
104.	Ⓣ Ⓕ	Ⓣ Ⓕ	◯
105.	Ⓣ Ⓕ	Ⓣ Ⓕ	◯
106.	Ⓣ Ⓕ	Ⓣ Ⓕ	◯
107.	Ⓣ Ⓕ	Ⓣ Ⓕ	◯
108.	Ⓣ Ⓕ	Ⓣ Ⓕ	◯
109.	Ⓣ Ⓕ	Ⓣ Ⓕ	◯
110.	Ⓣ Ⓕ	Ⓣ Ⓕ	◯
111.	Ⓣ Ⓕ	Ⓣ Ⓕ	◯
112.	Ⓣ Ⓕ	Ⓣ Ⓕ	◯
113.	Ⓣ Ⓕ	Ⓣ Ⓕ	◯
114.	Ⓣ Ⓕ	Ⓣ Ⓕ	◯
115.	Ⓣ Ⓕ	Ⓣ Ⓕ	◯

ANSWER SHEET
Practice Test 3

ON THE ACTUAL CHEMISTRY TEST, THE REMAINING QUESTIONS MUST BE ANSWERED BY RETURNING TO THE SECTION OF YOUR ANSWER SHEET YOU STARTED FOR CHEMISTRY.

24. Ⓐ Ⓑ Ⓒ Ⓓ Ⓔ
25. Ⓐ Ⓑ Ⓒ Ⓓ Ⓔ
26. Ⓐ Ⓑ Ⓒ Ⓓ Ⓔ
27. Ⓐ Ⓑ Ⓒ Ⓓ Ⓔ
28. Ⓐ Ⓑ Ⓒ Ⓓ Ⓔ
29. Ⓐ Ⓑ Ⓒ Ⓓ Ⓔ
30. Ⓐ Ⓑ Ⓒ Ⓓ Ⓔ
31. Ⓐ Ⓑ Ⓒ Ⓓ Ⓔ
32. Ⓐ Ⓑ Ⓒ Ⓓ Ⓔ
33. Ⓐ Ⓑ Ⓒ Ⓓ Ⓔ
34. Ⓐ Ⓑ Ⓒ Ⓓ Ⓔ
35. Ⓐ Ⓑ Ⓒ Ⓓ Ⓔ
36. Ⓐ Ⓑ Ⓒ Ⓓ Ⓔ
37. Ⓐ Ⓑ Ⓒ Ⓓ Ⓔ
38. Ⓐ Ⓑ Ⓒ Ⓓ Ⓔ
39. Ⓐ Ⓑ Ⓒ Ⓓ Ⓔ

40. Ⓐ Ⓑ Ⓒ Ⓓ Ⓔ
41. Ⓐ Ⓑ Ⓒ Ⓓ Ⓔ
42. Ⓐ Ⓑ Ⓒ Ⓓ Ⓔ
43. Ⓐ Ⓑ Ⓒ Ⓓ Ⓔ
44. Ⓐ Ⓑ Ⓒ Ⓓ Ⓔ
45. Ⓐ Ⓑ Ⓒ Ⓓ Ⓔ
46. Ⓐ Ⓑ Ⓒ Ⓓ Ⓔ
47. Ⓐ Ⓑ Ⓒ Ⓓ Ⓔ
48. Ⓐ Ⓑ Ⓒ Ⓓ Ⓔ
49. Ⓐ Ⓑ Ⓒ Ⓓ Ⓔ
50. Ⓐ Ⓑ Ⓒ Ⓓ Ⓔ
51. Ⓐ Ⓑ Ⓒ Ⓓ Ⓔ
52. Ⓐ Ⓑ Ⓒ Ⓓ Ⓔ
53. Ⓐ Ⓑ Ⓒ Ⓓ Ⓔ
54. Ⓐ Ⓑ Ⓒ Ⓓ Ⓔ
55. Ⓐ Ⓑ Ⓒ Ⓓ Ⓔ

56. Ⓐ Ⓑ Ⓒ Ⓓ Ⓔ
57. Ⓐ Ⓑ Ⓒ Ⓓ Ⓔ
58. Ⓐ Ⓑ Ⓒ Ⓓ Ⓔ
59. Ⓐ Ⓑ Ⓒ Ⓓ Ⓔ
60. Ⓐ Ⓑ Ⓒ Ⓓ Ⓔ
61. Ⓐ Ⓑ Ⓒ Ⓓ Ⓔ
62. Ⓐ Ⓑ Ⓒ Ⓓ Ⓔ
63. Ⓐ Ⓑ Ⓒ Ⓓ Ⓔ
64. Ⓐ Ⓑ Ⓒ Ⓓ Ⓔ
65. Ⓐ Ⓑ Ⓒ Ⓓ Ⓔ
66. Ⓐ Ⓑ Ⓒ Ⓓ Ⓔ
67. Ⓐ Ⓑ Ⓒ Ⓓ Ⓔ
68. Ⓐ Ⓑ Ⓒ Ⓓ Ⓔ
69. Ⓐ Ⓑ Ⓒ Ⓓ Ⓔ
70. Ⓐ Ⓑ Ⓒ Ⓓ Ⓔ

PRACTICE TEST 3

Note: For all questions involving solutions and/or chemical equations, assume that the system is in water unless otherwise stated.

Reminder: You may *not* use a calculator on these tests.

The following symbols have the meanings listed unless otherwise noted.

H	= enthalpy	g	= gram(s)
M	= molar	J	= joule(s)
n	= number of moles	kJ	= kilojoule(s)
P	= pressure	L	= liter(s)
R	= molar gas constant	mL	= milliliter(s)
S	= entropy	mm	= millimeter(s)
T	= temperature	mol	= mole(s)
V	= volume	V	= volt(s)
atm	= atmosphere(s)		

PART A

Directions: Every set of the given lettered choices below refers to the numbered statements or formulas immediately following it. Choose the one lettered choice that best fits each statement or formula and then fill in the corresponding oval on the answer sheet. Each choice may be used once, more than once, or not at all in each set.

Questions 1–4 refer to the following diagram:

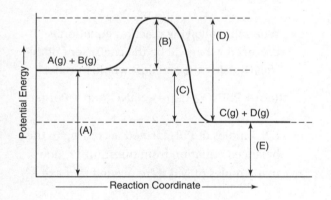

1. The activation energy of the forward reaction is shown by

2. The activation energy of the reverse reaction is shown by

3. The heat of the reaction for the forward reaction is shown by

4. The potential energy of the reactants is shown by

GO ON TO THE NEXT PAGE

Questions 5–7 refer to the following diagram and information:

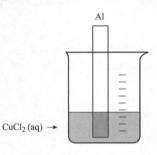

Al

CuCl$_2$ (aq) →

When a strip of aluminum is placed in a solution of copper(II) chloride, a reaction takes place. As time goes by, a brown solid forms on the strip of aluminum and the blue solution turns clearer.

5. One product of the reaction would be

   (A) AlCl$_2$
   (B) AlCl$_3$
   (C) AlCl$_4$
   (D) CuAl$_3$
   (E) CuAl$_2$

6. The aluminum is

   (A) being deprotonated
   (B) being disproportionated
   (C) being reduced
   (D) being oxidized
   (E) gaining mass

7. The copper(II) ion concentration in solution

   (A) is zero at the beginning of the reaction
   (B) has no effect on the rate of the reaction
   (C) stays the same and doesn't influence the color of the solution
   (D) increases, making the solution less blue
   (E) decreases, making the solution less blue

Questions 8–11 match the following equations to the appropriate descriptions:

   (A) $V/T = k$
   (B) $P/T = k$
   (C) $PV = k$
   (D) $P_T = P_1 + P_2 + P_3 \cdots$
   (E) $PT = k$

8. This equation shows the volume decreasing as the pressure is increased when the temperature is held constant. It is an example of Boyle's Law.

9. This equation shows the pressure increasing as the temperature is increased when the volume is held constant. It is an example of Gay-Lussac's Law.

10. This equation shows the volume increasing as the temperature is increased when the pressure is held constant. It is an example of Charles's Law.

11. This equation shows that the total pressure of a mixture of gases is equal to the sum of the partial pressures of the component gases.

Questions 12–14

   (A) 1
   (B) 2
   (C) 3
   (D) 4
   (E) 5

12. When the following reaction equation is balanced, what will be the coefficient of the NaNO$_3$ using the smallest whole number?

   NaI(aq) + Pb(NO$_3$)$_2$(aq) → NaNO$_3$(aq) + PbI$_2$(s)

13. If 0.5 moles of PbI$_2$ formed according to the balanced equation from question 12, how many moles of NaI were needed to make it?

14. If NaNO$_3$ goes into solution as ions, into how many ions would it dissociate?

GO ON TO THE NEXT PAGE

Questions 15–18

   (A) Ionic substance
   (B) Polar covalent substance
   (C) Nonpolar covalent substance
   (D) Amorphous substance
   (E) Metallic network

15. $MgCl_2(s)$

16. $HCl(g)$

17. $CH_3–CH_3(g)$

18. $Cu(s)$

Questions 19–23

   (A) Dissociation
   (B) Amphoteric
   (C) Phenolphthalein
   (D) Dehydration
   (E) Deliquescence

19. The reason why a blue crystal of $CuSO_4 \cdot 5H_2O$ turns white when heated

20. The reason why ionic substances dissolved in water exhibit conductivity

21. The reason there may be a pink color in a basic solution

22. The reason why a substance may act like an acid or like a base depending on the substance it is in the presence of

23. The reason why an ionic solid may dissolve in the moisture it absorbs from the air

---

PART B

ON THE ACTUAL CHEMISTRY TEST, THE FOLLOWING TYPE OF QUESTION MUST BE ANSWERED ON A SPECIAL SECTION (LABELED "CHEMISTRY") AT THE LOWER LEFT-HAND CORNER OF PAGE 2 OF YOUR ANSWER SHEET. THESE QUESTIONS WILL BE NUMBERED BEGINNING WITH 101 AND MUST BE ANSWERED ACCORDING TO THE FOLLOWING DIRECTIONS.

> **Directions:** Every question below contains two statements, I in the left-hand column and II in the right-hand column. For each question, decide if statement I is true or false <u>and</u> if statement II is true or false and fill in the corresponding T or F ovals on your answer sheet. *<u>Fill in oval CE only if statement II is a correct explanation of statement I.</u>

Sample Answer Grid:

CHEMISTRY * Fill in oval CE only if II is a correct explanation of I.

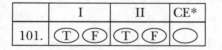

GO ON TO THE NEXT PAGE

101. Elements in the upper/left corner of the Periodic Table are active metals — BECAUSE — metals have larger ionic radii than their atomic radii.

102. A synthesis reaction that is nonspontaneous and has a negative value for its heat of reaction will not occur until some heat is added — BECAUSE — nonspontaneous exothermic reactions need enough activation energy to get them started.

103. Transition elements in a particular period may have the same oxidation number — BECAUSE — they have a complete outer energy level.

104. When a crystal is added to a supersaturated solution of itself, the crystal does not appear to change — BECAUSE — the supersaturated solution is holding more solute than its normal solubility.

105. Equilibrium is a static condition — BECAUSE — at equilibrium, the forward reaction rate equals the reverse reaction rate.

106. The ionic bond is the strongest bond — BECAUSE — ionic bonds have electrostatic attraction due to the loss and gain of electron(s).

107. In the equilibrium reaction $N_2(g) + 3H_2(g) \leftrightarrow 2NH_3(g) + heat$ when the pressure in the reaction chamber is increased, the reaction shifts to the right — BECAUSE — the increase in pressure causes the reaction to shift to the right to decrease the pressure since 4 volumes on the left become 2 volumes on the right.

108. If the forward reaction of an equilibrium is exothermic, adding heat to the system favors the reverse reaction — BECAUSE — additional heat causes a stress on the system, and the system moves in the direction that releases the stress.

109. An element that has an electron configuration of $1s^2 2s^2 2p^6 3s^2 3p^6 3d^3 4s^2$ is a transition element — BECAUSE — the transition elements from scandium to zinc are filling the $3d$ orbitals.

110. The most electronegative elements in the periodic chart are found among nonmetals — BECAUSE — electronegativity is a measure of the ability of an atom to draw valence electrons to itself.

111. Basic anhydrides react in water to form bases — BECAUSE — metallic oxides react with water to form solutions that have an excess of hydroxide ions.

112. There are 3 moles of atoms in 18 grams of water — BECAUSE — there are 6×10^{23} atoms in 1 mole.

113. Benzene is a good electrolyte — BECAUSE — a good electrolyte has charged ions that carry the electric current.

GO ON TO THE NEXT PAGE

	I		II
114.	Normal butyl alcohol and 2-butanol are isomers	BECAUSE	isomers vary in the number of neutrons in the nucleus of the atom.
115.	The reaction of CaCO₃ and HCl goes to completion	BECAUSE	reactions that form a precipitate tend to go to completion.

PART C

Directions: Every question or incomplete statement below is followed by five suggested answers or completions. Choose the one that is best in each case and then fill in the corresponding oval on the answer sheet.

24. What are the simplest whole-number coefficients that balance this equation?

$$\cdots C_4H_{10} + \cdots O_2 \rightarrow \cdots CO_2 + \cdots H_2O$$

   (A) 1, 6, 4, 2
   (B) 2, 13, 8, 10
   (C) 1, 6, 1, 5
   (D) 3, 10, 16, 20
   (E) 4, 26, 16, 20

25. How many atoms are present in the formula $KAl(SO_4)_2$?

   (A) 7
   (B) 9
   (C) 11
   (D) 12
   (E) 13

26. All of the following are compounds EXCEPT

   (A) copper sulfate
   (B) carbon dioxide
   (C) washing soda
   (D) air
   (E) lime

27. What volume of gas, in liters, would 2.0 moles of hydrogen occupy at STP?

   (A) 11.2
   (B) 22.4
   (C) 33.6
   (D) 44.8
   (E) 67.2

28. What is the maximum number of electrons held in the d orbitals?

   (A) 2
   (B) 6
   (C) 8
   (D) 10
   (E) 14

29. If an element has an atomic number of 11, it will combine most readily with an element that has an electron configuration of

   (A) $1s^2 2s^2 2p^6 3s^2 3p^1$
   (B) $1s^2 2s^2 2p^6 3s^2 3p^2$
   (C) $1s^2 2s^2 2p^6 3s^2 3p^3$
   (D) $1s^2 2s^2 2p^6 3s^2 3p^4$
   (E) $1s^2 2s^2 2p^6 3s^2 3p^5$

GO ON TO THE NEXT PAGE

30. An example of a physical property is

    (A) rusting
    (B) decay
    (C) souring
    (D) low melting point
    (E) high heat of formation

31. A gas at STP that contains 6.02×10^{23} atoms and forms diatomic molecules will occupy

    (A) 11.2 L
    (B) 22.4 L
    (C) 33.6 L
    (D) 67.2 L
    (E) 1.06 qt

32. When excited electrons cascade to lower energy levels in an atom,

    (A) visible light is always emitted
    (B) the potential energy of the atom increases
    (C) the electrons always fall back to the first energy level
    (D) the electrons fall indiscriminately to all levels
    (E) the electrons fall back to a lower unfilled energy level

33. Mass spectroscopy uses the concept that

    (A) charged particles are evenly deflected in a magnetic field
    (B) charged particles are deflected in a magnetic field inversely to the mass of the particles
    (C) particles of heavier mass are deflected in a magnetic field to a greater degree than lighter particles
    (D) particles are evenly deflected in a magnetic field

34. The bond that describes an interaction between two orbitals that is not symmetrical about a line between the two atoms' nuclei is called

    (A) a pi bond
    (B) a sigma bond
    (C) a hydrogen bond
    (D) a covalent bond
    (E) an ionic bond

35. What is the boiling point of water at the top of Pikes Peak? (Note: Pikes Peak is well above sea level.)

    (A) It is 100°C.
    (B) It is >100°C since the pressure is less than at ground level.
    (C) It is <100°C since the pressure is less than at ground level.
    (D) It is >100°C since the pressure is greater than at ground level.
    (E) It is <100°C since the pressure is greater than at ground level.

36. The atomic structure of the alkane series contains the hybrid orbitals designated as

    (A) sp
    (B) sp^2
    (C) sp^3
    (D) sp^3d^2
    (E) sp^4d^3

GO ON TO THE NEXT PAGE

37. Which of the following is (are) true for this reaction?

$$Cu + 4HNO_3 \rightarrow Cu(NO_3)_2 + 2H_2O + 2NO_2(g)$$

   I. It is an oxidation-reduction reaction.
  II. Copper is oxidized.
 III. The oxidation number of nitrogen goes from +5 to +4.

(A) I only
(B) III only
(C) I and II only
(D) II and III only
(E) I, II, and III

38. Which of the following properties can be attributed to water?

   I. It has a permanent dipole moment attributed to its molecular structure.
  II. It is a very good conductor of electricity.
 III. It has its polar covalent bonds with hydrogen on opposite sides of the oxygen atom, so that the molecule is linear.

(A) I only
(B) III only
(C) I and II only
(D) II and III only
(E) I, II, and III

39. All of the following statements are true for this reaction EXCEPT

$$HCl(g) + H_2O(\ell) \rightarrow H_3O^+(aq) + Cl^-(aq)$$

(A) H_3O^+ is the conjugate acid of H_2O.
(B) Cl^- is the conjugate base of HCl.
(C) H_2O is behaving as a Brønsted-Lowry base.
(D) HCl is a weaker Brønsted-Lowry acid than H_2O.
(E) The reaction essentially goes to completion.

40. What information about the subatomic particles in a particular carbon atom could be determined from the name carbon-13?

   I. The number of protons
  II. The number of neutrons
 III. The number of electrons

(A) I only
(B) III only
(C) I and II only
(D) I, II, and III
(E) II and III only

41. Which of the following salts will hydrolyze in water to form basic solutions?

   I. NaCl
  II. $CuSO_4$
 III. K_3PO_4

(A) I only
(B) III only
(C) I and II only
(D) II and III only
(E) I, II, and III

42. When 1 mole of NaCl is dissolved in 1,000 grams of water, the boiling point of the water is changed more than when 1 mole of which other substance is added to the same mass of water?

(A) $CaCl_2$
(B) C_2H_5OH
(C) $AlBr_3$
(D) MgF_2
(E) $FeCl_3$

43. What is the structure associated with the BF_3 molecule?

(A) Linear
(B) Trigonal planar
(C) Tetrahedron
(D) Trigonal pyramidal
(E) Bent or V-shaped

GO ON TO THE NEXT PAGE

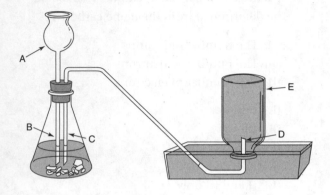

44. What letter designates an error in this laboratory setup?

    (A) A (upper part of tube)
    (B) B (lower part of tube)
    (C) C
    (D) D
    (E) E

45. If the reaction in question 44 created a gas, to where would the contents of the flask be expelled under these conditions?

    (A) A
    (B) B
    (C) C
    (D) D
    (E) E

46. The most active nonmetal has

    (A) a high electronegativity
    (B) a low electronegativity
    (C) a medium electronegativity
    (D) large atomic radii
    (E) a deliquescent property

47. In the reaction $Fe + S \rightarrow FeS$, which is true?

    (A) $Fe + 2e^- \rightarrow Fe^{2+}$
    (B) $Fe \rightarrow Fe^{2+} + 2e^-$
    (C) $Fe^{2+} \rightarrow Fe + 2e^-$
    (D) $S \rightarrow S^{2-} + 2e^-$
    (E) $S^{2-} + 2e^- \rightarrow S$

48. What is the pH of a solution with a hydroxide ion concentration of 0.00001 mole/liter?

    (A) -5
    (B) -1
    (C) 5
    (D) 9
    (E) 14

49. Hydrogen peroxide decomposes into water and oxygen according to the following equation.

$$2 \, H_2O_2(aq) \rightarrow 2 \, H_2O(\ell) + O_2(g)$$

The oxygen is being

    (A) oxidized only
    (B) reduced only
    (C) both oxidized and reduced
    (D) neither oxidized nor reduced
    (E) acidified

50. $\cdots C_2H_4(g) + \cdots O_2(g) \rightarrow \cdots CO_2(g) + \cdots H_2O(l)$

    If the equation for the above reaction is balanced with whole-number coefficients, what is the coefficient for oxygen gas?

    (A) 1
    (B) 2
    (C) 3
    (D) 4
    (E) 5

51. It is found that 273 L of a gas at 0.821 atm and a temperature of 273 K has a mass of 40.0 grams. What is the molar mass of the gas in g/mol?

    (A) 22.4
    (B) 28.0
    (C) 14.0
    (D) 4.00
    (E) 2.00

GO ON TO THE NEXT PAGE

52. A compound whose molecular mass is 90.0 grams contains 40.0% carbon, 6.67% hydrogen, and 53.33% oxygen. What is the true formula of the compound?

(A) $C_2H_2O_4$
(B) CH_2O_4
(C) C_3H_6O
(D) C_3HO_3
(E) $C_3H_6O_3$

53. How many moles of CaO are needed to react with an excess of water to form 370 grams of calcium hydroxide?

(A) 1.0
(B) 2.0
(C) 3.0
(D) 4.0
(E) 5.0

54. To what volume, in milliliters, must 50.0 milliliters of 3.50 M H_2SO_4 be diluted in order to make 2.00 M H_2SO_4?

(A) 25.0
(B) 60.1
(C) 87.5
(D) 93.2
(E) 101

55. A small value of K_{eq} indicates that equilibrium occurs

(A) at a low product concentration
(B) at a high product concentration
(C) after considerable time
(D) with the help of a catalyst
(E) with no forward reaction

56. A student measured 10.0 milliliters of an HCl solution into a beaker and titrated it with a standard NaOH solution that was 0.09 M. The initial NaOH burette reading was 34.7 milliliters while the final reading showed 49.2 milliliters.

What is the molarity of the HCl solution?

(A) 0.13
(B) 0.47
(C) 0.52
(D) 1.57
(E) 2.43

57. A student made the following observations in the laboratory:

(a) Sodium metal reacted vigorously with water while a strip of magnesium did not seem to react at all.
(b) The magnesium strip reacted with dilute hydrochloric acid faster than an iron strip.
(c) A copper rivet suspended in silver nitrate solution was covered with silver-colored stalactites in several days, and the resulting solution had a blue color.
(d) Iron filings dropped into the blue solution were coated with an orange color.

The order of *decreasing* strength as reducing agents is:

(A) Na, Mg, Fe, Ag, Cu
(B) Mg, Na, Fe, Cu, Ag
(C) Ag, Cu, Fe, Mg, Na
(D) Na, Fe, Mg, Cu, Ag
(E) Na, Mg, Fe, Cu, Ag

GO ON TO THE NEXT PAGE

58. A sample of ethyl alcohol and water that were previously mixed in the laboratory must now be separated. Which separation technique could accomplish this?

    (A) Filtration
    (B) Magnetism
    (C) Visual inspection
    (D) Distillation
    (E) Separating funnel

59. Which of these statements is NOT correct?

    (A) In an exothermic reaction, ΔH is negative and the enthalpy decreases.
    (B) In an endothermic reaction, ΔH is positive and the enthalpy increases.
    (C) In a reaction where ΔG is negative, the forward reaction is spontaneous.
    (D) In a reaction where ΔG is positive, ΔS may also be positive.
    (E) In a reaction where ΔH is positive and ΔS is negative, the forward reaction is spontaneous.

60. A student filled a eudiometer with 32. milliliters of oxygen and 4.0 milliliters of hydrogen over mercury. How much of which gas would be left uncombined after the mixture was sparked?

    (A) None of either
    (B) 3.0 mL H_2
    (C) 24 mL O_2
    (D) 28 mL O_2
    (E) 30. mL O_2

61. What would be the *total* volume, in milliliters, of gases in question 60 after sparking?

    (A) 16
    (B) 24
    (C) 34
    (D) 36
    (E) 40

62. How can the addition of a catalyst affect an exothermic reaction?

    I. Speed up the reaction.
    II. Slow down the reaction.
    III. Increase the amount of product formed.

    (A) I only
    (B) II only
    (C) I and II only
    (D) II and III only
    (E) I, II, and III

63. In which period of the periodic table is the most electronegative element found?

    (A) 1
    (B) 2
    (C) 3
    (D) 4
    (E) 5

64. What could be the equilibrium constant for this reaction: $a\text{A} + b\text{B} \rightleftharpoons c\text{C} + d\text{D}$, if A and D are solids?

    (A) $\dfrac{[\text{C}]^c[\text{D}]^d}{[\text{A}]^a[\text{B}]^b}$

    (B) $\dfrac{[\text{A}]^a[\text{B}]^b}{[\text{C}]^c[\text{D}]^d}$

    (C) $\dfrac{[\text{C}]^c}{[\text{B}]^b}$

    (D) $\dfrac{[\text{C}]^c[\text{D}]^d}{[\text{A}]^a}$

    (E) $[\text{A}]^a[\text{B}]^b[\text{C}]^c[\text{D}]^d$

65. Which of the following does NOT react with a dilute solution of sulfuric acid?

    (A) $NaNO_3$
    (B) Na_2S
    (C) Na_3PO_4
    (D) Na_2CO_3
    (E) $NaOH$

GO ON TO THE NEXT PAGE

66. Which of these statements is the best explanation for the sp^3 hybridization of carbon's electrons in methane, CH_4?

(A) The new orbitals are one s orbital and three p orbitals.
(B) The s electron is promoted to the p orbitals.
(C) The s orbital is deformed into a p orbital.
(D) Four new and equivalent orbitals are formed.
(E) The s orbital electron loses energy to fall back into a partially filled p orbital.

67. The intermolecular force that is most significant in explaining the variation of the boiling point of water from the boiling points of similarly structured molecules is

(A) hydrogen bonding
(B) van der Waals forces
(C) covalent bonding
(D) ionic bonding
(E) coordinate covalent bonding

68. If K for the reaction $H_2 + I_2 \rightleftharpoons 2HI$ is equal to 45.9 at 450°C, and 1 mole of H_2 and 1 mole of I_2 are introduced into a 1-liter box at that temperature, what will be the expression for K at equilibrium?

(A) $\dfrac{[x^2]^2}{[1-x][1-x]}$

(B) $\dfrac{[2x]^2}{[1-x][1-x]}$

(C) $\dfrac{[2x]^2}{[x][x]}$

(D) $\dfrac{[1-x][1-x]}{[2x]^2}$

(E) $\dfrac{[1-x][1-x]}{[x^2]^2}$

69. In the decomposition reaction of potassium chlorate, if 96.0 grams of oxygen are produced, the mass of potassium chloride produced would be

(A) 149.2 grams
(B) 74.2 grams
(C) 37.1 grams
(D) 96.0 grams
(E) unable to be determined

70. Alpha radiation

(A) bends away from atomic nuclei
(B) is the same as a stream of electrons
(C) is composed of hydrogen nuclei
(D) does not naturally occur
(E) has a neutral charge

STOP

If you finish before one hour is up, you may go back to check your work or complete unanswered questions.

1. **B**	7. **E**	13. **A**	19. **D**
2. **D**	8. **C**	14. **B**	20. **A**
3. **C**	9. **B**	15. **A**	21. **C**
4. **A**	10. **A**	16. **B**	22. **B**
5. **B**	11. **D**	17. **C**	23. **E**
6. **D**	12. **B**	18. **E**	

101. **T, F**	105. **F, T**	109. **T, T, CE**	113. **F, T**
102. **T, T, CE**	106. **F, T**	110. **T, T, CE**	114. **T, F**
103. **T, F**	107. **T, T, CE**	111. **T, T, CE**	115. **T, T**
104. **F, T**	108. **T, T, CE**	112. **T, T**	

24. **B**	36. **C**	48. **D**	60. **E**
25. **D**	37. **E**	49. **C**	61. **C**
26. **D**	38. **A**	50. **C**	62. **A**
27. **D**	39. **D**	51. **D**	63. **B**
28. **D**	40. **D**	52. **E**	64. **C**
29. **E**	41. **B**	53. **E**	65. **A**
30. **D**	42. **B**	54. **C**	66. **D**
31. **A**	43. **B**	55. **A**	67. **A**
32. **E**	44. **C**	56. **A**	68. **B**
33. **B**	45. **A**	57. **E**	69. **A**
34. **A**	46. **A**	58. **D**	70. **A**
35. **C**	47. **B**	59. **E**	

ANSWERS EXPLAINED

1. **(B)** The activation energy of the forward reaction is the energy needed to begin the reaction.

2. **(D)** For the reverse reaction to occur, activation energy equal to the sum of (B) + (C) is needed. This is shown by (D).

3. **(C)** The heat of the reaction is the heat liberated between the level of potential energy of the reactants and that of the products. This is quantity (C) on the diagram.

4. **(A)** The potential energy of the reactants is the total of the original potential energies of the reactants shown by (A).

5. **(B)** This is a single replacement reaction in which aluminum chloride forms. The formula for aluminum chloride is $AlCl_3$.

6. **(D)** The oxidation state of aluminum is changing from 0 to +3; therefore, the aluminum is being oxidized.

7. **(E)** The original solution of $CuCl_2$ is blue due to the presence of Cu^{2+} ions in solution. The copper(II) ions are being used in the reaction; therefore, the concentration of the copper(II) ion will be decreasing and the solution will become less blue as time goes by.

8. **(C)** This equation shows the volume decreasing as the pressure is increased when the temperature is held constant. It is an example of Boyle's Law.

9. **(B)** This equation shows the pressure increasing as the temperature is increased when the volume is held constant. It is an example of Gay-Lussac's Law.

10. **(A)** This equation shows the volume increasing as the temperature is increased when the pressure is held constant. It is an example of Charles's Law.

11. **(D)** This equation shows that the total pressure of a mixture of gases is equal to the sum of the partial pressures of the component gases.

12. **(B)** The balanced reaction equation is:

$$2NaI(aq) + Pb(NO_3)_2(aq) \rightarrow 2NaNO_3(aq) + PbI_2(s)$$

13. **(A)** The balanced reaction equation shows that 1 mole of PbI_2 is produced from 2 moles of NaI. Therefore, only 1 mole of NaI would be required to make 0.5 mole of PbI_2.

14. **(B)** $NaNO_3$ is an ionic substance that dissociates into two ions, Na^+ and NO_3^-.

15. **(A)** $MgCl_2$ is ionic because it is the product of an active metal (Mg) combining with a very active nonmetal (Cl).

16. **(B)** The electronegativity difference between H and Cl is between 0.5 and 1.7. This indicates an unequal sharing of electrons, which results in a polar covalent bond.

17. **(C)** Ethane, CH_3–CH_3(g), has symmetrically arranged C–H polar bonds so that the ethane molecule is nonpolar covalent.

18. **(E)** Cu(s) is a metal.

19. **(D)** When hydrated copper sulfate is heated, the crystal crumples as the water is forced out of the structure, and a white powder is the result.

20. **(A)** When an ionic substance dissolves in water, the ionic substance breaks up into the ions that make it up. This process is called dissociation.

21. **(C)** Phenolphthalein is used as an acid–base indicator because it turns pink in the presence of a base.

22. **(B)** A substance that can act as both an acid and a base depending on what it is in the presence of is said to be amphoteric.

23. **(E)** Hygroscopic materials have the ability to absorb moisture from the air. Some hygroscopic materials have such an ability to do so that they can dissolve into a puddle from that moisture. Such substances are said to be deliquescent.

101. **(T, F)** The assertion is true that active metals are found in the upper left corner of the Periodic Table because of their ability to lose their outer electron(s). These metals actually have smaller ionic radii than their atomic radii.

102. **(T, T, CE)** The assertion is explained by the reason. The graphic display of this is:

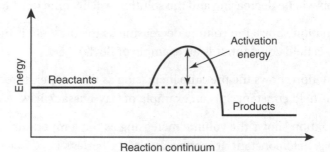

103. **(T, F)** The assertion is true but the reasoning is false. Transition elements have incomplete inner energy levels that are being filled with the additional electrons, thus leaving the outer energy level the same in most cases. As a result, these elements have common oxidation numbers.

104. **(F, T)** The assertion is false and the reason is true. A supersaturated solution is holding more than its normal solubility, and the addition of a crystal causes crystallization to occur.

105. **(F, T)** Equilibrium is a dynamic condition because of the reason stated. The assertion is false; the reason, true.

106. **(F, T)** Covalent bonds are stronger than ionic bonds, generally.

107. **(T, T, CE)** An equilibrium system must have a gaseous reactant or product for pressure to affect the equilibrium. Then increased pressure will cause the reaction to go in the direction that reduces the concentration of gaseous substances. Since four volumes of gases on the left became two volumes on the right, the reaction shifts to the right.

108. **(T, T, CE)** The assertion is explained by the reason; both are true.

109. **(T, T, CE)** Both the assertion and the reason are true; they explain that the element's orbital designation places it in the first transition series of filling the $3d$ orbitals.

110. **(T, T, CE)** The assertion is true; the reason is true and explains why nonmetals have the highest electronegativity.

111. **(T, T, CE)** The assertion is explained by the reason; both are true.

112. **(T, T)** There are 3 moles of atoms in 18 g of water because 18 g is 1 mol of water molecules and each molecule has three atoms. The reason does not explain the assertion but is also true.

113. **(F, T)** Benzene is a nonionizing substance and therefore a nonelectrolyte. The reason is a true statement.

114. **(T, F)** Isomers have the same empirical formula but vary in their structural formulas.

115. **(T, T)** The reaction does go to completion, but a gaseous product is formed, not a precipitate, so II does not explain I.

24. **(B)** The correct coefficients are 2, 13, 8, and 10.

25. **(D)** 1K + 1Al + 2S + 8O = 12 total

26. **(D)** Air is a mixture; all others are compounds. Washing soda (C) is sodium carbonate, and lime (E) is calcium oxide.

27. **(D)** One mole of a gas at STP occupies 22.4 L. So two moles of a gas at STP occupy 2.0 mol $\times$ 22.4 L = 44.8 L.

28. **(D)** The maximum number of electrons in each kind of orbital is:

$$s = 2 \text{ in one orbital}$$
$$p = 6 \text{ in three orbitals}$$
$$d = 10 \text{ in five orbitals}$$
$$f = 14 \text{ in seven orbitals}$$

29. **(E)** The element with atomic number 11 is sodium with 1 electron in the $3s$ orbital. It would readily combine with the element that has $3p^5$ as the outer orbital since it needs only 1 more electron to fill it.

30. **(D)** The only physical property named in the list is low melting point.

31. **(A)** If the gas is diatomic, then 6.02×10^{23} atoms will form $6.02 \times 10^{23}/2$ molecules. At STP, 6.02×10^{23} molecules occupy 22.4 L, so half that number will occupy 11.2 L.

32. **(E)** Cascading excited electrons can fall only to lower energy levels that are unfilled.

33. **(B)** Mass spectroscopy uses a magnetic field to separate isotopes by bending their path. The lighter ones are bent farther than the heavier ones.

34. **(A)** The pi bond is a bond between two *p* orbitals, like this:

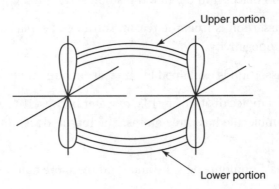

35. **(C)** At Pikes Peak (alt. approx. 14,000 ft) the pressure is lower than at ground level; therefore the vapor pressure at a lower temperature will equal the outside pressure and boiling will occur.

36. **(C)** The alkanes contain the sp^3 hybrid orbitals.

37. **(E)** I, II, and III are correct.

38. **(A)** Only I is correct.

39. **(D)** HCl is actually a stronger Brønsted-Lowry acid than H_2O, which is why the reaction occurs as shown.

40. **(D)** The name carbon-13 supplies enough information to determine the number of all three subatomic particles. The name carbon indicates 6 protons as carbon has only 6 protons. Since atoms are neutral, an equal number of electrons must be present; therefore, carbon-13 has 6 electrons. The number that follows the name, 13, indicates the mass number of the atom, which counts the number of protons and neutrons. Since the atom has 6 protons, it must contain 7 neutrons.

41. **(B)** III is a salt from a strong base and a weak acid, which hydrolyzes to form a basic solution with water.

42. **(B)** Since NaCl dissociates into two ions, it can change the boiling point of water more than the ethyl alcohol does, which does not dissociate. Colligative properties depend on the number of solute particles present in the solution. All the other substances dissociate into more particles per formula unit.

43. **(B)** The VSEPR model shows BF_3 is trigonal planar and so is related to the triangle shape on one plane.

44. **(C)** The delivery tube is below the fluid level in the flask and will cause liquid to be forced up the thistle tube when gas is evolved in the reaction.

45. **(A)** At first the fluid will be expelled up the thistle tube by the gas generated and exerting pressure in the reaction flask. When the level of the fluid falls below the end of the thistle tube, the gas will then be released through the thistle tube.

46. **(A)** The most active nonmetal has a high attraction for another electron—thus high electronegativity.

47. **(B)** Fe loses electrons to form the Fe^2 ion.

48. **(D)** The K_w of water = $[H^+][OH^-] = 10^{-14}$. If $[OH^-] = 10^{-5}$ mol/L, then

$$[H^+] = 10^{-14}/10^{-5} = 10^{-9}$$
$$pH = -\log[H^+] \text{ (by definition)}$$
$$pH = -[-9]$$
$$pH = 9$$

49. **(C)** The oxidation state of oxygen in hydrogen peroxide (H_2O_2) is –1. In the products the oxidation state is both –2 (in H_2O) and 0 (in O_2). Therefore, the oxygen is being both oxidized and reduced. This process is called disproportionation.

50. **(C)** The correctly balanced equation is

$$C_2H_4(g) + 3O_2(g) \rightarrow 2CO_2(g) + 2H_2O(\ell)$$

51. **(D)** The values provided should be plugged into the Ideal Gas Law to solve for the number of moles of gas present.

$$PV = nRT$$
$$(0.821\,\text{atm})(273\,\text{L}) = n\left(0.0821\frac{\text{L} \cdot \text{atm}}{\text{mol} \cdot \text{K}}\right)(273\,\text{K})$$
$$n = 10.0 \text{ mol}$$

To find the molar mass of the gas, the mass of the sample must be divided by the number of moles of the gas in the sample. Therefore, 40.0 g/10.0 moles gives a molar mass of 4.00 g/mol.

52. **(E)** To find the simple or empirical formula, divide each % by the element's atomic mass.

Carbon
$$40 \div 12 = 3.333$$

Hydrogen
$$6.67 \div 1 = 6.67$$

Oxygen
$$53.33 \div 16 = 3.33$$

Next, divide each quotient by the smallest quotient in an attempt to get small whole numbers.

$$3.33 \div 3.33 = 1 \text{ C}$$
$$6.67 \div 3.33 = 2 \text{ H}$$
$$3.33 \div 3.33 = 1 \text{ O}$$

The simplest formula is CH_2O, which has a molecular mass of 30. The true molecular mass is given as 90.0, which is three times the simplest. Therefore, the true formula is $C_3H_6O_3$.

53. **(E)** The reaction is:

$$x \text{ mol} \qquad 370 \text{ g}$$
$$CaO + H_2O \rightarrow Ca(OH)_2$$
$$Ca(OH)_2 \text{ molecular mass} = 74$$

370 g ÷ 74 = 5.0 mol of $Ca(OH)_2$ is wanted. The reaction equation shows 1 mol of CaO produces 1 mol of $Ca(OH)_2$, so the answer is 5.0 mol.

54. **(C)** In dilution problems, this formula can be used:

$$M_{before} \times V_{before} = M_{after} \times V_{after}$$

Substituting gives:

$$3.50 \times 50.0 = 2 \times (?\ x)$$

$x = 87.5$ mL, new volume after dilution

55. **(A)** For K_{eq} to be small, the numerator, which is made up of the concentration(s) of the product(s) at equilibrium, must be smaller than the denominator.

56. **(A)** The amount of NaOH used is

$$49.2 - 34.7 = 14.5 \text{ mL}$$
$$\text{Using } M_1 \times V_1 = M_2 \times V_2 \text{ gives}$$
$$0.09 \text{ M} \times 14.5 \text{ mL} = M_2 \times 10 \text{ mL}$$
$$M_2 = 0.13 \text{ M}$$

57. **(E)** The reactions recorded indicated that the ease of losing electrons is greater in sodium than magnesium, greater in magnesium than iron, greater in iron than copper, and finally greater in copper than silver.

58. **(D)** Distillation takes advantage of the boiling point differences between two liquids in a mixture. Since water and ethyl alcohol have different boiling points, they could be separated via this method.

59. **(E)** All the first four statements are correct.

The Gibbs free-energy equation is:

$$\Delta G = \Delta H - T \Delta S$$

In choice (E), if ΔH is positive and ΔS is negative, then ΔG will definitely be positive, which means that the forward reaction will not occur spontaneously.

60. **(E)** H_2 to O_2 ratio by volume is 2 : 1 in the formation of water. Therefore, 4.0 mL H_2 will react with 2.0 mL of O_2 to make 4.0 mL of steam.

$$2H_2(g) + O_2(g) \rightarrow 2H_2O(g)$$

This leaves 30. mL of O_2 uncombined.

61. **(C)** There will be 30. mL of O_2 + 4.0 mL of steam = 34 mL total.

62. **(A)** By definition, a catalyst can be used to speed up a reaction without itself being consumed, so I is correct.

63. **(B)** The most electronegative element is fluorine (F), found in period 2.

64. **(C)** Solids are incorporated into the K value and therefore do not appear in the equilibrium constant expression.

65. **(A)** Only $NaNO_3$ will not react because it is a neutral substance.

66. **(D)** When hybridization forms the sp^3 orbitals in methane, CH_4, four entirely new orbitals, different from but equivalent to the former s and p orbitals, result.

67. **(A)** Hydrogen bonding between water molecules causes the boiling point to be higher than would be expected.

68. **(B)** At the beginning of the reaction

$$[H_2] = 1 \text{ mol/L}$$
$$[I_2] = 1 \text{ mol/L}$$
$$[HI] = 0$$

At equilibrium

$$(H_2 + I_2 \rightleftharpoons 2HI)$$
(Let x = moles/liter of
H_2 and I_2 in HI form)

$$[H_2] = (1 - x) \text{ mol/L}$$
$$[I_2] = (1 - x) \text{ mol/L}$$
$$[HI] = 2x \text{ mol/L}$$

Then, substituting the above values into the equation, you get:

$$K = \frac{[HI]^2}{[H_2][I_2]}$$
$$= \frac{[2x]^2}{[1 - x][1 - x]} = 45.9$$

69. **(A)** The decomposition reaction of potassium chlorate is

$$2KClO_3(s) \rightarrow 2KCl(s) + 3O_2(g)$$

Since 96.0 grams of O_2 were produced, which is 3 moles of O_2, then 2 moles of KCl were also produced. Based on the Periodic Table, 2 moles of KCl has a mass of 149.2 grams.

70. **(A)** Alpha radiation is a type of radioactive decay that naturally and spontaneously occurs when the nucleus of an atom releases a composite particle of two protons and two neutrons (essentially the nucleus of a helium atom). Since protons are positive and neutrons are negative, the overall charge of an alpha particle is positive, and so it would bend away from other positively charged particles such as the nuclei of atoms.

CALCULATING YOUR SCORE

Your score on Practice Test 3 can now be computed manually. The actual test will be scored by machine, but the same method is used to arrive at the raw score. You get one point for each correct answer. For each wrong answer, you lose one-fourth of a point. Questions that you omit or that have more than one answer are not counted. On your answer sheet mark all correct answers with a "C" and all incorrect answers with an "X."

Determining Your Raw Test Score

Total the number of correct answers you have recorded on your answer sheet. It should be the same as the total of all the numbers you place in the block in the lower left corner of each area of the Subject Area summary in the next section.

- **A.** Enter the total number of correct answers here: _____
  Now count the number of wrong answers you recorded on your answer sheet.
- **B.** Enter the total number of wrong answers here: _____
  Multiply the number of wrong answers in B by 0.25.
- **C.** Enter that product here: _____
  Subtract the result in C from the total number of right answers in A.
- **D.** Enter the result of your subtraction here: _____
- **E.** Round the result in D to the nearest whole number: _____.
  This is your raw test score.

Conversion of Raw Scores to Scaled Scores

Your raw score is converted by the College Board into a scaled score. The College Board scores range from 200 to 800. This conversion is done to ensure that a score earned on any edition of a particular SAT Subject Test in Chemistry is comparable to the same scaled score earned on any other edition of the same test. Because some editions of the tests may be slightly easier or more difficult than others, scaled scores are adjusted so that they indicate the same level of performance regardless of the edition of the test taken and the ability of the group that takes it. Consequently, a specific raw score on one edition of a particular test will not necessarily translate to the same scaled score on another edition of the same test.

Because the practice tests in this book have no large population of scores with which they can be scaled, scaled scores cannot be determined.

Results from previous SAT Chemistry tests appear to indicate that the conversion of raw scores to scaled scores GENERALLY follows this pattern:

Raw Score	Scaled Score	Raw Score	Scaled Score
85–82	800–800	30–25	540–520
81–75	790–760	25–20	520–490
75–70	760–740	20–15	490–460
70–65	740–710	15–10	460–430
65–60	710–690	10–5	430–400
60–55	690–670	5–0	400–370
55–50	670–640	0 to –5	370–340
50–45	640–620	–5 to –10	340–310
45–40	620–590	–10 to –15	310–290
40–35	590–570	–15 to –20	290–270
35–30	570–540	–20 or lower	270–200

Note that this scale provides only a *general idea* of what a raw score may translate into on a scaled score range of 800–200. Scaling on every test is usually slightly different. Some students who had taken the SAT Subject Test in Chemistry after using this book had reported that they have scored slightly higher on the SAT test than on the practice tests in this book. They *all* reported that preparing well for the test paid off in a better score!

DIAGNOSING YOUR NEEDS

After taking Practice Test 3, check your answers against the correct ones. Then fill in the chart below.

In the space under each question number, place a check if you answered that question correctly.

➡ EXAMPLE

If your answer to question 5 was correct, place a check in the appropriate box.

Next, total the check marks for each section and insert the number in the designated block. Now do the arithmetic indicated and insert your percent for each area.

Subject Area*	(✓) Questions Answered Correctly								
I. **Atomic Theory and Structure,** including periodic relationships	110	112	70	28	32	33	34	40	66
☐ No. of checks ÷ 9 × 100 = _____%									
II. **Chemical Bonding and Molecular Structure**	15	16	17	18	106	25	29	38	43	67
No. of checks ÷ 10 × 100 = _____%										
III. **States of Matter and Kinetic Molecular Theory of Gases**	8	9	10	11	19	23	27	31	35
☐ No. of checks ÷ 9 × 100 = _____%									

The subject areas have been expanded to identify specific areas in the text.

Subject Area*	(✓) Questions Answered Correctly						
IV. Solutions, including concentration units, solubility, and colligative properties	20	104	42	54			
☐ No. of checks ÷ 4 × 100 = _____%							
V. Acids and Bases	21	22	113	115	39	41	48
☐ No. of checks ÷ 7 × 100 = _____%							
VI. Oxidation-Reduction	5	6	7	37	47	49	57
☐ No. of checks ÷ 11 × 100 = _____%							

Subject Area*	(✓) Questions Answered Correctly										
VII. Stoichiometry	12	13	14	24	50	51	52	53	60	61	69
☐ No. of checks ÷ 11 × 100 = _____%											

Subject Area*	(✓) Questions Answered Correctly	
VIII. Reaction Rates	1	108
☐ No. of checks ÷ 2 × 100 = _____%		

Subject Area*	(✓) Questions Answered Correctly				
IX. Equilibrium	105	107	55	64	68
☐ No. of checks ÷ 5 × 100 = _____%					
X. Thermodynamics: energy changes in chemical reactions, randomness, and criteria for spontaneity	2	3	4	102	59	62
☐ No. of checks ÷ 6 × 100 = _____%					

Subject Area*	(✓) Questions Answered Correctly				
XI. Descriptive Chemistry: physical and chemical properties of elements and their familiar compounds; organic chemistry; periodic properties	101	103	109	111	114
	26	30	36	46	63	65
☐ No. of checks ÷ 11 × 100 = _____%					
XII. Laboratory: equipment, procedures, observations, safety, calculations, and interpretation of results	44	45	56	57	58
☐ No. of checks ÷ 5 × 100 = _____%					

The subject areas have been expanded to identify specific areas in the text.

To develop a study plan, refer to pages 31–33.

Determine the correct answer for each question. Then, using a No. 2 pencil, blacken completely the oval containing the letter of your choice.

1. Ⓐ Ⓑ Ⓒ Ⓓ Ⓔ
2. Ⓐ Ⓑ Ⓒ Ⓓ Ⓔ
3. Ⓐ Ⓑ Ⓒ Ⓓ Ⓔ
4. Ⓐ Ⓑ Ⓒ Ⓓ Ⓔ
5. Ⓐ Ⓑ Ⓒ Ⓓ Ⓔ
6. Ⓐ Ⓑ Ⓒ Ⓓ Ⓔ
7. Ⓐ Ⓑ Ⓒ Ⓓ Ⓔ
8. Ⓐ Ⓑ Ⓒ Ⓓ Ⓔ
9. Ⓐ Ⓑ Ⓒ Ⓓ Ⓔ
10. Ⓐ Ⓑ Ⓒ Ⓓ Ⓔ
11. Ⓐ Ⓑ Ⓒ Ⓓ Ⓔ
12. Ⓐ Ⓑ Ⓒ Ⓓ Ⓔ
13. Ⓐ Ⓑ Ⓒ Ⓓ Ⓔ
14. Ⓐ Ⓑ Ⓒ Ⓓ Ⓔ
15. Ⓐ Ⓑ Ⓒ Ⓓ Ⓔ
16. Ⓐ Ⓑ Ⓒ Ⓓ Ⓔ

17. Ⓐ Ⓑ Ⓒ Ⓓ Ⓔ
18. Ⓐ Ⓑ Ⓒ Ⓓ Ⓔ
19. Ⓐ Ⓑ Ⓒ Ⓓ Ⓔ
20. Ⓐ Ⓑ Ⓒ Ⓓ Ⓔ
21. Ⓐ Ⓑ Ⓒ Ⓓ Ⓔ
22. Ⓐ Ⓑ Ⓒ Ⓓ Ⓔ
23. Ⓐ Ⓑ Ⓒ Ⓓ Ⓔ

ON THE ACTUAL CHEMISTRY TEXT. THE FOLLOWING TYPE OF QUESTION MUST BE ANSWERED ON A SPECIAL SECTION (LABELED "CHEMISTRY") AT THE LOWER LEFT-HAND CORNER OF PAGE 2 OF YOUR ANSWER SHEET. THESE QUESTIONS WILL BE NUMBERED BEGINNING WITH 101 AND MUST BE ANSWERED ACCORDING TO THE DIRECTIONS.

CHEMISTRY* Fill in oval CE only if II is a correct explanation of I.

	I	II	CE*
101.	Ⓣ Ⓕ	Ⓣ Ⓕ	◯
102.	Ⓣ Ⓕ	Ⓣ Ⓕ	◯
103.	Ⓣ Ⓕ	Ⓣ Ⓕ	◯
104.	Ⓣ Ⓕ	Ⓣ Ⓕ	◯
105.	Ⓣ Ⓕ	Ⓣ Ⓕ	◯
106.	Ⓣ Ⓕ	Ⓣ Ⓕ	◯
107.	Ⓣ Ⓕ	Ⓣ Ⓕ	◯
108.	Ⓣ Ⓕ	Ⓣ Ⓕ	◯
109.	Ⓣ Ⓕ	Ⓣ Ⓕ	◯
110.	Ⓣ Ⓕ	Ⓣ Ⓕ	◯
111.	Ⓣ Ⓕ	Ⓣ Ⓕ	◯
112.	Ⓣ Ⓕ	Ⓣ Ⓕ	◯
113.	Ⓣ Ⓕ	Ⓣ Ⓕ	◯
114.	Ⓣ Ⓕ	Ⓣ Ⓕ	◯
115.	Ⓣ Ⓕ	Ⓣ Ⓕ	◯

ANSWER SHEET
Practice Test 4

ON THE ACTUAL CHEMISTRY TEST, THE REMAINING QUESTIONS MUST BE ANSWERED BY RETURNING TO THE SECTION OF YOUR ANSWER SHEET YOU STARTED FOR CHEMISTRY.

23. Ⓐ Ⓑ Ⓒ Ⓓ Ⓔ	39. Ⓐ Ⓑ Ⓒ Ⓓ Ⓔ	55. Ⓐ Ⓑ Ⓒ Ⓓ Ⓔ
24. Ⓐ Ⓑ Ⓒ Ⓓ Ⓔ	40. Ⓐ Ⓑ Ⓒ Ⓓ Ⓔ	56. Ⓐ Ⓑ Ⓒ Ⓓ Ⓔ
25. Ⓐ Ⓑ Ⓒ Ⓓ Ⓔ	41. Ⓐ Ⓑ Ⓒ Ⓓ Ⓔ	57. Ⓐ Ⓑ Ⓒ Ⓓ Ⓔ
26. Ⓐ Ⓑ Ⓒ Ⓓ Ⓔ	42. Ⓐ Ⓑ Ⓒ Ⓓ Ⓔ	58. Ⓐ Ⓑ Ⓒ Ⓓ Ⓔ
27. Ⓐ Ⓑ Ⓒ Ⓓ Ⓔ	43. Ⓐ Ⓑ Ⓒ Ⓓ Ⓔ	59. Ⓐ Ⓑ Ⓒ Ⓓ Ⓔ
28. Ⓐ Ⓑ Ⓒ Ⓓ Ⓔ	44. Ⓐ Ⓑ Ⓒ Ⓓ Ⓔ	60. Ⓐ Ⓑ Ⓒ Ⓓ Ⓔ
29. Ⓐ Ⓑ Ⓒ Ⓓ Ⓔ	45. Ⓐ Ⓑ Ⓒ Ⓓ Ⓔ	61. Ⓐ Ⓑ Ⓒ Ⓓ Ⓔ
30. Ⓐ Ⓑ Ⓒ Ⓓ Ⓔ	46. Ⓐ Ⓑ Ⓒ Ⓓ Ⓔ	62. Ⓐ Ⓑ Ⓒ Ⓓ Ⓔ
31. Ⓐ Ⓑ Ⓒ Ⓓ Ⓔ	47. Ⓐ Ⓑ Ⓒ Ⓓ Ⓔ	63. Ⓐ Ⓑ Ⓒ Ⓓ Ⓔ
32. Ⓐ Ⓑ Ⓒ Ⓓ Ⓔ	48. Ⓐ Ⓑ Ⓒ Ⓓ Ⓔ	64. Ⓐ Ⓑ Ⓒ Ⓓ Ⓔ
33. Ⓐ Ⓑ Ⓒ Ⓓ Ⓔ	49. Ⓐ Ⓑ Ⓒ Ⓓ Ⓔ	65. Ⓐ Ⓑ Ⓒ Ⓓ Ⓔ
34. Ⓐ Ⓑ Ⓒ Ⓓ Ⓔ	50. Ⓐ Ⓑ Ⓒ Ⓓ Ⓔ	66. Ⓐ Ⓑ Ⓒ Ⓓ Ⓔ
35. Ⓐ Ⓑ Ⓒ Ⓓ Ⓔ	51. Ⓐ Ⓑ Ⓒ Ⓓ Ⓔ	67. Ⓐ Ⓑ Ⓒ Ⓓ Ⓔ
36. Ⓐ Ⓑ Ⓒ Ⓓ Ⓔ	52. Ⓐ Ⓑ Ⓒ Ⓓ Ⓔ	68. Ⓐ Ⓑ Ⓒ Ⓓ Ⓔ
37. Ⓐ Ⓑ Ⓒ Ⓓ Ⓔ	53. Ⓐ Ⓑ Ⓒ Ⓓ Ⓔ	69. Ⓐ Ⓑ Ⓒ Ⓓ Ⓔ
38. Ⓐ Ⓑ Ⓒ Ⓓ Ⓔ	54. Ⓐ Ⓑ Ⓒ Ⓓ Ⓔ	70. Ⓐ Ⓑ Ⓒ Ⓓ Ⓔ

PRACTICE TEST 4

Note: For all questions involving solutions and/or chemical equations, assume that the system is in water unless otherwise stated.

Reminder: You may *not* use a calculator on these tests.

The following symbols have the meanings listed unless otherwise noted.

H	= enthalpy	T	= temperature	L	= liter(s)
M	= molar	V	= volume	mL	= milliliter(s)
n	= number of moles	atm	= atmosphere(s)	mm	= millimeter(s)
P	= pressure	g	= gram(s)	mol	= mole(s)
R	= molar gas constant	J	= joule(s)	V	= volt(s)
S	= entropy	kJ	= kilojoule(s)		

PART A

Directions: Every set of the given lettered choices below refers to the numbered statements or formulas immediately following it. Choose the one lettered choice that best fits each statement or formula and then fill in the corresponding oval on the answer sheet. Each choice may be used once, more than once, or not at all in each set.

Questions 1–3 refer to the following graphs:

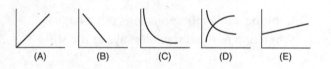

(A)   (B)   (C)   (D)   (E)

1. The graph that best shows the relationship of gas volume to temperature, with pressure held constant

2. The graph that best shows the relationship of gas volume to pressure, with temperature held constant

3. The graph that best shows the relationship of the number of grams of a solid that are soluble in 100 grams of H_2O at varying temperatures if the solubility begins at a small quantity and increases at a slow, steady pace as the temperature is increased

Questions 4–7

(A) A molecule
(B) A mixture of compounds
(C) An isotope
(D) An isomer
(E) An acid salt

4. The simplest unit of water that retains its properties

5. A commercial cake mix

6. An atom with the same number of protons as another atom of the same element but a different number of neutrons

7. Classification of $Ca(HCO_3)_2$

GO ON TO THE NEXT PAGE

Questions 8–10

(A) 1
(B) 6
(C) 9
(D) 10
(E) 14

8. The atomic number of an atom with an electron dot arrangement similiar to $\cdot \ddot{\text{I}} \cdot$

9. The number of atoms represented in the formula Na_2CO_3

10. The number that represents the most acid pH

Questions 11–14

(A) Density
(B) Equilibrium constant
(C) Molar mass
(D) Freezing point
(E) Molarity

11. Can be expressed in moles of solute per liter of solution

12. Can be expressed in grams per liter of a gas

13. Will NOT be affected by changes in temperature and pressure

14. At STP, can be used to determine the molecular mass of a pure gas

Questions 15–18

(A) Buffer
(B) Indicator
(C) Arrhenius acid
(D) Arrhenius base
(E) Neutral condition

15. Resists a rapid change of pH

16. Exhibits different colors in acidic and basic solutions

17. At 25°C, the aqueous solution has a pH < 7.

18. At 25°C, the aqueous solution has a pH > 7.

Questions 19–23

(A) O_2
(B) SO_2
(C) CO
(D) CO_2
(E) O_3

19. In the stratosphere, screens out a large fraction of the ultraviolet rays of the sun

20. Is a product of the incomplete combustion of hydrocarbons

21. A gas produced by the heating of potassium chlorate

22. A gas that is slightly soluble in water and gives a weakly acid solution

23. The gas with the slowest effusion rate

GO ON TO THE NEXT PAGE

ON THE ACTUAL CHEMISTRY TEST, THE FOLLOWING TYPE OF QUESTION MUST BE ANSWERED ON A SPECIAL SECTION (LABELED "CHEMISTRY") AT THE LOWER LEFT-HAND CORNER OF PAGE 2 OF YOUR ANSWER SHEET. THESE QUESTIONS WILL BE NUMBERED BEGINNING WITH 101 AND MUST BE ANSWERED ACCORDING TO THE FOLLOWING DIRECTIONS.

Directions: Every question below contains two statements, I in the left-hand column and II in the right-hand column. For each question, decide if statement I is true or false <u>and</u> if statement II is true or false and fill in the corresponding T or F ovals on your answer sheet. *<u>Fill in oval CE only if statement II is a correct explanation of statement I.</u>

Sample Answer Grid:

CHEMISTRY * Fill in oval CE only if II is a correct explanation of I.

	I	II	CE*
101.	Ⓣ Ⓕ	Ⓣ Ⓕ	◯

<table>
<tr><th>I</th><th></th><th>II</th></tr>
<tr><td>101. According to the Kinetic Molecular Theory, the particles of a gas are in random motion above absolute zero</td><td>BECAUSE</td><td>the degree of random motion of gas molecules varies inversely with the temperature of the gas.</td></tr>
<tr><td>102. An electron has wave properties as well as particle properties</td><td>BECAUSE</td><td>the design of a particular experiment determines which properties are verified.</td></tr>
<tr><td>103. The alkanes are considered a homologous series</td><td>BECAUSE</td><td>homologous series have the same functional group but differ in formula by the addition of a fixed group of atoms.</td></tr>
<tr><td>104. When an atom of an active metal becomes an ion, the radius of the ion is less than that of the atom</td><td>BECAUSE</td><td>the nucleus of an active metallic ion has less positive charge than the electron "cloud."</td></tr>
<tr><td>105. When the heat of formation for a compound is negative, ΔH is negative</td><td>BECAUSE</td><td>a negative heat of formation indicates that a reaction is exothermic with a negative enthalpy change.</td></tr>
<tr><td>106. Water is a polar substance</td><td>BECAUSE</td><td>the sharing of the bonding electrons in water is unequal.</td></tr>
<tr><td>107. A catalyst accelerates a chemical reaction</td><td>BECAUSE</td><td>a catalyst lowers the activation energy of the reaction.</td></tr>
</table>

PRACTICE TEST 4

GO ON TO THE NEXT PAGE

	I		**II**

108. Copper is an oxidizing agent in the reaction with silver nitrate solution

BECAUSE copper loses electrons in a reaction with silver ions.

109. The rate of diffusion (or effusion) of hydrogen gas compared with that of helium gas is 1:4

BECAUSE the rate of diffusion (or effusion) of gases varies inversely as the square root of the molecular mass.

110. A gas heated from 10°C to 100°C at constant pressure will increase in volume

BECAUSE as Charles's Law states, if the pressure remains constant, the volume varies directly as the absolute temperature varies.

111. The Gibbs free-energy equation cannot be used to predict the spontaneity of a reaction

BECAUSE both enthalpy change and entropy change are part of the Gibbs free energy equation.

112. The complete electrolysis of 45 grams of water will yield 40 grams of H_2 and 5 grams of O_2

BECAUSE water is composed of hydrogen and oxygen in a ratio of 8:1 by mass.

113. 320 calories or 1.34×10^3 joules of heat will melt 4 grams of ice at 0°C

BECAUSE the heat of fusion of water is 80 calories per gram or 3.34×10^2 joules per gram.

114. When 2 liters of oxygen gas react with 2 liters of hydrogen completely, the limiting factor is the volume of the oxygen

BECAUSE the coefficients in balanced equations of gaseous reactions give the volume relationships of the involved gases.

115. Water is a good solvent for ionic and/or polar covalent substances

BECAUSE water shows hydrogen bonding between oxygen atoms.

GO ON TO THE NEXT PAGE

Directions: Every question or incomplete statement below is followed by five suggested answers or completions. Choose the one that is best in each case and then fill in the corresponding oval on the answer sheet.

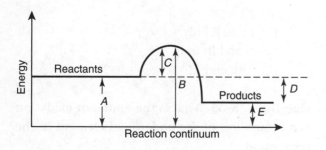

24. In this graphic representation of a chemical reaction, which arrow depicts the activation energy of the forward reaction?

   (A) *A*
   (B) *B*
   (C) *C*
   (D) *D*
   (E) *E*

25. How many liters (STP) of O_2 can be produced by completely decomposing 2.00 moles of $KClO_3$?

   (A) 11.2
   (B) 22.4
   (C) 33.6
   (D) 44.8
   (E) 67.2

26. Which of the following statements is true?

   (A) A catalyst cannot lower the activation energy.
   (B) A catalyst can lower the activation energy.
   (C) A catalyst affects only the activation energy of the forward reaction.
   (D) A catalyst affects only the activation energy of the reverse reaction.
   (E) A catalyst is permanently changed after the activation energy is reached.

27. Which of the following is the correct structural representation of sodium?

   (A) $11\,p$
       $11\,n$ Nucleus and electron configuration:

       $1s^2 2s^2 2p^6 3s^2 3p^6 4s^2 4p^3$

   (B) $11\,p$
       $12\,n$ Nucleus and electron configuration:

       $1s^2 2s^2 2p^6 3s^2 3p^6 4s^2 3d^1 4p^2$

   (C) $23\,p$
       $23\,n$ Nucleus and electron configuration:

       $1s^2 2s^2 2p^6 3s^1$

   (D) $23\,p$
       $23\,n$ Nucleus and electron configuration:

       $1s^2 2s^2 2p^6 3s^2 3p^6 3d^3 4s^2$

   (E) $11\,p$
       $12\,n$ Nucleus and electron configuration:

       $1s^2 2s^2 2p^6 3s^1$

28. If the molecular mass of NH_3 is 17, what is the density of this compound at STP?

   (A) 0.25 g/L
   (B) 0.76 g/L
   (C) 1.52 g/L
   (D) 3.04 g/L
   (E) 9.11 g/L

GO ON TO THE NEXT PAGE

29. Which bond(s) is (are) ionic?

    I. H—Cl(g)
    II. S—Cl(g)
    III. Cs—F(s)

(A) I only
(B) III only
(C) I and II only
(D) II and III only
(E) I, II, and III

30. Aromatic hydrocarbons are represented by which of the following?

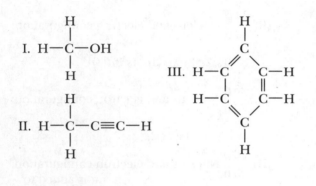

(A) I only
(B) III only
(C) I and II only
(D) II and III only
(E) I, II, and III

31. According to placement in the Periodic Table, which statement(s) regarding the first ionization energies of certain elements should be true?

    I. Li has a higher value than Na.
    II. K has a higher value than Cs.
    III. Na has a higher value than Al.

(A) I only
(B) III only
(C) I and II only
(D) II and III only
(E) I, II, and III

32. Redox reactions include which of the following?

    I. $Na_2S(aq) + PbCl_2(aq) \rightarrow NaCl(s) + PbS(s)$
    II. $CH_4(g) + O_2(g) \rightarrow CO_2(g) + H_2O(\ell)$
    III. $H_2(g) + O_2(g) \rightarrow H_2O(\ell)$

(A) I only
(B) III only
(C) I and II only
(D) II and III only
(E) I, II, and III

Questions 33–35. What is the apparent oxidation state (number) of the underlined element in the compound

33. K$\underline{H}$CO$_3$?

(A) +1
(B) +2
(C) +3
(D) +4
(E) +5

34. M$\underline{g}$SO$_4$?

(A) +1
(B) −1
(C) +2
(D) −2
(E) +3

35. $\underline{C}$O$_2$?

(A) +2
(B) −2
(C) +4
(D) −4
(E) +5

GO ON TO THE NEXT PAGE

36. An atom with an electron configuration of $1s^22s^22p^63s^23p^4$ will probably exhibit which oxidation state?

   (A) +2
   (B) −2
   (C) +3
   (D) −3
   (E) +5

37. In the Lewis dot structure X:, what is the predictable oxidation number?

   (A) +1
   (B) −1
   (C) +2
   (D) −2
   (E) +3

Questions 38–40.

   (A) Balance
   (B) Barometer
   (C) Condenser
   (D) Funnel
   (E) Pipette

38. Commonly used in the laboratory to transfer an exact volume of liquid from one container to another

39. Commonly used in the laboratory in a distillation setup

40. Commonly used in the laboratory in a filtration setup

41. If you collected hydrogen gas by the displacement of water and under the conditions shown:

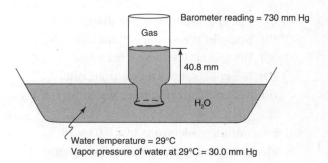

Barometer reading = 730 mm Hg
Gas
40.8 mm
H₂O
Water temperature = 29°C
Vapor pressure of water at 29°C = 30.0 mm Hg

which of the following would give you the pressure of the hydrogen in the bottle?

   (A) 730. mm − 40.8 mm
   (B) 730. mm − 30.0 mm
   (C) 730. mm − 30.0 mm/13.6 + 40.8 mm
   (D) 730. mm − 30.0 mm/13.6 − 40.8 mm
   (E) 730. mm − 40.8 mm/13.6 − 30.0 mm

42. What occurs when a reaction is at equilibrium and more reactant is added to the container?

   (A) The equilibrium remains unchanged.
   (B) The forward reaction rate increases.
   (C) The reverse reaction rate increases.
   (D) The forward reaction rate decreases.
   (E) The reverse reaction rate decreases.

43. How much heat energy is released when 8 grams of hydrogen are burned? The thermal equation is $2H_2(g) + O_2(g) \rightarrow 2H_2O(g) + 483.6$ kJ.

   (A) 241.8 kJ
   (B) 483.6 kJ
   (C) 967.2 kJ
   (D) 1,934 kJ
   (E) 3,869 kJ

GO ON TO THE NEXT PAGE

44. Would the reaction shown below be spontaneous? Why or why not?

$$ZnCl_2(aq) + Au(s) \rightarrow Zn(s) + AuCl_3(aq)$$

(A) Yes, gold is more active than zinc.
(B) No, gold is less active than zinc.
(C) Yes, gold is less active than zinc.
(D) No, gold is more active than zinc.
(E) No, gold and zinc have the same activity.

45. Four moles of electrons ($4 \times 6.02 \times 10^{23}$ electrons) would electroplate how many grams of silver from a silver nitrate solution?

(A) 108
(B) 216
(C) 324
(D) 432
(E) 540

46. A 5.0 M solution of HCl has how many moles of H^+ ion in 1 liter?

(A) 0.50
(B) 1.0
(C) 2.0
(D) 2.5
(E) 5.0

47. What is the K_{sp} for silver acetate if a saturated solution contains 2×10^{-3} moles of silver ion/liter of solution?

(A) 2×10^{-3}
(B) 2×10^{-6}
(C) 4×10^{-3}
(D) 4×10^{-6}
(E) 4×10^{6}

48. The following data were obtained for H_2O and H_2S:

	Formula Mass	Freezing Point (°C)	Boiling Point (°C)
H_2O	18	0	100
H_2S	34	−83	−60

What is the best explanation for the variation of physical properties between these two compounds?

(A) The H_2S has stronger bonds between molecules.
(B) The H_2O has a great deal of hydrogen bonding.
(C) The bond angles differ by about 15°.
(D) The formula mass is of prime importance.
(E) The oxygen atom has a smaller radius and thus cannot bump into other molecules as often as the sulfur.

49. What is the pOH of a solution that has 0.00001 mole of H_3O^+/liter of solution?

(A) 2
(B) 3
(C) 4
(D) 5
(E) 9

50. How many grams of sulfur are present in 1 mole of H_2SO_4?

(A) 2
(B) 32
(C) 49
(D) 64
(E) 98

51. What is the approximate mass, in grams, of 1 liter of nitrous oxide, N_2O, at STP?

(A) 1
(B) 2
(C) 11.2
(D) 22
(E) 44

GO ON TO THE NEXT PAGE

52. If the simplest formula of a substance is CH_2 and its molecular mass is 56, what is its true formula?

    (A) CH_2
    (B) C_2H_4
    (C) C_3H_4
    (D) C_4H_8
    (E) C_5H_{10}

Questions 53 and 54 refer to the following diagrams of two methods of collecting gases:

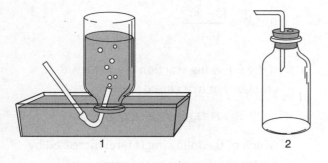

53. Method 1 is best suited to collect

    (A) a gas denser than air
    (B) a gas less dense than air
    (C) a gas that is insoluble in water
    (D) a gas that is soluble in water
    (E) a gas that has a distinct color

54. Which of these gases, because of its density and solubility, should be collected by Method 2?

    (A) NH_3
    (B) H_2
    (C) HCl
    (D) CO_2
    (E) He

55. What is the molar mass of $CaCO_3$?

    (A) 68 g/mol
    (B) 75 g/mol
    (C) 82 g/mol
    (D) 100 g/mol
    (E) 116 g/mol

56. What volume, in liters, will be occupied at STP by 4 grams of H_2?

    (A) 11.2
    (B) 22.4
    (C) 33.6
    (D) 44.8
    (E) 56.0

57. How many moles of KOH are needed to neutralize 196 grams of sulfuric acid? (H_2SO_4 = 98 amu)

    (A) 1.0
    (B) 1.5
    (C) 2.0
    (D) 4.0
    (E) 6.0

58. What volume, in liters, of $NH_3(g)$ is produced when 22.4 liters of $N_2(g)$ are made to combine completely with a sufficient quantity of $H_2(g)$ under appropriate conditions?

    (A) 11.2
    (B) 22.4
    (C) 44.8
    (D) 67.2
    (E) 89.6

59. What volume, in liters, of SO_2 will result from the complete burning of 64 grams of sulfur?

    (A) 2.00
    (B) 11.2
    (C) 44.8
    (D) 126
    (E) 158

GO ON TO THE NEXT PAGE

60. The amount of energy required to melt 5.00 grams of ice at 0°C would also heat 1 gram of water at 4°C to what condition? (Heat of fusion = 80 cal/g or 3.34×10^2 J/g; heat of vaporization = 540 cal/g or 2.26×10^3 J/g)

   (A) water at 90°C
   (B) water at 100°C
   (C) steam at 100°C
   (D) Part of the water would be vaporized to steam.
   (E) All of the water would be vaporized to steam.

61. How many moles of electrons are needed to electroplate a deposit of 0.5 mole of silver from a silver nitrate solution?

   (A) 0.5
   (B) 1
   (C) 27
   (D) 54
   (E) 108

62. All of the following statements about carbon dioxide are true EXCEPT:

   (A) It can be prepared by the action of acid on $CaCO_3$.
   (B) It is used in fire extinguishers.
   (C) It dissolves slightly in water at room temperature.
   (D) It sublimes rather than melts at 20°C and 1 atm pressure.
   (E) It is a product of photosynthesis in plants.

63. Three moles of H_2 and 3 moles of I_2 are introduced into a liter box at a temperature of 490°C. What will the K expression be for this reaction? ($K = 45.9$)

   (A) $K = \dfrac{[H_2][I_2]}{[HI]}$

   (B) $K = \dfrac{[HI]}{[H_2][I_2]}$

   (C) $K = \dfrac{2x}{(x)(x)}$

   (D) $K = \dfrac{(2x)^2}{(3-x)^2}$

   (E) $K = \dfrac{(3-x)^2}{(2x)^2}$

64. If the following reaction has achieved equilibrium in a closed system:

   $$N_2O_4(g) \rightleftharpoons 2\,NO_2(g)$$

   which of the following is (are) increased by decreasing the size of the container?

   I. The value of K
   II. The concentration of $N_2O_4(g)$
   III. The rate of the reverse reaction

   (A) I only
   (B) III only
   (C) I and II only
   (D) II and III only
   (E) I, II, and III

65. Which of the following correctly completes this nuclear reaction: $^{14}_{7}N + {}^{4}_{2}He \rightarrow \cdots + {}^{1}_{1}H$?

   (A) $^{17}_{8}O$
   (B) $^{16}_{9}O$
   (C) $^{17}_{8}N$
   (D) $^{17}_{7}N$
   (E) $^{16}_{8}O$

GO ON TO THE NEXT PAGE

PRACTICE TEST 4

66. How many grams of NaCl will be needed to make 100. milliliters of 2 M solution?

(A) 5.85
(B) 11.7
(C) 29.2
(D) 58.5
(E) 117

67. How many grams of H_2SO_4 are in 1,000. grams of a 10.% solution? (1 mol of H_2SO_4 = 98 g)

(A) 1.0
(B) 9.8
(C) 10.
(D) 98
(E) 100.

68. If 1 mole of ethyl alcohol in 1,000 grams of water depresses the freezing point by 1.86° Celsius, what will be the freezing point of a solution of 1 mole of ethyl alcohol in 500 grams of water?

(A) −0.93°C
(B) −1.86°C
(C) −2.79°C
(D) −3.72°C
(E) −5.58°C

69. Which nuclear reaction shows the release of a beta particle?

(A) $^{235}_{92}U + ^{1}_{0}n \rightarrow ^{93}_{36}Kr + ^{140}_{56}Ba + 3\,^{1}_{0}n$
(B) $^{210}_{84}Po \rightarrow ^{206}_{82}Pb + ^{4}_{2}He$
(C) $^{14}_{6}C \rightarrow ^{14}_{7}N + ^{0}_{-1}e$
(D) $^{106}_{47}Ag + ^{0}_{-1}e \rightarrow ^{106}_{46}Pd$
(E) $^{38}_{19}K \rightarrow ^{38}_{18}Ar + ^{0}_{+1}e$

70. The density of ammonia gas, NH_3, at STP is closest to which value below?

(A) 0.1 g/mL
(B) 1.0 g/mL
(C) 2.0 g/mL
(D) 1.0 g/L
(E) 2.0 g/L

If you finish before one hour is up, you may go back to check your work or complete unanswered questions.

ANSWER KEY
Practice Test 4

1. **A**	7. **E**	13. **C**	19. **E**
2. **C**	8. **C**	14. **A**	20. **C**
3. **E**	9. **B**	15. **A**	21. **A**
4. **A**	10. **A**	16. **B**	22. **D**
5. **B**	11. **E**	17. **C**	23. **B**
6. **C**	12. **A**	18. **D**	

101. **T, F**	105. **T, T, CE**	109. **F, T**	113. **T, T, CE**
102. **T, T, CE**	106. **T, T, CE**	110. **T, T, CE**	114. **F, T**
103. **T, T, CE**	107. **T, T, CE**	111. **F, T**	115. **T, F**
104. **T, F**	108. **F, T**	112. **F, F**	

24. **C**	36. **B**	48. **B**	60. **D**
25. **E**	37. **C**	49. **E**	61. **A**
26. **B**	38. **E**	50. **B**	62. **E**
27. **E**	39. **C**	51. **B**	63. **D**
28. **B**	40. **D**	52. **D**	64. **D**
29. **B**	41. **E**	53. **C**	65. **A**
30. **B**	42. **B**	54. **C**	66. **B**
31. **C**	43. **C**	55. **D**	67. **E**
32. **D**	44. **B**	56. **D**	68. **D**
33. **A**	45. **D**	57. **D**	69. **C**
34. **C**	46. **E**	58. **C**	70. **D**
35. **C**	47. **D**	59. **C**	

ANSWERS EXPLAINED

1. **(A)** The volume of a gas increases as temperature increases provided that pressure remains constant. This is a direct proportion. Heating a balloon is a good example.

2. **(C)** The volume of a gas decreases as the pressure is increased provided that the temperature is held constant. This is shown by the inversely proportional curve in (C). Pressure increase on a closed cylinder is a good example.

3. **(E)** The graph shows that there is a starting quantity in solution, and a slight positive slope to the right indicates a directly proportional change in the solubility as temperature rises.

4. **(A)** This is the definition of any molecule.

5. **(B)** A commercial cake mix is a mixture of ingredients.

6. **(C)** This is the definition of an isotope.

7. **(E)** An acid salt contains one or more H atoms in the salt formula separating a positive ion and the hydrogen-bearing negative ion. For example, Na_2SO_4 is a *normal* salt and $NaHSO_4$ is an *acid* salt because of the presence of H in the hydrogen sulfate ion. In $Ca(HCO_3)_2$, the same is true. This is classified as an acid salt.

8. **(C)** An atom with atomic number 9 would have a 2,7 electron configuration, which matches the outer energy level of iodine.

9. **(B)** There are 2 Na, 1 C, and 3 O, which add to 6 atoms.

10. **(A)** pH from 0 to 6 is acid, 7 neutral, 8 to 14 basic. Most acid is 1.

11. **(E)** Molarity is defined as moles of solute/liter of solution.

12. **(A)** Gas densities can be expressed in grams/liter.

13. **(C)** Molar mass is not affected by pressure and temperature.

14. **(A)** If the density of a gas is known, the mass of 1 L can be multiplied by 22.4 to find the molecular mass because 1 mol occupies 22.4 L at STP.

15. **(A)** Buffers resist changes in pH.

16. **(B)** Color change is the function of indicators.

17. **(C)** On the pH scale, from 0 to 6 is acid and 7 is neutral.

18. **(D)** On the pH scale, from 8 to 14 is basic.

19. **(E)** The ozone (O_3) in the stratosphere absorbs ultraviolet rays from the sun.

20. **(C)** When hydrocarbons containing C and H do not have enough oxygen to combust with $O_2(g)$ completely, the product will be CO, carbon monoxide.

21. **(A)** $2KClO_3 \rightarrow 2KCl + 3O_2(g)$ is the reaction that occurs.

22. **(D)** CO_2 is slightly soluble in water, forming carbonic acid, H_2CO_3, which is a weak acid.

23. **(B)** SO_2 has the highest molar mass.

101. **(T, F)** The assertion is true, but the degree of motion of gas molecules is directly related to the temperature.

102. **(T, T, CE)** Assertion and reason are true; an electron can be treated as either an electromagnetic wave or a bundle of negative charge.

103. **(T, T, CE)** A homologous series increases each member by a constant number of carbons and hydrogens. Examples are the alkane, alkene, and alkyne series, which each increase the chain by a CH_2 group. The reason is true and does explain the assertion.

104. **(T, F)** The nuclear charge of an active metallic ion is greater than that of the electron cloud. The reason is false.

105. **(T, T, CE)** A negative heat of formation indicates that the reaction is exothermic and the enthalpy change is negative.

106. **(T, T, CE)** Water is a polar molecule because there is unequal sharing of bonding electrons.

107. **(T, T, CE)** This is a function of a catalyst—to speed up a reaction without permanently changing itself. Assertion and reason are true.

108. **(F, T)** The Cu is losing electrons and thus being oxidized; the assertion is false. It is furnishing electrons and thus is a reducing agent; the reason is true.

109. **(F, T)** $H_2 = 2$, He = 4 (molecular mass); then inversely $\sqrt{4}:\sqrt{2} = \sqrt{2}:\sqrt{2}$ is the rate of diffusion of hydrogen to helium. The assertion is false; the reason, true.

110. **(T, T, CE)** Since the gas is being heated at constant pressure, it expands. The temperatures are converted to kelvins (K) by adding 273° to the Celsius readings. The fraction must be $\frac{373}{273}$ and this will increase the volume.

111. **(F, T)** Gibbs free energy is useful in indicating the conditions under which a chemical reaction will occur. Therefore, the equation can be used to predict the spontaneity of a reaction. Change in enthalpy and change in entropy are both part of the Gibbs free energy equation: $\Delta G = \Delta H - T\Delta S$.

112. **(F, F)** Water is $\frac{1}{9}$ hydrogen and $\frac{8}{9}$ oxygen by weight. Both assertion and reason are false.

113. **(T, T, CE)** Four grams of ice would require 4×80 cal/g = 320 cal or $4 \times 3.34 \times 10^2$ J/g = 1.34×10^3 J to melt the ice.

114. **(F, T)** The reaction is: $2H_2 + O_2 \rightarrow 2H_2O$. The coefficients of this gaseous reaction show that 2 liters of hydrogen react with 1 liter of oxygen. This would leave 1 liter of unreacted oxygen. The limiting factor is the hydrogen.

115. **(T, F)** The reason why water is a good solvent is false.

24. **(C)** The energy necessary to get the reaction started, which is the activation energy, is shown at *C*.

25. **(E)** $2KClO_3 \rightarrow 2KCl + 3O_2(g)$ shows that 2.00 mol of $KClO_3$ yield 3 mol of O_2.

$$3 \, mol \times \frac{22.4 \, L}{1 \, mol} = 67.2 \, L$$

26. **(B)** A catalyst can speed up a reaction by lowering the activation energy needed to start the reaction and then keep it going.

27. **(E)** The atomic number gives the number of protons in the nucleus and the total number of electrons. The mass number indicates the total number of protons and neutrons in the nucleus—for Na, 23 (11 protons + 12 neutrons).

28. **(B)** Density $= \dfrac{Mass}{Volume}$. For gases this is expressed as grams per liter. Since 1 molar mass of a gas occupies 22.4 L, 17 g/22.4 L = 0.76 g/L.

29. **(B)** Choice III is made up of elements from extreme sides of the Periodic Table and will therefore form ionic bonds.

30. **(B)** Only III is a ring hydrocarbon of the aromatic series.

31. **(C)** Since Li is higher in Group 1 than Na, and K is higher than Cs, they have smaller radii and hence higher ionization energies. Al is to the right of Na and therefore has a higher ionization energy.

32. **(D)** Only II and III are redox reactions. The first reaction is a precipitation reaction. PbS is the precipitate, and the oxidation states of the substances in the reaction are not changing. The second equation is a combustion reaction. The oxidation states of carbon and oxygen change from –4 to +4 and 0 to –2, respectively. The third reaction is a synthesis reaction. The oxidation states of hydrogen and oxygen change from 0 to +1 and 0 to –2, respectively.

33. **(A)**
34. **(C)** These answers are based on the fact that the total of the assigned oxidation numbers times their occurrence for all the atoms in a compound is zero.
35. **(C)**

36. **(B)** This orbital configuration shows 6 electrons in the third energy level. The atom would like to gain $2e^-$ to fill the $3p$ and thereby gain a –2 oxidation number.

37. **(C)** With this structure, the atom would tend to lose these electrons and get a +2 charge.

38. **(E)** The pipette is used to transfer liquid from one container to another.

39. **(C)** The condenser tube is used in distillation.

40. **(D)** The funnel is used to hold the filter paper.

41. **(E)** The pressure in the bottle would be less than atmospheric pressure by the Hg equivalent height of the 30 mm of water above the level in the collecting pan. This is calculated as 40.8 mm water/(13.6 mm water/1 mm Hg) and must be subtracted from atmospheric pressure. The other adjustment is to subtract the vapor pressure of water that is in the hydrogen gas since it was collected over water. This pressure is given as

30.0 mm Hg. Subtracting each of these from 730.mm Hg, the given atmospheric pressure, you have 730. mm − 40.8 mm/13.6 − 30.0 mm.

42. **(B)** The equilibrium shifts in the direction that tends to relieve the stress and thus regain equilibrium.

43. **(C)** The thermal reaction shows 2 mol of hydrogen reacting, or 4 g. Therefore, 8 g would release twice the amount of energy:

$$2 \times 483.6 \text{ kJ} = 967.2 \text{ kJ}.$$

44. **(B)** The reaction would not be spontaneous because gold is less active than zinc. This could be determined by looking at the activity series for metals. Gold is the least active of all the metals and will not replace another metal in a compound spontaneously.

45. **(D)** Since $Ag^+ + 1e^- \rightarrow Ag^0$, 1 mol of electrons yields 1 mol of silver; 1 mol silver = 6.02×10^{23} atoms, and 4×108 g/mol = 432 g

46. **(E)** $5.0 \text{ M} = \dfrac{5.0 \text{ mol}}{\text{L}}$, and since HCl ionizes completely there would be 5.0 mol of H^+ and 5.0 mol of Cl^- ions.

47. **(D)** $K_{sp} = [Ag^+][C_2H_3O_2^-] = [2 \times 10^{-3}][2 \times 10^{-3}]$

(Since

$$AgC_2H_3O_2 \rightleftharpoons Ag^+ + C_2H_3O_2^-$$

the silver ion and acetate ion concentrations are equal.)

$$K_{sp} = 4 \times 10^{-6}$$

48. **(B)** It is the explanation for the observed high boiling point and high freezing point of water compared with hydrogen sulfide.

49. **(E)** $pH = -\log[H^+]$
$= -\log[10^{-5}] = -[-5] = 5$
Since pH + pOH = 14, pOH = 14 − 5 = 9.

50. **(B)** 1 mol H_2SO_4 contains 1 molar mass of sulfur, that is, 32 g.

51. **(B)** N_2O = 44 g/mol

$$(2 \times 14 + 16 = 44)$$

1 mol of a gas occupies 22.4 L, so 44 g/22.4 L = 1.99 g/L.

52. **(D)** CH_2 = 14

$$(12 + 2 = 14 \text{ molecular mass})$$

$$56 \div 14 = 4$$

Then $4 \times CH_2 = C_4H_8$.

53. **(C)** Only insoluble gases can be collected in this way.

54. **(C)** HCl is very soluble in water and denser than air, so it is suited to the No. 2 collection method.

55. **(D)** Ca = 40

   C = 12

   3O = 48

   $\overline{}$

   100 g/mol

56. **(D)** Gram-molecular mass of H_2 is 2 g. 4 g is 2 mol, and each mole occupies 22.4 L. $2 \times 22.4 = 44.8$ L.

57. **(D)** $\overset{x\,mol}{2KOH} + \overset{196\,g}{H_2SO_4} \rightarrow K_2SO_4 + 2H_2O$
   $\;\;\;\;\;$ 2 mol $\;\;$ 98 g

   $$\frac{x\,\text{mol}}{2\,\text{mol}} = \frac{196\,\text{g}}{98\,\text{g}}$$

   Then $x = 4.0$ mol.

58. **(C)** $\overset{22.4\,L}{\underset{1\,L}{N_2(g)}} + 3H_2(g) \rightarrow \overset{x\,L}{\underset{2\,L}{2NH_3(g)}}$

   $$\frac{22.4\,\text{L}}{1\,\text{L}} = \frac{x\,\text{L}}{2\,\text{L}}$$

   Then $x = 44.8$ L.

59. **(C)** $\overset{64\,g}{\underset{32\,g}{S}} + O_2 \rightarrow \overset{x\,L}{\underset{22.4\,L}{SO_2}}$

   $$\frac{64\,\text{g}}{32\,\text{g}} = \frac{x\,\text{L}}{22.4\,\text{L}}$$

   or

   $$64\,\text{g} \times \frac{1\,\cancel{\text{mol S}}}{32\,\text{g}} \times \frac{1\,\text{mol SO}_2}{1\,\cancel{\text{mol S}}} = 2\,\text{mol SO}_2$$

   $$2\,\cancel{\text{mol SO}_2} \times \frac{22.4\,\text{L SO}_2}{1\,\cancel{\text{mol SO}_2}} = 44.8\,\text{L SO}_2$$

   Then $x = 44.8$ L.

60. **(D)** 5.00 g ice to water = 5.00×80 cal = 400 cal. 1 g at 4° can go to 100°C as water and absorb 1 cal/°C. Then 400 cal − (100° − 4°) = 400 − 96 = 304 cal. 304 cal can change $\dfrac{304\,\text{cal}}{540\,\text{cal/g}}$ or 0.56 g of water to steam. There obviously is not enough heat to vaporize all the water. If done in joules, 5.00 g of ice to water = $5.00 \times 3.34 \times 10^2$ g/J = 1.67×10^3 J.

   $$1\,\text{g at 4°C to 100°C} = \Delta\text{temp} = 96°$$

   $$96° \times 4.18\,\text{J/g/°C} = 17.3\,\text{J/g} = 1.73 \times 10^1\,\text{J} = 0.0173 \times 10^3\,\text{J}$$

   1.67×10^3 J − 0.0173×10^3 J = 1.65×10^3 J left for vaporization. Since 1 g requires 2.26×10^3 J for vaporization, (D) is the answer.

61. **(A)** Since Ag^+ gains $1e^-$ to become Ag^0, 0.5 mol requires 0.5 mol of electrons.

62. **(E)** CO_2 is a reactant in photosynthesis, not a product. The reaction is

$$6CO_2 + 6H_2O \rightarrow \underset{\text{simple sugar}}{C_6H_{12}O_6} + 6O_2(g)$$

or

$$6CO_2 + 5H_2O \rightarrow \underset{\text{cellulose}}{C_6H_{12}O_5} + 6O_2(g)$$

63. **(D)** $K = \dfrac{[HI]^2}{[H_2][I_2]}$

Let x = moles of H_2 and also of I_2 that combine to form HI.

Then at equilibrium $[H_2] = 3 - x$, $[I_2] = 3 - x$, $[HI] = 2x$.

$$\text{Then } K = \frac{(2x)^2}{(3-x)(3-x)} \text{ or } = \frac{(2x)^2}{(3-x)^2}$$

64. **(D)** In a closed system, decreasing the size of the container will cause the pressure to increase. When pressure is applied to an equilibrium involving gases, the reaction that lowers the pressure by decreasing the number of molecules will increase in rate. In this reaction, the rate of the reverse reaction, in which 2 molecules are decreased to 1, increases, thus reducing pressure while also increasing the concentration of N_2O_4. Thus, II and III are true.

65. **(A)** This is Rutherford's famous artificial transmutation experiment, done in 1919.

66. **(B)** $2\,M = \dfrac{2\,mol}{1,000\,mL}$

$$2 \text{ mol of NaCl} = 2 \times 58.5\,g$$
$$= 117.0\,g$$

$$2\,M = 117\,g/1,000\,mL,$$

$$\text{so } \frac{117\,g}{1,000\,mL} = \frac{x\,g}{100.\,mL},$$
$$x = 11.7\,g$$

67. **(E)** Percent is by mass, so 10.% is $0.10 \times 1,000.$ g or 100. g.

68. **(D)** Since colligative properties, like freezing point, are related to the concentration of the solute particles and since the decrease in volume of the solution causes the concentration of the solution to be doubled, the colligative effect will be doubled.

69. **(C)** The nuclear reactions shown release: (A) a neutron, (B) an alpha particle, (C) a beta particle, (D) no particles, (E) a positron.

70. **(C)** The density of a gas at STP can be determined by dividing the molar mass of the gas by 22.4 L (the molar volume of a gas at STP). The molar mass of ammonia is 17.0 grams per mole. Divide 17.0 g per mol by 22.4 L per mole to calculate the density of approximately 1 g/L.

CALCULATING YOUR SCORE

Your score on Practice Test 4 can now be computed manually. The actual test will be scored by machine, but the same method is used to arrive at the raw score. You get one point for each correct answer. For each wrong answer, you lose one-fourth of a point. Questions that you omit or that have more than one answer are not counted. On your answer sheet mark all correct answers with a "C" and all incorrect answers with an "X."

Determining Your Raw Test Score

Total the number of correct answers you have recorded on your answer sheet. It should be the same as the total of all the numbers you place in the block in the lower left corner of each area of the Subject Area summary in the next section.

A. Enter the total number of correct answers here: _____

Now count the number of wrong answers you recorded on your answer sheet.

B. Enter the total number of wrong answers here: _____

Multiply the number of wrong answers in B by 0.25.

C. Enter that product here: _____

Subtract the result in C from the total number of right answers in A.

D. Enter the result of your subtraction here: _____

E. Round the result in D to the nearest whole number: _____.

This is your raw test score.

Conversion of Raw Scores to Scaled Scores

Your raw score is converted by the College Board into a scaled score. The College Board scores range from 200 to 800. This conversion is done to ensure that a score earned on any edition of a particular SAT Subject Test in Chemistry is comparable to the same scaled score earned on any other edition of the same test. Because some editions of the tests may be slightly easier or more difficult than others, scaled scores are adjusted so that they indicate the same level of performance regardless of the edition of the test taken and the ability of the group that takes it. Consequently, a specific raw score on one edition of a particular test will not necessarily translate to the same scaled score on another edition of the same test.

Because the practice tests in this book have no large population of scores with which they can be scaled, scaled scores cannot be determined.

Results from previous SAT Chemistry tests appear to indicate that the conversion of raw scores to scaled scores GENERALLY follows this pattern:

Raw Score	Scaled Score	Raw Score	Scaled Score
85–82	800–800	30–25	540–520
81–75	790–760	25–20	520–490
75–70	760–740	20–15	490–460
70–65	740–710	15–10	460–430
65–60	710–690	10–5	430–400
60–55	690–670	5–0	400–370
55–50	670–640	0 to –5	370–340
50–45	640–620	–5 to –10	340–310
45–40	620–590	–10 to –15	310–290
40–35	590–570	–15 to –20	290–270
35–30	570–540	–20 or lower	270–200

Note that this scale provides only a *general idea* of what a raw score may translate into on a scaled score range of 800–200. Scaling on every test is usually slightly different. Some students who had taken the SAT Subject Test in Chemistry after using this book had reported that they have scored slightly higher on the SAT test than on the practice tests in this book. They *all* reported that preparing well for the test paid off in a better score!

DIAGNOSING YOUR NEEDS

After taking Practice Test 4, check your answers against the correct ones. Then fill in the chart below.

In the space under each question number, place a check if you answered that question correctly.

➥ Example _____

If your answer to question 5 was correct, place a check in the appropriate box.

Next, total the check marks for each section and insert the number in the designated block. Now do the arithmetic indicated, and insert your percent for each area.

Subject Area*	(✓) Questions Answered Correctly											
I. Atomic Theory and Structure, including periodic relationships			6	8	102	107	26	31	36	37	65	69
No. of checks ÷ 10 × 100 = _____%												
II. Chemical Bonding and Molecular Structure				4	9	106	29	33	34	35	46	48
No. of checks ÷ 9 × 100 = _____%												
III. States of Matter and Kinetic Molecular Theory of Gases					1	2	14	23	101	109	110	28
No. of checks ÷ 8 × 100 = _____%												
IV. Solutions, including concentration units, solubility, and colligative properties							3	11	66	67	68	
No. of checks ÷ 5 × 100 = _____%												
V. Acids and Bases							7	10	15	16	17	18
No. of checks ÷ 6 × 100 = _____%												
VI. Oxidation-Reduction							108	112	32	44	45	61
No. of checks ÷ 6 × 100 = _____%												
VII. Stoichiometry	113	114	70	25	50	51	52	55	56	57	58	59
No. of checks ÷ 12 × 100 = _____%												
VIII. Reaction Rates											107	27
No. of checks ÷ 2 × 100 = _____%												
IX. Equilibrium								42	47	49	63	64
No. of checks ÷ 5 × 100 = _____%												
X. Thermodynamics: energy changes in chemical reactions, randomness, and criteria for spontaneity								105	111	24	43	60
☐ No. of checks ÷ 6 × 100 = _____%												

The subject areas have been expanded to identify specific areas in the text.

Subject Area*	(✓) Questions Answered Correctly					
XI. Descriptive Chemistry: physical and chemical properties of elements and their familiar compounds; organic chemistry; periodic properties	5	12	13	19	20	21
	22	103	115	30	62	

☐ No. of checks ÷ 11 × 100 = _____%

Subject Area*	(✓) Questions Answered Correctly					
XII. Laboratory: equipment, procedures, observations, safety, calculations, and interpretation of results	38	39	40	41	53	54

☐ No. of checks ÷ 6 × 100 = _____%

The subject areas have been expanded to identify specific areas in the text.

To develop a study plan, refer to pages 31–33.

PRACTICE TEST 4

Appendixes

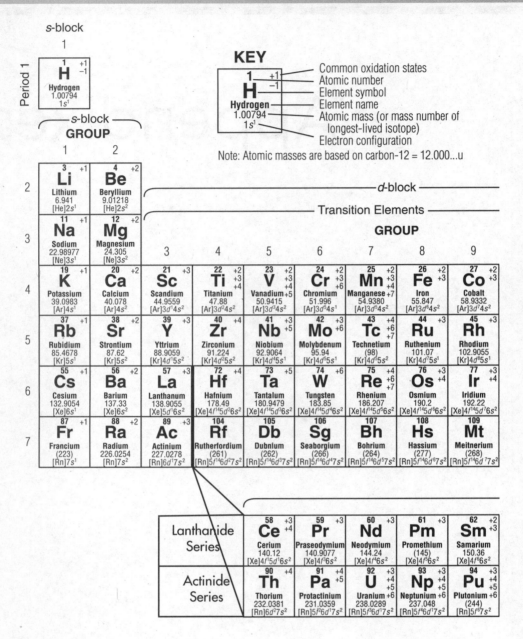

2	0
He	
Helium	
4.00260	
$1s^2$	

p-block
GROUP

13	14	15	16	17	18
5 +3	6 −4	7 −3	8 −2	9 −1	10 0
B	**C** +2	**N** −2	**O**	**F**	**Ne**
	+4	−1			
		+1			
Boron	Carbon	+2	Oxygen	Fluorine	Neon
10.81	12.011	Nitrogen +3	15.9994	18.998403	20.1797
$[He]2s^22p^1$	$[He]2s^22p^2$	14.0067 +4	$[He]2s^22p^4$	$[He]2s^22p^5$	$[He]2s^22p^6$
		$[He]2s^22p^3$ +5			
13 +3	14 −4	15 −3	16 −2	17 +1	18 0
Al	**Si** +2	**P** +3	**S** +4	**Cl** +5	**Ar**
	+4	+5	+6	+3 +7	
Aluminum	Silicon	Phosphorus	Sulfur	Chlorine	Argon
26.98154	28.0855	30.97376	32.066	35.453	39.948
$[Ne]3s^23p^1$	$[Ne]3s^23p^2$	$[Ne]3s^23p^3$	$[Ne]3s^23p^4$	$[Ne]3s^23p^5$	$[Ne]3s^23p^6$

10	11	12					
28 +2	29 +1	30 +2	31 +3	32 −4	33 −3	34 −2	35 −1	36 0
Ni +3	**Cu** +2	**Zn**	**Ga**	**Ge** +2	**As** +3	**Se** +4	**Br** +5	**Kr** +2
				+4	+5	+6		
Nickel	Copper	Zinc	Gallium	Germanium	Arsenic	Selenium	Bromine	Krypton
58.69	63.546	65.39	69.72	72.61	74.9216	78.96	79.904	83.80
$[Ar]3d^84s^2$	$[Ar]3d^{10}4s^1$	$[Ar]3d^{10}4s^2$	$[Ar]3d^{10}4s^24p^1$	$[Ar]3d^{10}4s^24p^2$	$[Ar]3d^{10}4s^24p^3$	$[Ar]3d^{10}4s^24p^4$	$[Ar]3d^{10}4s^24p^5$	$[Ar]3d^{10}4s^24p^6$
46 +2	47 +1	48 +2	49 +3	50 +2	51 −3	52 −2	53 −1	54 0
Pd +4	**Ag**	**Cd**	**In**	**Sn** +4	**Sb** +3	**Te** +4	**I** +5	**Xe** +4
					+5	+6	+7	+6
Palladium	Silver	Cadmium	Indium	Tin	Antimony	Tellurium	Iodine	Xenon
106.42	107.8682	112.41	114.82	118.710	121.757	127.60	126.9045	131.29
$[Kr]4d^{10}$	$[Kr]4d^{10}5s^1$	$[Kr]4d^{10}5s^2$	$[Kr]4d^{10}5s^25p^1$	$[Kr]4d^{10}5s^25p^2$	$[Kr]4d^{10}5s^25p^3$	$[Kr]4d^{10}5s^25p^4$	$[Kr]4d^{10}5s^25p^5$	$[Kr]4d^{10}5s^25p^6$
78 +2	79 +1	80 +1	81 +1	82 +2	83 +3	84 +2	85	86 0
Pt +4	**Au** +3	**Hg** +2	**Tl** +3	**Pb** +4	**Bi** +5	**Po** +4	**At**	**Rn**
Platinum	Gold	Mercury	Thallium	Lead	Bismuth	Polonium	Astatine	Radon
195.08	196.9665	200.59	204.383	207.2	208.9804	(209)	(210)	(222)
$[Xe]4f^{14}5d^96s^1$	$[Xe]4f^{14}5d^{10}6s^1$	$[Xe]4f^{14}5d^{10}6s^2$	$[Xe]4f^{14}5d^{10}6s^26p^1$	$[Xe]4f^{14}5d^{10}6s^26p^2$	$[Xe]4f^{14}5d^{10}6s^26p^3$	$[Xe]4f^{14}5d^{10}6s^26p^4$	$[Xe]4f^{14}5d^{10}6s^26p^5$	$[Xe]4f^{14}5d^{10}6s^26p^6$
110	111	112	113	114	115	116	117	118
Ds	**Rg**	**Cn**	**Nh**	**Fl**	**Mc**	**Lv**	**Ts**	**Og**
Darmstadtium	Roentgenium	Copernicium	Nihonium	Flerovium	Moscovium	Livermorium	Tennessine	Oganesson
(281)	(272)	(285)	(286)	(289)	(289)	(293)	(294)	(294)
$[Rn]5f^{14}6d^87s^2$	$[Rn]5f^{14}6d^97s^2$	$[Rn]5f^{14}6d^{10}7s^2$	$[Rn]5f^{14}6d^{10}7s^27p^1$	$[Rn]5f^{14}6d^{10}7s^27p^2$	$[Rn]5f^{14}6d^{10}7s^27p^3$	$[Rn]5f^{14}6d^{10}7s^27p^4$	$[Rn]5f^{14}6d^{10}7s^27p^5$	$[Rn]5f^{14}6d^{10}7s^27p^6$

f-block

63 +2	64 +3	65 +3	66 +3	67 +3	68 +3	69 +3	70 +2	71 +3
Eu +3	**Gd**	**Tb**	**Dy**	**Ho**	**Er**	**Tm**	**Yb** +3	**Lu**
Europium	Gadolinium	Terbium	Dysprosium	Holmium	Erbium	Thulium	Ytterbium	Lutetium
151.96	157.25	158.9254	162.50	164.9304	167.26	168.9342	173.04	174.967
$[Xe]4f^76s^2$	$[Xe]4f^75d^16s^2$	$[Xe]4f^96s^2$	$[Xe]4f^{10}6s^2$	$[Xe]4f^{11}6s^2$	$[Xe]4f^{12}6s^2$	$[Xe]4f^{13}6s^2$	$[Xe]4f^{14}6s^2$	$[Xe]4f^{14}5d^16s^2$
95 +3	96 +3	97 +3	98 +3	99	100	101	102	103
Am +4	**Cm**	**Bk** +4	**Cf**	**Es**	**Fm**	**Md**	**No**	**Lr**
+5								
Americium +6	Curium	Berkelium	Californium	Einsteinium	Fermium	Mendelevium	Nobelium	Lawrencium
(243)	(247)	(247)	(251)	(252)	(257)	(258)	(259)	(262)
$[Rn]5f^77s^2$	$[Rn]5f^76d^17s^2$	$[Rn]5f^97s^2$	$[Rn]5f^{10}7s^2$	$[Rn]5f^{11}7s^2$	$[Rn]5f^{12}7s^2$	$[Rn]5f^{13}7s^2$	$[Rn]5f^{14}7s^2$	$[Rn]5f^{14}6d^17s^2$

SOME IMPORTANT EQUATIONS

Density (d)

$$\text{Density} = \frac{\text{Mass}}{\text{Volume}}$$

$$d = \frac{m}{v}$$

Percent Error

Percent error

$$= \frac{\text{Measured value} - \text{Accepted value}}{\text{Accepted value}} \times 100\%$$

Percent Yield

$$\text{Percent yield} = \frac{\text{Actual yield}}{\text{Expected yield}} \times 100\%$$

Percentage Composition

Percentage composition by mass

$$= \frac{\text{Mass of element}}{\text{Mass of compound}} \times 100\%$$

Boyle's Law

$$P_1 V_1 = P_2 V_2$$

$$PV = k$$

Charles's Law

$$\frac{V_1}{T_1} = \frac{V_2}{T_2}$$

Dalton's Law of Partial Pressures

$$P_T = p_a + p_b + p_c + \cdots$$

Ideal Gas Law

$$PV = nRT$$

Combined Gas Law

$$\frac{P_1 V_1}{T_1} = \frac{P_2 V_2}{T_2}$$

Graham's Law of Effusion

$$\frac{r_1}{r_2} = \sqrt{\frac{M_2}{M_1}}$$

General Equation for Equilibrium Constant of the Equation

$$aA + bB \rightleftharpoons cC + dD$$

$$K = \frac{[C]^c [D]^d}{[A]^a [B]^b} = \text{equilibrium constant}$$

Equilibrium Constant for Water

$$K_w = [H^+][OH^-] = 1 \times 10^{-14}$$

Molarity (M)

$$\text{Molarity} = \frac{\text{Moles of solute}}{\text{Liters of solution}}$$

Titration

$$M_a V_a = M_b V_b \ (a = \text{acid}, \ b = \text{base})$$

Molality (m)

$$\text{Molality} = \frac{\text{Moles of solute}}{\text{Kilograms of solvent}}$$

Dilution

$$M_c V_c = M_d V_d$$

Rate of Reaction

$$\text{Rate} = k[A]^x [B]^y$$

where [A] and [B] are molar concentrations of reactants, and k is a rate constant

Hess's Law

$$\Delta H_{\text{net}} = \Delta H_1 + \Delta H_2 + \cdots$$

Entropy Change

$$\Delta S = S_{\text{products}} - S_{\text{reactants}}$$

Gibbs Free Energy

$$\Delta G = \Delta H - T\Delta S$$

SOME USEFUL TABLES

A

Standard Energies of Formation of Compounds at 1 atm and 298 K				
	Heat (Enthalpy) of Formation* ($\Delta H_f°$)		Free Energy of Formation ($\Delta G_f°$)	
Compound	kJ/mol	kcal/mol	kJ/mol	kcal/mol
Aluminum oxide Al_2O_3(s)	–1676.5	–400.5	–1583.1	–378.2
Ammonia NH_3(g)	–46.0	–11.0	–16.3	–3.9
Barium sulfate $BaSO_4$(s)	–1473.9	–352.1	–1363.0	–325.6
Calcium hydroxide $Ca(OH)_2$(s)	–986.6	–235.7	–899.2	–214.8
Carbon dioxide CO_2(g)	–393.9	–94.1	–394.7	–94.3
Carbon monoxide CO(g)	–110.5	–26.4	–137.3	–32.8
Copper(II) sulfate $CuSO_4$(s)	–771.9	–184.4	–662.2	–158.2
Ethane C_2H_6(g)	–84.6	–20.2	–33.1	–7.9
Ethene (ethylene) C_2H_4(g)	52.3	12.5	68.2	16.3
Ethyne (acetylene) C_2H_2(g)	226.9	54.2	209.3	50.0
Hydrogen fluoride HF(g)	–271.3	–64.8	–273.3	–65.3
Hydrogen iodide HI(g)	26.4	6.3	1.7	0.4
Iodine chloride ICl(g)	18.0	4.3	–5.4	–1.3
Lead(II) oxide PbO(s)	–215.6	–51.5	–188.4	–45.0
Methane CH_4(g)	–74.9	–17.9	–50.7	–12.1
Magnesium oxide MgO(s)	–601.9	–143.8	–569.7	–136.1
Nitrogen(II) oxide NO(g)	90.4	21.6	86.7	20.7
Nitrogen(IV) oxide NO_2(g)	33.1	7.9	51.5	12.3
Potassium chloride KCl(s)	–437.0	–104.4	–409.4	–97.8
Sodium chloride NaCl(s)	–411.5	–98.3	–384.3	–91.8
Sulfur dioxide SO_2(g)	–296.8	–70.9	–300.1	–71.7
Water H_2O(g)	–242.0	–57.8	–228.6	–54.6
Water H_2O(ℓ)	–285.9	–68.3	–237.3	–56.7

*Minus sign indicates an exothermic reaction.
Sample equations:

$$2Al(s) + \frac{3}{2}O_2(g) \rightarrow Al_2O_3(s) + 400.5 \text{ kcal}$$

$$2Al(s) + \frac{3}{2}O_2(g) \rightarrow Al_2O_3(s) \quad \Delta H = -400.5 \text{ kcal/mol}$$

B

Heats of Reaction at 1 atm and 298 K		
Reaction	ΔH	
	kJ	kcal
$CH_4(g) + 2O_2(g) \rightarrow CO_2(g) + 2H_2O(\ell)$	−890.8	−212.8
$C_3H_8(g) + 5O_2(g) \rightarrow 3CO_2(g) + 4H_2O(\ell)$	−2,221.1	−530.6
$CO_3OH(\ell) + \frac{3}{2}O_2(g) \rightarrow CO_2(g) + 2H_2O(\ell)$	−726.7	−173.6
$C_6H_{12}O_6(s) + 6O_2(g) \rightarrow 6CO_2(g) + 6H_2O(\ell)$	−2,804.2	−669.9
$CO(g) + \frac{1}{2}O_2(g) \rightarrow CO_2(g)$	−283.4	−67.7
$C_8H_{18}(\ell) + \frac{25}{2}O_2(g) \rightarrow 8CO_2(g) + 9H_2O(\ell)$	−5,453.1	−1,302.7
$KNO_3(s) \xrightarrow{H_2O} K^+(aq) + NO_3^-(aq)$	34.7	8.3
$NaOH(s) \xrightarrow{H_2O} Na^+(aq) + OH^-(aq)$	−44.4	−10.6
$NH_4Cl(s) \xrightarrow{H_2O} NH_4^+(aq) + Cl^-(aq)$	14.7	3.5
$NH_4NO_3(s) \xrightarrow{H_2O} NH_4^+(aq) + NO_3^-(aq)$	25.5	6.1
$NaCl(s) \xrightarrow{H_2O} Na^+(aq) + Cl^-(aq)$	3.8	0.9
$KClO_3(s) \xrightarrow{H_2O} K^+(aq) + ClO_3^-(aq)$	41.4	9.9
$LiBr(s) \xrightarrow{H_2O} Li^+(aq) + Br^-(aq)$	−49.0	−11.7
$H^+(aq) + OH^-(aq) \rightarrow H_2O(\ell)$	−57.8	−13.8

C

Symbols Used In Nuclear Chemistry		
alpha particle	^4_2He	α
beta particle (electron)	$^0_{-1}e$	β^-
gamma radiation		γ
neutron	^1_0n	n
proton	^1_1H	p
deuteron	^2_1H	
triton	^3_1H	
positron	$^0_{+1}e$	β^+

D

Physical Constants and Conversion Factors			
Name	Symbol	Value(s)	Units
Avogadro number	N_A	6.02×10^{23} mol	
Charge of electron	e	1.60×10^{-19} C	coulomb
Electron-volt	eV	1.60×10^{-19} J	joule
Speed of light	c	3.00×10^8 m/s	meters/second
Planck's constant	h	6.63×10^{-34} J · s	joule-second
		1.58×10^{-37} kcal · s	kilocalorie-second
Universal gas constant	R	0.0821 L · atm/mol · K	liter-atmosphere/mole-kelvin
Atomic mass unit	μ	1.66×10^{-24} g	gram
Volume standard, liter	L	1×10^3 cm^3 = 1 dm^3	cubic centimeters, cubic decimeter
Standard pressure, atmosphere	atm	101.3 kPa	kilopascals
		760 mm Hg	millimeters of mercury
		760 torr	torr
Heat equivalent, kilocalorie	kcal	4.18×10^3 J	joules

Physical Constants for H_2O

Specific heat	4.18 J/g K
Molal freezing point depression	1.86°C/m
Molal boiling point elevation	0.52°C/m
Heat of fusion	79.72 cal/g, 333.6 J/g
Heat of vaporization	539.4 cal/g, 2,259 J/g

E

Standard (SI) Units		
Symbol	**Name**	**Quantity**
m	meter	length
kg	kilogram	mass
Pa	pascal	pressure
K	kelvin	thermodynamic temperature
mol	mole	amount of substance
J	joule	energy, work, quantity of heat
s	second	time
C	coulomb	quantity of electricity
V	volt	electric potential, potential difference
L	liter	volume

F

Prefixes Used with SI Units			
Prefix	**Symbol**	**Meaning**	**Scientific Notation**
exa-	E	1,000,000,000,000,000,000	10^{18}
peta-	P	1,000,000,000,000,000	10^{15}
tera-	T	1,000,000,000,000	10^{12}
giga-	G	1,000,000,000	10^{9}
mega-	M	1,000,000	10^{6}
kilo-	k	1,000	10^{3}
hecto-	h	100	10^{2}
deka-	da	10	10^{1}
—	—	1	10^{0}
deci-	d	0.1	10^{-1}
centi-	c	0.01	10^{-2}
milli-	m	0.001	10^{-3}
micro-	μ	0.000 001	10^{-6}
nano-	n	0.000 000 001	10^{-9}
pico-	p	0.000 000 000 001	10^{-12}
femto-	f	0.000 000 000 000 001	10^{-15}
atto-	a	0.000 000 000 000 000 001	10^{-18}

THE CHEMICAL ELEMENTS

(Atomic masses in this table are based on the atomic mass of carbon-12 being exactly 12.)

Name	Symbol	Atomic Number	Atomic Mass	Name	Symbol	Atomic Number	Atomic Mass
Actinium	Ac	89	(227)	Hassium	Hs	108	(269)
Aluminum	Al	13	26.98	Helium	He	2	4.00
Americium	Am	95	(243)	Holmium	Ho	67	164.93
Antimony	Sb	51	121.75	Hydrogen	H	1	1.008
Argon	Ar	18	39.95	Indium	In	49	114.82
Arsenic	As	33	74.92	Iodine	I	53	126.90
Astatine	At	85	(210)	Iridium	Ir	77	192.2
Barium	Ba	56	137.34	Iron	Fe	26	55.85
Berkelium	Bk	97	(247)	Krypton	Kr	36	83.80
Beryllium	Be	4	9.01	Lanthanum	La	57	138.91
Bismuth	Bi	83	208.98	Lawrencium	Lr	103	(262)
Bohrium	Bh	107	(264)	Lead	Pb	82	207.19
Boron	B	5	10.81	Lithium	Li	3	6.94
Bromine	Br	35	79.90	Livermorium	Lv	116	(293)
Cadmium	Cd	48	112.40	Lutetium	Lu	71	174.97
Caesium	Cs	55	132.91	Magnesium	Mg	12	24.31
Calcium	Ca	20	40.08	Manganese	Mn	25	54.94
Californium	Cf	98	(251)	Meitnerium	Mt	109	(268)
Carbon	C	6	12.01	Mendelevium	Md	101	(258)
Cerium	Ce	58	140.12	Mercury	Hg	80	200.59
Chlorine	Cl	17	35.45	Molybdenum	Mo	42	95.94
Chromium	Cr	24	52.00	Moscovium	Mc	115	(289)
Cobalt	Co	27	58.93	Neodymium	Nd	60	144.24
Copper	Cu	29	63.55	Neon	Ne	10	20.18
Curium	Cm	96	(247)	Neptunium	Np	93	237.05
Darmstadium	Ds	110	(281)	Nickel	Ni	28	58.71
Dubnium	Db	105	(262)	Nihonium	Nh	113	(286)
Dysprosium	Dy	66	162.50	Niobium	Nb	41	92.90
Einsteinium	Es	99	(254)	Nitrogen	N	7	14.01
Erbium	Er	68	167.26	Nobelium	No	102	(259)
Europium	Eu	63	151.96	Oganesson	Og	118	(294)
Fermium	Fm	100	(257)	Osmium	Os	76	190.2
Flerovium	Fl	114	(289)	Oxygen	O	8	16.00
Fluorine	F	9	19.00	Palladium	Pd	46	106.4
Francium	Fr	87	(223)	Phosphorus	P	15	30.97
Gadolinium	Gd	64	157.25	Platinum	Pt	78	195.09
Gallium	Ga	31	69.72	Plutonium	Pu	94	(244)
Germanium	Ge	32	72.59	Polonium	Po	84	(210)
Gold	Au	79	196.97	Potassium	K	19	39.10
Hafnium	Hf	72	178.49	Praseodymium	Pr	59	140.90

A number in parentheses is the mass number of the most stable isotope.

Name	Symbol	Atomic Number	Atomic Mass
Promethium	Pm	61	(145)
Protactinium	Pa	91	231.04
Radium	Ra	88	(226)
Radon	Rn	86	(222)
Rhenium	Re	75	186.2
Rhodium	Rh	45	102.91
Roentgenium	Rg	111	(280)
Rubidium	Rb	37	85.47
Ruthenium	Ru	44	101.07
Rutherfordium	Rf	104	(261)
Samarium	Sm	62	150.35
Scandium	Sc	21	44.95
Seaborgium	Sg	106	(266)
Selenium	Se	34	78.96
Silicon	Si	14	28.09
Silver	Ag	47	107.89
Sodium	Na	11	22.99
Strontium	Sr	38	87.62
Sulfur	S	16	32.06

Name	Symbol	Atomic Number	Atomic Mass
Tantalum	Ta	73	180.95
Technetium	Tc	43	(99)
Tellurium	Te	52	127.60
Tennessine	Ts	117	(294)
Terbium	Tb	65	158.92
Thallium	Tl	81	204.37
Thorium	Th	90	232.03
Thulium	Tm	69	168.93
Tin	Sn	50	118.69
Titanium	Ti	22	47.90
Tungsten	W	74	183.85
Uranium	U	92	238.03
Vanadium	V	23	50.94
Xenon	Xe	54	131.30
Ytterbium	Yb	70	173.04
Yttrium	Y	39	88.91
Zinc	Zn	30	65.37
Zirconium	Zr	40	91.22

Glossary

(See Index for additional references)

absolute temperature Temperature measured on the absolute scale, which has its origin at absolute zero. *See also* **Kelvin scale**.

absorption The process of taking up by capillary, osmotic, chemical, or solvent action, as a sponge absorbs water.

acid A water solution that has an excess of hydrogen ions; an acid turns litmus paper pink or red, has a sour taste, and neutralizes bases to form salts.

acidic anhydride A nonmetallic oxide that, when placed in water, reacts to form an acid solution.

acid salt A salt formed by replacing part of the hydrogen ions of a dibasic or tribasic acid with metallic ions. Examples: $NaHSO_4$, NaH_2PO_4.

actinide series The series of radioactive elements starting with actinium, No. 89, and ending with lawrencium, No. 103.

activated charcoal A specially treated and finely divided form of carbon, which possesses a high degree of adsorption.

activation energy The minimum energy necessary to start a reaction.

adsorption The adhesion (in an extremely thin layer) of the molecules of gases, of dissolved substances, or of liquids to the surfaces of solid or liquid bodies with which they come into contact.

alcohol An organic hydroxyl compound formed by replacing one or more hydrogen atoms of a hydrocarbon with an equal number of hydroxyl (OH) groups.

aldehyde An organic compound formed by dehydrating oxidized alcohol; contains the characteristic –CHO group.

alkali Usually, a strong base, such as sodium hydroxide or potassium hydroxide.

alkaline Referring to any substance that has basic properties.

alkyl A substitutent obtained from a saturated hydrocarbon by removing one hydrogen atom. Examples: methyl (CH_3^-), ethyl ($C_2H_5^-$).

allotropic forms Forms of the same element that differ in their crystalline structures.

alloy A substance composed of two or more metals, which are intimately mixed; usually made by melting the metals together.

alpha particles Positively charged helium nuclei.

amine A compound such as CH_3NH_2, derived from ammonia by substituting one or more hydrocarbon radicals for hydrogen atoms.

amino acid One of the "building blocks" of proteins; contains one or more NH_2^- groups that have replaced the same number of hydrogen atoms in an organic acid.

amorphous Having no definite crystalline structure.

amphoteric Referring to a hydroxide that may have either acidic or basic properties, depending on the substance with which it reacts.

analysis The breaking down of a compound into two or more simpler substances.

anhydride A compound derived from another compound by the removal of water; it will combine with water to form an acid (acidic anhydride) or a base (basic anhydride).

anhydrous Containing no water.

anion An ion or particle that has a negative charge and thus is attracted to a positively charged anode.

anode The electrode in an electrolytic cell that has a positive charge and attracts negative ions.

antichlor A substance used to remove the excess of chlorine in the bleaching process.

aromatic compound A compound whose basic structure contains the benzene ring; it usually has an odor.

atmosphere The layer of gases surrounding Earth; also, a unit of pressure (1 atm = approx. 760 mm of Hg or torr).

atom The smallest particle of an element that retains the properties of that element and can enter into a chemical reaction.

atomic energy *See* **nuclear energy**, a more accurate term.

atomic mass (relative atomic mass or atomic weight) The average mean value of the isotopic masses of the atoms of an element. It indicates the relative mass of the element as compared with that of carbon-12, which is assigned a mass of exactly 12 atomic mass units.

atomic mass unit One twelfth of the mass of a carbon-12 atom; equivalent to 1.660531×10^{-27} kilogram (abbreviation: amu or μ).

atomic number The number that indicates the order of an element in the periodic system; numerically equal to the number of protons in the nucleus of the atom, or the number of negative electrons located outside the nucleus of the atom.

atomic radius One-half the distance between adjacent nuclei in the crystalline or solid phase of an element; the distance from the atomic nucleus to the valence electrons.

atomic weight *See* **atomic mass**.

Aufbau Principle The principle that states that an electron occupies the lowest energy orbital that can receive it.

Avogadro's hypothesis *See under* **laws**.

Avogadro's number The number of molecules in 1 gram-molecular volume of a substance, or the number of atoms in 1 gram-atomic mass of an element; equal to 6.022169×10^{23}. *See also* **mole**.

barometer An instrument, invented by Torricelli in 1643, used for measuring atmospheric pressure.

base A water solution that contains an excess of hydroxide ions; a proton acceptor; a base turns litmus paper blue and neutralizes acids to form salts.

basic anhydride A metallic oxide that forms a base when placed in water.

beta particles High-speed, negatively charged electrons $_{-1}^{0}e$ or $_{-1}^{0}\beta$ emitted in radiation.

binary Referring to a compound composed of two elements, such as H_2O.

boiling point The temperature at which the vapor pressure of a liquid equals the atmospheric pressure.

bond energy The energy needed to break a chemical bond and form a neutral atom.

bonding The union of atoms to form compounds or molecules by filling their outer shells of electrons. This can be done through giving and taking electrons (ionic) or by sharing electrons (covalent).

Boyle's Law *See under* **laws**.

brass An alloy of copper and zinc.

breeder reactor A nuclear reactor in which more fissionable material is produced than is used up during operation.

Brownian movement Continuous zigzagging movement of colloidal particles in a dispersing medium, as viewed through an ultramicroscope.

buffer A substance that, when added to a solution, makes changing the pH of the solution more difficult.

calorie A unit of heat; the amount of heat needed to raise the temperature of 1 gram of water 1 degree on the Celsius scale.

calorimeter An instrument used to measure the amount of heat liberated or absorbed during a change.

carbonated water Water containing dissolved carbon dioxide.

carbon dating The use of radioactive carbon-14 to estimate the ages of ancient materials, such as archeological or paleontological specimens.

catalyst A substance that speeds up or slows down a reaction without being permanently changed itself.

cathode The electrode in an electrolytic cell that is negatively charged and attracts positive ions.

cathode rays Streams of electrons given off by the cathode of a vacuum tube.

cation An ion that has a positive charge.

Celsius scale A temperature scale divided into 100 equal divisions and based on water freezing at 0° and boiling at 100°. Synonymous with centigrade.

chain reaction A reaction produced during nuclear fission when at least one neutron from each fission produces another fission, so that the process becomes self-sustaining without additional external energy.

Charles's Law *See under* **laws**.

chemical change A change that alters the atomic structures of the substances involved and results in different properties.

chemical property A property that determines how a substance will behave in a chemical reaction.

chemistry The science concerned with the compositions of substances and the changes that they undergo.

colligative property A property of a solution that depends primarily on the concentration, not the type, of particles present.

colloids Particles larger than those found in a solution but smaller than those in a suspension.

Combining Volumes *See Gay-Lussac's, under* **laws**.

combustion A chemical action in which both heat and light are given off.

compound A substance composed of elements chemically united in definite proportions by weight.

condensation (a) A change from gaseous to liquid state; (b) the union of like or unlike molecules with the elimination of water, hydrogen chloride, or alcohol.

Conservation of Energy *See under* **laws**.

Conservation of Matter *See under* **laws**.

control rod In a nuclear reactor, a rod of a certain metal such as cadmium, which controls the speed of the chain reaction by absorbing neutrons.

coordinate covalence Covalence in which both electrons in a pair come from the same atom.

covalent bonding Bonding accomplished through the sharing of electrons so that atoms can fill their outer shells.

critical mass The smallest amount of fissionable material that will sustain a chain reaction.

critical temperature The temperature above which no gas can be liquefied, regardless of the pressure applied.

crystalline Having a definite molecular or ionic structure.

crystallization The process of forming definitely shaped crystals when water is evaporated from a solution of the substance.

cyclotron A device used to accelerate charged particles to high energies for bombarding the nuclei of atoms.

Dalton's Law of Partial Pressures *See under* **laws**.

decomposition The breaking down of a compound into simpler substances or into its constituent elements.

Definite Composition *See under* **laws**.

dehydrate To take water from a substance.

dehydrating agent A substance able to withdraw water from another substance, thereby drying it.

deliquescence The absorption by a substance of water from the air, so that the substance becomes wet.

denatured alcohol Ethyl alcohol that has been "poisoned" in order to produce (by avoiding federal tax) a cheaper alcohol for industrial purposes.

density The mass per unit volume of a substance; the mathematical formula is $D = m/V$, where $D =$ density, $m =$ mass, and $V =$ volume.

destructive distillation The process of heating an organic substance, such as coal, in the absence of air to break it down into solid and volatile products.

deuterium An isotope of hydrogen, sometimes called heavy hydrogen, with an atomic weight of 2.

dew point The highest temperature at which water vapor condenses out of the air.

dialysis The process of separation of a solution by diffusion through a semipermeable membrane.

diffusion The process whereby gases or liquids intermingle freely of their own accord.

dipole–dipole attraction A relatively weak force of attraction between polar molecules; a component of van der Waals forces.

displacement A change by which an element takes the place of another element in a compound.

dissociation (ionic) The separation of the ions of an ionic compound due to the action of a solvent.

distillation The process of first vaporizing a liquid and then condensing the vapor back into a liquid, leaving behind the nonvolatile impurities.

double bond A bond between atoms involving two electron pairs. In organic chemistry: unsaturated.

double displacement A reaction in which two chemical substances exchange ions with the formation of two new compounds.

dry ice Solid carbon dioxide.

ductile Capable of being drawn into thin wire.

effervescence The rapid escape of excess gas that has been dissolved in a liquid.

efflorescence The loss by a substance of its water of hydration on exposure to air at ordinary temperatures.

effusion The flow of a gas through a small aperture.

Einstein equation $E = mc^2$, which relates mass to energy; E = energy in ergs, m = mass in grams, and c = velocity of light, 3×10^{10} centimeters/second.

electrode A terminal of an electrolytic cell.

electrode potential The difference in potential between an electrode and the solution in which it is immersed.

electrolysis The process of separating the ions in a compound by means of electrically charged poles.

electrolyte A liquid that will conduct an electric current.

electrolytic cell A cell in which electrolysis is carried out.

electron A negatively charged particle found outside the nucleus of the atom; it has a mass of 9.109×10^{-28} gram.

electron dot symbol *See* **Lewis dot symbol**.

electronegativity The numerical expression of the relative strength with which the atoms of an element attract valence electrons to themselves; the higher the number, the greater the attraction.

electron volt A unit for expressing the kinetic energy of subatomic particles; the energy acquired by an electron when it is accelerated by a potential difference of 1 volt; equals 1.6×10^{-12} erg or 23.1 kilocalories/mole (abbreviation: eV).

electroplating Depositing a thin layer of (usually) a metallic element on the surface of another metal by electrolysis.

element One of the more than 100 "building blocks" of which all matter is composed. An element consists of atoms of only one kind and cannot be decomposed further by ordinary chemical means.

empirical formula A formula that shows only the simplest ratio of the numbers and kinds of atoms, such as CH_4.

emulsifying agent A colloidal substance that forms a film about the particles of two immiscible liquids, so that one liquid remains suspended in the other.

emulsion A suspension of fine particles or droplets of one liquid in another, the two liquids being immiscible in each other; the droplets are surrounded by a colloidal (emulsifying) agent.

endothermic Referring to a chemical reaction that results in an overall absorption of heat from its surroundings.

energy The capacity to do work. In every chemical change energy is either given off or taken in. Forms of energy are heat, light, motion, sound, and electrical, chemical, and nuclear energy.

enthalpy The heat content of a chemical system.

entropy The measure of the randomness or disorder that exists in a system.

equation A shorthand method of showing the changes that take place in a chemical reaction.

equilibrium The point in a reversible reaction at which the forward reaction is occurring at the same rate as the opposing reaction.

erg A unit of energy or work done by a force of 1 dyne (1/980 g of force) acting through a distance of 1 centimeter; equals 2.4×10^{-11} kilocalorie.

ester An organic salt formed by the reaction of an alcohol with an organic (or inorganic) acid.

esterification A chemical reaction between an alcohol and an acid, in which an ester is formed.

ether An organic compound containing the –O– group.

eudiometer A graduated glass tube into which gases are placed and subjected to an electric spark; used to measure the individual volumes of combining gases.

evaporation The process in which molecules of a liquid (or a solid) leave the surface in the form of vapor.

exothermic Referring to a chemical reaction that results in the giving off of heat to its surroundings.

Fahrenheit scale The temperature scale that has 32° as the freezing point of water and 212° as the boiling point.

fallout The residual radioactivity from an atmospheric nuclear test, which eventually settles on the surface of Earth.

Faraday's Law *See under* **laws**.

filtration The process by which suspended matter is removed from a liquid by passing the liquid through a porous material.

First Law of Thermodynamics *See under* **laws**.

fission A nuclear reaction that releases energy because of the splitting of large nuclei into smaller ones.

fixation of nitrogen Any process for converting atmospheric nitrogen into compounds, such as ammonia and nitric acid.

flame The glowing mass of gas and luminous particles produced by the burning of a gaseous substance.

flammable Capable of being easily set on fire; combustible (same as *inflammable*).

fluorescence Emission by a substance of electromagnetic radiation, usually visible, as the immediate result of (and only during) absorption of energy from another source.

fluoridation Addition of small amounts of fluoride (usually NaF) to drinking water to help prevent tooth decay.

flux In metallurgy: a substance that helps to melt and remove the solid impurities as slag. In soldering: a substance that cleans the surface of the metal to be soldered. In nucleonics: the concentration of nuclear particles or rays.

formula An expression that uses the symbols for elements and subscripts to show the basic makeup of a substance.

formula mass The sum of the atomic mass units of all the atoms (or ions) contained in a formula.

fractional crystallization The separation of the components in a mixture of dissolved solids by evaporation according to individual solubilities.

fractional distillation The separation of the components in a mixture of liquids having different boiling points by vaporization.

free energy *See* **Gibbs free energy**.

freezing point The specific temperature at which a given liquid and its solid form are in equilibrium.

fuel Any substance used to furnish heat by combustion. *See also* **nuclear fuel**.

fuel cell A device for converting an ordinary fuel such as hydrogen or methane directly into electricity.

functional group A group of atoms that characterizes certain types of organic compounds, such as –OH for alcohols, and that reacts more or less independently.

fusion A nuclear reaction that releases energy because of the union of smaller nuclei to form larger ones.

fusion melting Changing a solid to the liquid state by heating.

galvanizing Applying a coating of zinc to iron or steel to protect the latter from rusting.

gamma rays A type of radiation consisting of high-energy waves that can pass through most materials. Symbol: γ

gas A phase of matter that has neither definite shape nor definite volume.

Gay-Lussac's Law *See under* **laws**.

Gibbs free energy Changes in Gibbs free energy, ΔG, are useful in indicating the conditions under which a chemical reaction will occur. The equation is $\Delta G = \Delta H - T\Delta S$, where ΔH = change in enthalpy and ΔS = change in entropy. If ΔG is negative, the reaction will proceed spontaneously to equilibrium.

glass An amorphous, usually translucent substance consisting of a mixture of silicates. Ordinary glass is made by fusing together silica and sodium carbonate and lime; the various forms of glass contain many other silicates.

Graham's Law *See under* **laws**.

gram A unit of weight in the metric system; the weight of 1 milliliter of water at 4°C (abbreviation: g).

gram-atomic mass The atomic mass, in grams, of an element.

gram-formula weight The formula weight, in grams, of a substance.

group A vertical column of elements in the periodic table that generally has similar properties.

half-life The time required for half of the mass of a radioactive substance to disintegrate.

half-reaction One of the two parts, either the reduction part or the oxidation part, of a redox reaction.

halogen Any of the five nonmetallic elements (fluorine, chlorine, bromine, iodine, astatine) that form part of Group 17 of the Periodic Table.

heat A form of molecular energy; it passes from a warmer body to a cooler one.

heat capacity (specific heat) The quantity of heat, in calories, needed to raise the temperature of 1 gram of a substance 1 degree on the Celsius scale.

heat of formation The quantity of heat either given off or absorbed in the formation of 1 mole of a substance from its elements.

heat of fusion The amount of heat, in calories, required to melt 1 gram of a solid; for water, 80 calories.

heat of vaporization The quantity of heat needed to vaporize 1 gram of a liquid at constant temperature and pressure; for water at 100°C, 540 calories.

heavy water (deuterium oxide, D_2O) Water in which the hydrogen atoms are replaced by atoms of the isotope of hydrogen, deuterium.

Henry's Law *See under* **laws**.

Hess's Law *See under* **laws**.

homogeneous Uniform; having every portion exactly like every other portion.

homologous Alike in structure; referring to series of organic compounds, such as hydrocarbons, in which each member differs from the next by the addition of the same group.

humidity The amount of moisture in the air.

hybridization The combination of two or more orbitals to form new orbitals.

hydrate A compound that has water molecules included in its crystalline makeup.

hydride Any binary compound containing hydrogen, such as HCl.

hydrogenation A process in which hydrogen is made to combine with another substance, usually organic, in the presence of a catalyst.

hydrogen bond A weak chemical linkage between the hydrogen of one polar molecule and the oppositely charged portion of a closely adjacent molecule.

hydrolysis Of carbohydrates: the action of water in the presence of a catalyst upon one carbohydrate to form simpler carbohydrates. Of salts: a reaction involving the splitting of water into its ions by the formation of a weak acid, a weak base, or both.

hydronium ion A hydrated ion, $H_2O \cdot H^+$ or H_3O^+.

hydroponics Growing plants without the use of soil, as in nutrient solution or in sand irrigated with nutrient solution.

hydroxyl Referring to the –OH radical.

hygroscopic Referring to the ability of a substance to draw water vapor from the atmosphere to itself and become wet.

hypothesis A possible explanation of the nature of an action or phenomenon; a hypothesis is not as completely developed as a theory.

Ideal Gas Law *See under* **laws**.

immiscible Referring to the inability of two liquids to mix.

indicator A dye that shows one color in the presence of the hydrogen ion (acid) and a different color in the presence of the hydroxyl ion (base).

inertia The property of matter whereby it remains at rest or, if in motion, remains in motion in a straight line unless acted upon by an outside force.

ion An atom or a group of combined atoms that carries one or more electric charges. Examples: NH_4^+, OH^-.

ionic bonding The bonding of ions due to their opposite charges.

ionic equation An equation showing a reaction among ions.

ionization The process in which ions are formed from neutral atoms.

ionization equation An equation showing the ions set free from an electrolyte.

isomerization The rearrangement of atoms in a molecule to form isomers.

isomers Two or more compounds having the same percentage composition but different arrangements of atoms in their molecules and hence different properties.

isotopes Two or more forms of an element that differ only in the number of neutrons in the nucleus and hence in their mass numbers.

IUPAC International Union of Pure and Applied Chemistry, an organization that establishes standard rules for naming compounds.

joule The SI unit of work or of energy equal to work done; 1 joule = 0.2388 calorie; 1 calorie = 4.18 joule.

Kelvin scale A temperature scale based on water freezing at 273 and boiling at 373 Kelvin units; its origin is absolute zero. Synonymous with *absolute scale*.

kernel (atomic) The nucleus and all the electron shells of an atom except the outer one; usually designated by the symbol for the atom.

ketone An organic compound containing the –CO– group.

kilocalorie A unit of heat; the amount of heat needed to raise the temperature of 1 kilogram of water 1 degree on the Celsius scale.

kindling temperature The temperature to which a given substance must be raised before it ignites.

Kinetic-Molecular Theory The theory that all molecules are in motion; this motion is most rapid in gases, less rapid in liquids, and very slow in solids.

lanthanide series The "rare earth" series of elements starting with lanthanum, No. 57, and ending with lutetium, No. 71.

law (in science) A generalized statement about the uniform behavior in natural processes.

laws

> *Avogadro's* Equal volumes of gases under identical conditions of temperature and pressure contain equal numbers of particles (atoms, molecules, ions, or electrons).
>
> *Boyle's* The volume of a confined gas is inversely proportional to the pressure to which it is subjected, provided that the temperature remains the same.
>
> *Charles's* The volume of a confined gas is directly proportional to the absolute temperature, provided that the pressure remains the same.
>
> *Combining Volumes* See *Gay-Lussac's* under **laws**.
>
> *Conservation of Energy* Energy can be neither created nor destroyed, so that the energy of the universe is constant.
>
> *Conservation of Matter* Matter can be neither created nor destroyed (or weight remains constant in an ordinary chemical change).
>
> *Dalton's* When a gas is made up of a mixture of different gases, the pressure of the mixture is equal to the sum of the partial pressures of the components.
>
> *Definite Composition* A compound is composed of two or more elements chemically combined in a definite ratio by weight.
>
> *Faraday's* During electrolysis, the weight of any element liberated is proportional (1) to the quantity of electricity passing through the cell, and (2) to the equivalent weight of the element.
>
> *First Law of Thermodynamics* The total energy of the universe is constant and cannot be created or destroyed.
>
> *Gay-Lussac's* The ratio between the combining volumes of gases and the product, if gaseous, can be expressed in small whole numbers.
>
> *Graham's* The rate of diffusion (or effusion) of a gas is inversely proportional to the square root of its molecular mass.
>
> *Henry's* The solubility of a gas (unless the gas is very soluble) is directly proportional to the pressure applied to the gas.

Hess's If a series of reactions are added together, the enthalpy change for the total reaction is the sum of the enthalpy changes for the individual steps.

Ideal Gas Any gas that obeys the gas laws perfectly. No such gas actually exists.

Multiple Proportions When any two elements, A and B, combine to form more than one compound, the different masses of B that unite with a fixed mass of A bear a small whole-number ratio to each other.

Periodic The chemical properties of elements vary periodically with their atomic numbers.

Second Law of Thermodynamics Heat cannot, of itself, pass from a cold body to a hot body.

Le Châtelier's Principle If a stress is placed on a system in equilibrium, the system will react in the direction that relieves the stress.

lepton An elementary particle; the electron and neutrino are believed to consist of leptons.

Lewis dot symbol The chemical symbol (kernel) for an atom, surrounded by dots to represent its outer level electrons. Examples: K·, Sr:.

liquid A phase of matter that has a definite volume but takes the shape of the container.

liquid air Air that has been cooled and compressed until it liquefies.

litmus An organic substance, obtained from the lichen plant and used as an indicator; it turns red in acidic solution and blue in basic solution.

London force The weakest of the van der Waals forces between molecules. These weak, attractive forces become apparent only when the molecules approach one another closely (usually at low temperatures and high pressure). They are due to the way the positive charges of one molecule attract the negative charges of another molecule because of the charge distribution at any one instant.

luminous Emitting a steady, suffused light.

malleable Capable of being hammered or pounded into thin sheets.

manometer A U-tube (containing mercury or some other liquid) used to measure the pressure of a confined gas.

mass The quantity of matter that a substance possesses; it can be measured by its resistance to a change in position or motion, and is not related to the force of gravity.

mass number The nearest whole number to the combined atomic mass of the individual atoms of an isotope when that mass is expressed in atomic mass units.

mass spectograph A device for determining the masses of electrically charged particles by separating them into distinct streams by means of magnetic deflection.

matter A substance that occupies space, has mass, and cannot be created or destroyed easily.

melting The change in phase of a substance from solid to liquid.

melting point The specific temperature at which a given solid changes to a liquid.

meson Any unstable, elementary nuclear particle having a mass between that of an electron and that of a proton.

metal (a) An element whose oxide combines with water to form a base; (b) an element that readily loses electrons and acquires a positive valence.

metallurgy The process involved in obtaining a metal from its ores.

meter The basic unit of length in the metric system; defined as 1,650,763.73 times the wavelength of krypton-86 when excited to give off an orange-red spectral line.

MeV A unit for expressing the kinetic energy of subatomic particles; equals 10^6 electron volts.

micron One thousandth of a millimeter (abbreviation: μ).

mineral An inorganic substance of definite composition found in nature.

miscible Referring to the ability of two liquids to mix with one another.

mixture A substance composed of two or more components, each of which retains its own properties.

moderator A substance such as graphite, paraffin, or heavy water used in a nuclear reactor to slow down neutrons.

molal solution A solution containing 1 mole of solute in 1,000 grams of solvent (indicated by m).

molar mass The mass arrived at by the addition of the atomic masses of the units that make up a molecule of an element or compound. Expressed in grams/mole, the molar mass of a gaseous substance at STP occupies a molar volume equal to 22.4 liters.

molar solution A solution containing 1 mole of solute in 1,000 milliliters of solution (indicated by M).

mole A unit of quantity that consists of 6.02×10^{23} particles.

molecular mass The sum of the masses of all the atoms in a molecule of a substance.

molecular theory *See* **Kinetic-Molecular Theory**.

molecule The smallest particle of a substance that retains the physical and chemical properties of that substance.

Example: He, Br_2, H_2O

monobasic acid An acid having only one hydrogen atom that can be replaced by a metal or a positive radical.

mordant A chemical, such as aluminum sulfate, used for fixing colors on textiles.

multiple proportions *See under* **laws**.

nascent (atomic) Referring to an element in the atomic form as it has just been liberated in a chemical reaction.

neutralization The union of the hydrogen ion of an acid and the hydroxyl ion of a base to form water.

neutron A subatomic particle found in the nucleus of the atom; it has no charge and has the same mass as the proton.

neutron capture A nuclear reaction in which a neutron attaches itself to a nucleus; a gamma ray is usually emitted simultaneously.

nitriding A process in which ammonia or a cyanide is used to produce case-hardened steel; a nitride is formed instead of a carbide.

nitrogen fixation Any process by which atmospheric nitrogen is converted into a compound such as ammonia or nitric acid.

noble gas A gaseous element that has a complete outer level of electrons; any of a group of rare gases (helium, neon, argon, krypton, xenon, and radon) that exhibit great stability and very low reaction rates.

noble gas structure The outer energy level electron configuration characteristic of the inert gases—two electrons for helium; eight electrons for all others.

nonelectrolyte A substance whose solution does not conduct a current of electricity.

nonmetal (a) An element whose oxide reacts with water to form an acid; (b) an element that takes on electrons and acquires a negative valence.

nonpolar compound A compound in whose molecules the atoms are arranged symmetrically so that the electric charges are uniformly distributed.

normal salt A salt in which all the hydrogen of the acid has been displaced by a metal.

normal solution A solution that contains 1 gram of H^+ (or its equivalent: 17 g of OH^-, 23 g of Na^+, 20 g of Ca^{2+}, etc.) in 1 liter of solution (indicated by N).

nuclear energy The energy released by spontaneously or artificially produced fission, fusion, or disintegration of the nuclei of atoms.

nuclear fuel A substance that is consumed during nuclear fission or fusion.

nuclear reaction Any reaction involving a change in nuclear structure.

nuclear reactor A device in which a controlled chain reaction of fissionable material can be produced.

nucleonics The science that deals with the constituents and all the changes in the atomic nucleus.

nucleus The center of the atom, which contains protons and neutrons.

nuclide A species of atom characterized by the constitution of its nucleus.

orbital A subdivision of a nuclear shell; it may contain none, one, or two electrons.

ore A natural mineral substance from which an element, usually a metal, may be obtained with profit.

organic acid An organic compound that contains the –COOH group.

organic chemistry The branch of chemistry dealing with carbon compounds, usually those found in nature.

oxidation The chemical process by which oxygen is attached to a substance; the process of losing electrons.

oxidation number (state) A positive or negative number representing the charge that an ion has or an atom appears to have when its electrons are counted according to arbitrarily accepted rules: (1) electrons shared by two unlike atoms are counted with the more electronegative atom; (2) electrons shared by two like atoms are divided equally between the atoms.

oxidation potential An electrode potential associated with the oxidation half-reaction.

oxidizing agent A substance that (a) gives up its oxygen readily, (b) removes hydrogen from a compound, (c) takes electrons from an element.

ozone An allotropic and very active form of oxygen, having the formula O_3.

paraffin series The methane series of hydrocarbons.

pascal The SI unit of pressure, equal to 1 newton per square meter.

pasteurization Partial sterilization of a substance, such as milk, by heating to approximately 65°C for ½ hour.

Pauli Exclusion Principle Each electron orbital of an atom can be filled by only two electrons, which have opposite spins.

period A horizontal row of elements in the Periodic Table.

Periodic Law *See under* **laws.**

petroleum (meaning "oil from stone") A complex mixture of gaseous, liquid, and solid hydrocarbons obtained from Earth.

pH A numerical expression of the hydrogen or hydronium ion concentration in a solution; defined as $-\log [H^+]$, where $[H^+]$ is the concentration of hydrogen ions, in moles per liter.

phenolphthalein An organic indicator; it is colorless in acid solution and red in the presence of OH^- ions.

photosynthesis The reaction taking place in all green plants that produces glucose from carbon dioxide and water under the catalytic action of chlorophyll in the presence of light.

physical change A change that does not involve any alteration in chemical composition.

physical property A property of a substance arrived at through observation of its smell, taste, color, density, and so on, which does not relate to chemical activity.

pi bond A bond between *p* orbitals.

pile A general term for a nuclear reactor; specifically, a graphite-moderated reactor in which uranium fuel is distributed throughout a "pile" of graphite blocks.

pitchblende A massive variety of uraninite that contains a small amount of radium.

plasma Very hot ionized gases.

polar covalent bond A bond in which electrons are closer to one atom than to another. *See also* **polar molecule.**

polar dot structure Representation of the arrangements of electrons around the atoms of a molecule in which the polar characteristics are shown by placing the electrons closer to the more electronegative atom.

polar molecule A molecule that has differently charged areas because of unequal sharing of electrons.

polyatomic ion A group of chemically united atoms that react as a unit and have an electric charge.

polymerization The process of combining several molecules to form one large molecule (polymer). (a) Additional polymerization: The addition of unsaturated molecules to each other. (b) Condensation polymerization: The reaction of two molecules by loss of a molecule of water.

positron A positively charged particle of electricity with about the same weight as the electron.

potential energy Energy due to the position of a body or to the configuration of its particles.

precipitate An insoluble compound formed in the chemical reaction between two or more substances in solution.

proteins Large, complex organic molecules, with nitrogen an essential part, found in plants and animals.

proton A subatomic particle found in the nucleus that has a positive charge.

qualitative analysis A term applied to the methods and procedures used to determine any or all of the constituent parts of a substance.

quantitative analysis A term applied to the methods and procedures used to determine the definite quantity or percentage of any or all of the constituent parts of a substance.

quenching Cooling a hot piece of metal rapidly, as in water or oil.

radiation The emission of particles and rays from a radioactive source; usually alpha and beta particles and gamma rays.

radioactive Referring to substances that have the ability to emit radiations (alpha or beta particles or gamma rays).

radioisotope An isotope that is radioactive, such as uranium-235.

reactant A substance involved in a reaction.

reaction A chemical transformation or change. The four basic types are combination (synthesis), decomposition (analysis), single replacement or single displacement, and double replacement or double displacement.

reaction potential The sum of the oxidation potential and reduction potential for a particular reaction.

reagent Any chemical taking part in a reaction.

recrystallization A series of crystallizations, repeated for the purpose of greater purification.

redox A shortened name for a reaction that involves reduction and oxidation.

reducing agent From an electron standpoint, a substance that loses its valence electrons to another element; a substance that is readily oxidized.

reduction A chemical reaction that removes oxygen from a substance; a gain of electrons.

reduction potential An electrode potential associated with a reduction half-reaction.

refraction (of light) The bending of light rays as they pass from one material into another.

relative humidity The ratio, expressed in percent, between the amount of water vapor in a given volume of air and the amount the same volume can hold when saturated at the same temperature.

resonance The phenomenon in a molecular structure that exhibits properties between those of a single bond and those of a double bond and thus possesses two or more alternative structures.

reversible reaction Any reaction that reaches an equilibrium, or that can be made to proceed from right to left as well as from left to right.

roasting Heating an ore (usually a sulfide) in an excess of air to convert the ore to an oxide, which can then be reduced.

salt A compound, such as NaCl, made up of a positive metallic ion and a negative nonmetallic ion or radical.

saturated solution A solution that contains the maximum amount of solute under the existing temperature and pressure.

Second Law of Thermodynamics *See under* **laws**.

sigma bond A bond between *s* orbitals or between an *s* orbital and another kind of orbital.

significant figures All the certain digits, that is, those recorded in a measurement, plus one uncertain digit.

slag The product formed when the flux reacts with the impurities of an ore in a metallurgical process.

solid A phase of matter that has a definite size and shape.

solubility A measure of the amount of solute that will dissolve in a given quantity of solvent at a given temperature.

solute The material that is dissolved to make a solution.

solution A uniform mixture of a solute in a solvent.

solvent The dispersing substance that allows the solute to go into solution.

specific gravity (mass) The ratio between the mass of a certain volume of a substance and the mass of an equal volume of water (or, in the case of gases, an equal volume of air); expressed as a single number.

specific heat The ratio between the number of calories needed to raise the temperature of a certain mass of a substance 1 degree on the Celsius scale and the number of calories needed to raise the temperature of the same mass of water 1 degree on the Celsius scale.

spectroscope An instrument used to analyze light by separating it into its component wavelengths.

spectrum The image formed when radiant energy is dispersed by a prism or grating into its various wavelengths.

spinthariscope A device for viewing through a microscope the flashes of light made by particles from radioactive materials against a sensitized screen.

spontaneous combustion (ignition) The process in which slow oxidation produces enough heat to raise the temperature of a substance to its kindling temperature.

stable Referring to a substance not easily decomposed or dissociated.

standard conditions An atmospheric pressure of 760 millimeters or torr or 1 atmosphere (mercury pressure) and a temperature of 0°C (273 K) (abbreviation: STP).

stratosphere The upper portion of the atmosphere, in which the temperature changes but little with altitude, and clouds of water never form.

strong acid (or base) An acid (or a base) capable of a high degree of ionization in water solution. Example: sulfuric acid (sodium hydroxide).

structural (graphic) formula A pictorial representation of the atomic arrangement of a molecule.

sublime To vaporize directly from the solid to the gaseous state, and then condense back to the solid.

substance A single kind of matter, element, or compound.

substitution product A product formed by the substitution of other elements or radicals for hydrogen atoms in hydrocarbons.

sulfation An accumulation of lead sulfate on the plates and at the bottom of a (lead) storage cell.

supersaturated solution A solution that contains a greater quantity of solute than is normally possible at a given temperature.

suspension A mixture of finely divided solid material in a liquid, from which the solid settles on standing.

symbol A letter or letters representing an element of the periodic table. Examples: O, Mn.

synthesis The chemical process of forming a substance from its individual parts.

Système International d'Unités The modernized metric system of measurements universally used by scientists. There are seven base units: kilogram, meter, second, ampere, kelvin, mole, and candela.

temperature The intensity or the degree of heat of a body, measured by a thermometer.

tempering The heating and then rapid cooling of a metal to increase its hardness.

ternary Referring to a compound composed of three different elements, such as H_2SO_4.

theory An explanation used to interpret the "mechanics" of nature's actions; a theory is more fully developed than a hypothesis.

thermochemical equation An equation that includes values for the calories absorbed or evolved.

thermoplastic Capable of being softened by heat; may be remolded.

thermosetting Capable of being permanently hardened by heat and pressure; resistant to the further effects of heat.

tincture An alcoholic solution of a substance, such as a tincture of iodine.

torr A unit of pressure defined as 1 millimeter of mercury; 1 torr equals 133.32 pascals.

tracer A minute quantity of radioactive isotope used in medicine and biology to study chemical changes within living tissues.

transmutation Conversion of one element into another, either by bombardment or by radioactive disintegration.

tribasic acid An acid that contains three replaceable hydrogen atoms in its molecule, such as H_3PO_4.

tritium A very rare, unstable, "triple-weight" hydrogen isotope (H^3) that can be made synthetically.

Tyndall effect The scattering of a beam of light as it passes through a colloidal material.

ultraviolet light The portion of the spectrum that lies just beyond the violet; therefore of short wavelength.

U.S.P. (United States Pharmacopeia) chemicals Chemicals certified as having a standard of purity that demonstrates their fitness for use in medicine.

valence The combining power of an element; the number of electrons gained, lost, or borrowed in a chemical reaction.

valence electrons The electrons in the outermost level or levels of an atom that determine its chemical properties.

van der Waals forces Weak attractive forces existing between molecules.

vapor The gaseous phase of a substance that normally exists as a solid or liquid at ordinary temperatures.

vapor pressure The pressure exerted by a vapor given off by a confined liquid or solid when the vapor is in equilibrium with its liquid or solid form.

volatile Easily changed to a gas or a vapor at relatively low pressure.

volt A unit of electrical potential or voltage, equal to the difference of potential between two points in a conducting wire carrying a constant current of 1 ampere when the power dissipated between these two points is equal to 1 watt (abbreviation: V).

volume The amount of three-dimensional space occupied by a substance.

VSEPR The valence shell electron pair repulsion model. It expresses the non-90° variations in bond angles for p orbitals in the outer energy levels of atoms in molecules because of electron repulsions.

water of hydration Water that is held in chemical combination in a hydrate and can be removed without essentially altering the composition of the substance. *See also* **Hydrate**.

weak acid (or base) An acid (or base) capable of being only slightly ionized in an aqueous solution. Example: acetic acid (ammonium hydroxide).

weak electrolyte A substance that, when dissolved in water, ionizes only slightly and hence is a poor conductor of electricity.

weight The measure of the force with which a body is attracted toward Earth by gravity.

work The product of the force exerted on a body and the distance through which the force acts; expressed mathematically by the equation $W = Fs$, where W = work, F = force, and s = distance.

X-rays Penetrating radiations, of extremely short wavelength, emitted when a stream of electrons strikes a solid target in a vacuum tube.

zeolite A natural or synthesized silicate used to soften water.

Index

density and, 164–165
   gas volumes and, 163–164
Molar volume, 163–164
Molarity, 195
Mole, 161
Molecular crystals, 110
Molecular geometry
   definition of, 102
   VSEPR and, 104–105
Molecular mass, 124
Molecule
   definition of, 38, 95, 142
   monoatomic, 95
Monoatomic molecules, 95
Moseley, Henry, 63
Multivalent metals, 118

N

Natural gas, 269
Negative ions, 314
Net ionic equation, 129
Neutralization, 248
Neutralization reaction, 254, 257
Neutrons, 62, 86
Newlands, John, 76
Nickel, 315
Nitric acid, 278
Nitrogen
   fixation of, 278
   properties of, 278
Nitrogen-fixing bacteria, 278
Noble gas notation, 75
Nonionizing product, 210
Nonmetals, 116
Nonpolar covalent bond, 97
Normal atmospheric pressure, 140
Normal salts, 275
Nuclear fission reactions, 88–89
Nuclear fusion reactions, 88–89
Nucleons, 63
Nucleus, 62

O

Octahedral, 103, 105
Open-hearth furnace, 279
Orbitals
   description of, 69, 71–72
   hybrid, 105–108
Ore, 279
Organic acids, 299–300
Organic chemistry
   definition of, 287
   hydrocarbons. *See* Hydrocarbon(s)
Ortho-, 296
Oxidation, 264, 268–269
Oxidation numbers, 65, 265
Oxidation states
   definition of, 265
   designation of, 265

redox reactions and, 267–268
   rules for assigning, 265–267
Oxidizer, 267
Oxidizing agent, 267
Oxygen, 121, 136–138, 313
Oxygen derivatives, 300
Oxygen furnace, 279
Ozone, 135–136, 138

P

Paper chromatography, 180
Para-, 296
Paradichlorobenzene, 186
Partial pressures, law of, 149
Pascal, 141
Paschen series, 66
Pauli Exclusion Principle, 69
Percent by mass, 195
Percent yield, 170
Percentage composition, 125–126
Perchlorate, 121
Periodic law, 78
Periodic table of elements
   description of, 78
   history of, 76–77
   properties related to, 78–83
Permanent dipole–dipole interactions, 179
pH, 251–252
pH meters, 309
Phenolphthalein, 247, 252–254
Phenyl, 296
Phlogiston theory, 136
Photographic plate, 85
Photons, 66
Photosynthesis, 287
Physical changes, 39
Pi bond, 109, 294
Pig iron, 279
Planck, Max, 68
Polar covalent bond, 97
Polyatomic ions, 116, 120–121, 123, 267
Polycrystalline solids, 178
Positive ions, 314
Positron, 86
Potassium, 314
Potassium dihydrogen phosphate, 122
Potential energy, 41
Potentials in electrochemical cells, 312
Precision, 48
Predicting of reactions, 207–211
Pressure
   air, 140
   atmospheric, 140
   barometric, 140
   corrections of, 149–150
   critical, 188
   equilibrium affected by changes in, 235
   temperature vs., 148–149
   units of, 140